WITHDRAWN

Environmental Radon

ENVIRONMENTAL SCIENCE RESEARCH

Editorial Board

Alexander Hollaender
*Council for Research Planning
in Biological Sciences, Inc.
Washington, D.C.*

Ronald F. Probstein
*Massachusetts Institute of Technology
Cambridge, Massachusetts*

Bruce L. Welch
*Environmental Biomedicine Research, Inc.
and
The Johns Hopkins University School of Medicine
Baltimore, Maryland*

Recent Volumes in this Series

Volume 29 – APPLICATION OF BIOLOGICAL MARKERS TO CARCINOGEN TESTING
Edited by Harry A. Milman and Stewart Sell

Volume 30 – INDIVIDUAL SUSCEPTIBILITY TO GENOTOXIC AGENTS IN THE HUMAN POPULATION
Edited by Frederick J. de Serres and Ronald W. Pero

Volume 31 – MUTATION, CANCER, AND MALFORMATION
Edited by Ernest H. Y. Chu and Walderico M. Generoso

Volume 32 – SHORT-TERM BIOASSAYS IN THE ANALYSIS OF COMPLEX ENVIRONMENTAL MIXTURES IV
Edited by Michael D. Waters, Shahbeg S. Sandhu, Joellen Lewtas, Larry Claxton, Gary Strauss, and Stephen Nesnow

Volume 33 – ACID RAIN: Economic Assessment
Edited by Paulette Mandelbaum

Volume 34 – ARCTIC AND ALPINE MYCOLOGY II
Edited by Gary A. Laursen, Joseph F. Ammirati, and Scott A. Redhead

Volume 35 – ENVIRONMENTAL RADON
Edited by C. Richard Cothern and James E. Smith, Jr.

Volume 36 – SHORT-TERM BIOASSAYS IN THE ANALYSIS OF COMPLEX ENVIRONMENTAL MIXTURES V
Edited by Shahbeg S. Sandhu, David M. DeMarini, Marc J. Mass, Martha M. Moore, and Judy L. Mumford

A Continuation Order Plan is available for this series. A continuation order will bring delivery of each new volume immediately upon publication. Volumes are billed only upon actual shipment. For further information please contact the publisher.

Environmental Radon

Edited by
C. RICHARD COTHERN
U.S. Environmental Protection Agency
Washington, D.C.

and
JAMES E. SMITH, Jr.
U.S. Environmental Protection Agency
Cincinnati, Ohio

PLENUM PRESS • NEW YORK AND LONDON

Library of Congress Cataloging in Publication Data

Environmental radon.

(Environmental science research; v. 35)
Bibliography: p.
Includes index.
1. Radon—Environmental aspects. 2. Air—Pollution, Indoor—Hygienic aspects. 3. Environmental chemistry. I. Cothern, C. Richard. II. Smith, James E., 1941- III. Series.
TD885.5.R33E58   1987   628.5′35   87-29108
ISBN 0-306-42707-9

© 1987 Plenum Press, New York
A Division of Plenum Publishing Corporation
233 Spring Street, New York, N.Y. 10013

All rights reserved

No part of this book may be reproduced, stored in a retrieval system, or transmitted in any form or by any means, electronic, mechanical, photocopying, microfilming, recording, or otherwise, without written permission from the Publisher

Printed in the United States of America

Preface

This volume is intended for the professional who is a newcomer to the area of environmental radon. It marks the first time that chapters on these subjects have been brought together in a single volume, and it is arranged so that anyone with some basic university-level chemistry and physics can develop a clear understanding of the different aspects involved. The volume is intended to serve as a supplementary textbook in public health, environmental, and health physics courses. It also can be used by the professional to get "up to speed" in this rapidly evolving field. The chapters are not necessarily a discussion of the latest research in this fast-moving field, but are intended to bring the reader to a level at which he can easily understand the current literature.

At the back of this volume the reader will find the references for the individual chapters, a general list of reading materials, a glossary, an appendix describing the equations for radioactive decay for a series of progeny, a table of often used conversion factors, and the addresses and brief biographies of the authors and editors. Both historical and SI (International System) units are used throughout the book to provide information for the widest range of readers.

Thanks go to Tom Hess for the idea for this volume and to Jessica Barron for help in editing. Thanks also to Robin Connor, Robley Evans, Naomi Harley, John Harley, Bruce Henschel, Werner Hofmann, Ron Kathren, C. Y. King, Ed Landa, Henry Lucas, Wayne Lowder, Ed Martel, Neal Nelson, Dan Strom, and James E. Watson, Jr., for review comments and suggestions.

The ideas and thoughts in this book are those of the individual authors and editors and do not necessarily reflect the policies of the U.S. Environmental Protection Agency.

C. Richard Cothern
Washington, D.C.
James E. Smith, Jr.
Cincinnati, Ohio

Contents

Chapter 1
Properties
C. Richard Cothern

1.1	Physical Properties	1
1.2	General Nuclear Properties	1
1.3	Radioactive Series	6
1.4	Radioactive Decay Properties	11
1.5	Units	12
	1.5.1 Activity	14
	1.5.2 Potential Alpha Energy	16
	1.5.3 Potential Alpha Energy Concentration	18
	1.5.4 Equilibrium-Equivalent Concentration	18
	1.5.5 Working Level	20
	1.5.6 Potential Alpha Energy Exposure	25
	1.5.7 Absorbed Dose	26
	1.5.8 Dose Equivalent	27

Chapter 2
History and Uses
C. Richard Cothern

2.1	History	31
	2.1.1 Historical Chronology of Radon	32
	2.1.2 Early Miner Experience	32
	2.1.3 Discovery of Radioactivity	33
	2.1.4 Early Studies	35
	2.1.5 Thorium Series and Thoron	39
	2.1.6 Other Properties of Radioactive Series	41
	2.1.7 Transmutation	43
	2.1.8 More on Characteristics	43

	2.1.9	Group Displacement Laws	45
	2.1.10	The 1920s, 1930s, and 1940s	46
	2.1.11	Recent Events	48
2.2	Uses		50
	2.2.1	Medical Applications	50
	2.2.2	Earthquake Prediction	51
	2.2.3	Atmospheric Transport	53
	2.2.4	Mines, Spas, and Questionable Medical Applications	54
	2.2.5	Other Uses	57

Chapter 3

Measurement

Douglas J. Crawford-Brown and Jacqueline Michel

3.1	Measurement of Radon and Progeny in Air		59
	3.1.1	Measurement of the Radon Alone	63
	3.1.2	Measurement of the Radon Progeny	71
	3.1.3	Measuring Progeny in the Body	74
	3.1.4	Calibration	75
3.2	Measurement of Radon in Water		77
	3.2.1	Introduction	77
	3.2.2	Sampling Methods	78
	3.2.3	Analytical Methods of Analysis	79

Chapter 4

Sources

Jacqueline Michel

4.1	Introduction		81
4.2	The Scale of Radon Transport Processes		83
4.3	Radon Transport Mechanisms		86
	4.3.1	Mechanisms of Transfer from Solids	86
	4.3.2	Transport by Groundwater	92
	4.3.3	Radon in the Atmosphere	93
4.4	Sources of Radon in Indoor Air		97
	4.4.1	Introduction	97
	4.4.2	Radon in Soils	98
	4.4.3	Radon in Surface Waters	108
	4.4.4	Radon in Groundwater	110
		4.4.4.1 Igneous and Metamorphic Rock Aquifers	111

CONTENTS

		4.4.4.2 Sedimentary Aquifers	113
		4.4.4.3 Radon Distribution in Groundwater	115
	4.4.5	Radon Contribution from Building Materials	118
	4.4.6	Radon from Fuels	123
	4.4.7	Radon Associated with Uranium and Phosphate Mining Activities	123
	4.4.8	Summary of Contribution of Radon to Indoor Air	126

Chapter 5

Human Exposure

Geoffrey G. Eichholz

5.1	Introduction		131
	5.1.1	Dietary Uptake of Radium	131
	5.1.2	Uptake of Radon Progeny	132
	5.1.3	Atmospheric Radon and its Progeny	133
5.2	Mining and Milling Activities		136
	5.2.1	General Characteristics	136
	5.2.2	Surface Mining	138
	5.2.3	Underground Mines	140
		5.2.3.1 Surface Effects of Uranium Mines	140
		5.2.3.2 Solution Mining	142
	5.2.4	Uranium Milling	143
	5.2.5	Phosphate Mining and Milling	147
		5.2.5.1 Phosphate Mine Sites	148
		5.2.5.2 Indoor Radon Levels near Phosphate Mines	149
		5.2.5.3 Use of Phosphate Slag in Building Materials	150
	5.2.6	Other Mineral Extraction Facilities	150
	5.2.7	Thoron Production in Mines and Mills	151
5.3	Radon Impact from Fossil Fuel Combustion		151
	5.3.1	Natural Gas	151
	5.3.2	Oil	153
	5.3.3	Coal-Fired Power Plants	153
5.4	Airborne Radon		154
	5.4.1	The Ambient Background	154
	5.4.2	Indoor Radon	157
		5.4.2.1 Introduction	157
		5.4.2.2 Soil Radon Seepage to Indoor Air	160
		5.4.2.3 Water Sources	166
		5.4.2.4 Building Materials	172

Chapter 6
Dosimetry
Douglas J. Crawford-Brown

6.1	Introduction	173
6.2	Anatomy and Morphology of the Lung	176
6.3	Reducing the Anatomy to a Mathematical Form	180
6.4	Changes in Anatomy with Age	182
6.5	Modeling Deposition	185
6.6	Movement on the Mucociliary Blanket	189
6.7	Dose Calculations	192
6.8	Critical Cells	199
6.9	Physiological Parameters	201
6.10	Previous Calculations of Dose from Radon and Progeny	202
6.11	Miscellaneous Factors Affecting Calculations of Lung Doses	206
6.12	A Comparison of Model Predictions and Experimental Data	207
6.13	Dosimetry of Ingested Radon	210
6.14	Discussion and Conclusions	212

Chapter 7
Health Effects
Fred T. Cross

7.1	Introduction	215
7.2	Multistage Theory of Carcinogenesis	216
7.3	Human Data	219
	7.3.1 Statistical Projection Models	221
	7.3.2 Lifetime Lung Cancer Risk Coefficient	224
	7.3.3 Lung Cancer Modifying Factors	225
	7.3.3.1 Smoking Effect	225
	7.3.3.2 Sex and Age Effect	226
	7.3.3.3 Time-Related Factors	227
7.4	Animal Studies	229
	7.4.1 Introduction	229
	7.4.2 Radon Inhalation Studies at the University of Rochester	232
	7.4.2.1 Introduction	232
	7.4.2.2 Experimental Results	232
	7.4.2.3 Discussion and Conclusions	234

CONTENTS

	7.4.3 Radon Inhalation Studies in France	234
	7.4.3.1 Introduction	234
	7.4.3.2 Experimental Results	235
	7.4.3.3 Discussion and Conclusions	237
	7.4.4 Radon Inhalation Studies at the Pacific Northwest Laboratory	237
	7.4.4.1 Introduction	237
	7.4.4.2 Exposure Parameters	238
	7.4.4.3 Experimental Results	238
	7.4.4.4 Discussion and Conclusions	241
	7.4.5 Summary of Radon Inhalation Studies in Animals	243
7.5	Comparison of Human and Animal Radon-Exposure Data	244

Chapter 8

Mitigation

Judith E. Cook and Daniel J. Egan, Jr.

8.1	Introduction	249
	8.1.1 Overview of Mitigation Techniques	250
	8.1.1.1 Ventilating Indoor Concentrations	250
	8.1.1.2 Reducing Radon Entry	250
	8.1.1.3 Removing the Radon Source	251
	8.1.1.4 Removing Radon and Radon Decay Products from Indoor Air	251
	8.1.2 Short-Term Mitigation Actions	251
	8.1.3 Limitations of Radon Mitigation	252
8.2	Evaluation of Sources and Entry Mechanisms	252
	8.2.1 House Construction Types	252
	8.2.2 Possible Entry Points	253
	8.2.3 Possible Depressurization Mechanisms	254
	8.2.4 Water and Building Materials	254
8.3	Options for Radon Reduction	255
	8.3.1 Ventilation—Diluting and Replacing Radon-Laden Indoor Air	255
	8.3.1.1 Natural Ventilation	256
	8.3.1.2 Forced Air Ventilation	257
	8.3.1.3 Forced Air Ventilation with Heat Recovery	258
	8.3.2 Preventing Radon Entry	258
	8.3.2.1 Reducing Entry Points	258

		8.3.2.2	Venting Radon from Soil Surrounding the House	260

		8.3.2.2	Venting Radon from Soil Surrounding the House	260
		8.3.2.3	Reducing Pressure Differentials	268
		8.3.2.4	Removing Radon from Water	269
	8.3.3	Air Cleaning		269
8.4	Evaluation and Maintenance of Radon Mitigation Systems			270
8.5	Developing a Mitigation Strategy			271

Chapter 9

Risk Assessment and Policy

William A. Mills and Daniel J. Egan, Jr.

9.1	Risk Assessment (William A. Mills)			273
	9.1.1	Introduction		273
	9.1.2	Source of Radon		274
	9.1.3	Derivation of Estimates of Radon Risk		275
	9.1.4	Uncertainties in Risk Assessment		279
	9.1.5	Estimates of Risk from Exposure to Indoor Radon		279
	9.1.6	Comparative Lung Cancer Risk		281
	9.1.7	Summary		283
9.2	Policy (Daniel J. Egan, Jr.)			283
	9.2.1	The Evolution of Current Concern		284
	9.2.2	Government Initiatives to Date		285
		9.2.2.1	Pennsylvania	286
		9.2.2.2	New Jersey	287
		9.2.2.3	Florida	288
		9.2.2.4	U.S. Environmental Protection Agency	288
	9.2.3	Policy Issues		289
		9.2.3.1	Assessing the Health Risk	290
		9.2.3.2	Assessing the Extent of Elevated Levels	291
		9.2.3.3	Developing Mitigation Methods	291
		9.2.3.4	Public Information and Guidance	291
		9.2.3.5	Consumer Protection	292
		9.2.3.6	Assisting or Subsidizing Mitigation	292
	9.2.4	Are Standards Needed?		292
		9.2.4.1	New Construction Issues	293
		9.2.4.2	Real Estate Transactions	293

                9.2.4.3   The Implications of Indoor Radon
                                Standards 294
        9.2.5   Summary 295

Glossary 297

Appendix A: Radioactive Decay 307

Appendix B: Conversion Factors 317

References 319

General Reading List 353

About the Authors and Editors 355

Index 359

1

Properties

C. RICHARD COTHERN

1.1 PHYSICAL PROPERTIES

Radon is a naturally occurring, colorless, odorless, almost chemically inert, and radioactive gas. Some of its properties are shown in Table 1.1. Compared to the other noble gases, radon is the heaviest and has the highest melting point, boiling point, critical temperature, and critical pressure. It is soluble in cold water, and its solubility decreases with increasing temperature as shown in Figure 1.1. This characteristic of radon causes it to be released during water-related activities in the home, such as washing clothes and dishes, taking showers or baths, flushing toilets, and general cleaning. Radon is not perfectly inert and is less so than lighter noble gases.

Although radon is an inert gas, it does form some compounds such as clathrates and complex fluorides.[1,2] Efforts have been unsuccessful to form oxides and other halides with radon.

1.2 GENERAL NUCLEAR PROPERTIES

An atom consists of a heavy concentration of mass at the center (the nucleus) surrounded by shells of electrons in different orbits. The primary constituents of the nucleus are neutrons and protons. The

---

C. RICHARD COTHERN • Office of the Administrator (A101F), U.S. Environmental Protection Agency, Washington, D.C.

TABLE 1.1. Some Physical Properties of ^{222}Rn[a]

Property	Value
Boiling point	−61.8°C
Melting point	−71°C
Critical temperature	104°C
Critical pressure	62 atmospheres
Density at normal temperature and pressure	9.96 kg m^{-3}
Volume of 27.03 pCi at normal temperature and pressure	1.6×10^{-20} m^3
Vapor pressure at	
−144.0°C	0.13 kPa
−126.3°C	1.3 kPa
−111.3°C	5.3 kPa
− 99.0°C	13 kPa
− 71.0°C	53 kPa
−61.8°C	100 kPa
Coefficients of solubility[b] at atmospheric pressure in water at:	
0°C	0.507
10°C	0.340
20°C	0.250
30°C	0.195
37°C	0.167
50°C	0.138
75°C	0.114
100°C	0.106

Coefficient of solubility in:	at 37°C	at 18°C	at 0°C
Absolute alcohol	—	6.17	8.28
Acetone	—	6.30	7.99
Amyl alcohol	—	10.6	—
Aniline	—	3.80	4.45
Animal fats	5.5–6.5	—	—
Benzene	—	12.82	—
Carbon disulfide	—	23.14	33.4
Chloroform	—	15.08	20.5
Ether	—	15.08	20.09
Fatty acids	3.6–7.3	—	—
Ethyl acetate	—	7.35	9.41
Formic acid	0.96	—	—
Glycerine	—	0.21	—
Hexane	—	16.56	23.4
Human blood	0.43	—	—
Human fat	6.33	—	—
Linoleic acid	6.3	—	—
Oleic acid	6.7	—	—
Olive oil	—	29.00	—
Petroleum (liquid paraffin)	—	9.20	12.6
Toluene	—	13.24	18.4
Xylene	—	12.75	—

[a] From References (3) and (4).
[b] The solubility coefficient or partition coefficient in water is defined as the ratio C_w/C_a, where C_w and C_a are the radon concentrations in water and air, respectively. The Bunsen coefficient is the ratio of volume of absorbed gas to that of the absorbing liquid at STP and the Oswald coefficient is the same ratio, but is not converted to STP. For more details, see Reference (4).

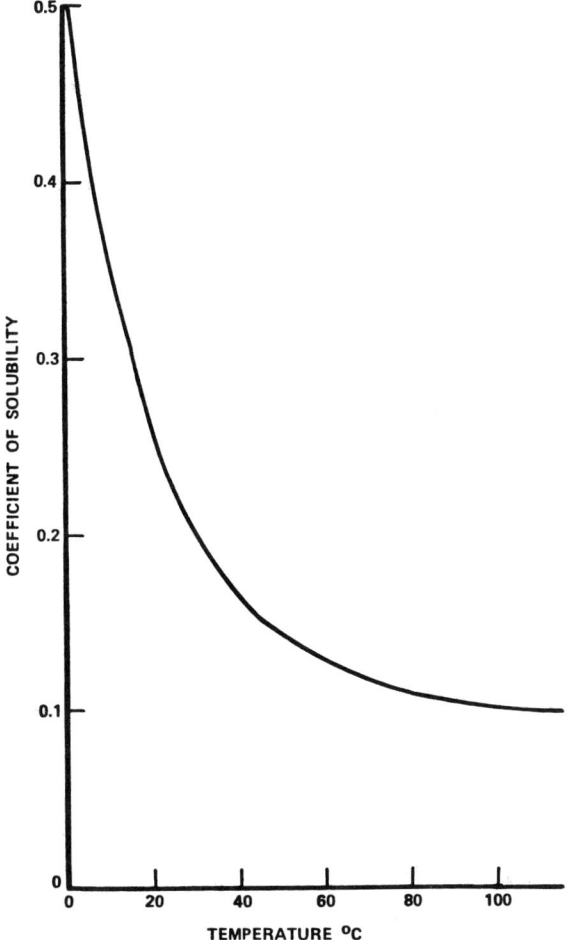

FIG. 1.1. Plot of the coefficient of solubility for radon as a function of water temperature. The solubility coefficient or partition coefficient in water is defined as the ratio C_w/C_a, where C_w and C_a are the radon concentrations in water and air, respectively.

neutrons have no charge, and the protons have a positive charge. The orbital electrons have a negative charge and are equal in number to the protons, making the atom neutral in overall charge. Each orbit that an electron can occupy has a maximum number of electrons that it can hold. How atoms interact with each other (i.e., their chemistry) depends upon how many electrons are in the outermost orbit. Due to the energy requirements of the atom, electrons tend to fall into lower orbits first until the maximum number for that orbit is achieved. Higher orbits are then generally filled in succession. With an input of energy, electrons can be moved to outer orbits. They will spontaneously "fall" to lower orbits, much like water flows downhill, until the maximum number for that orbit is reached. The energy lost in this process is emitted as light or X rays.

For example, the characteristics of the noble gases can be understood using the idea of electron orbits. They all possess filled outer electron orbits. When the orbits are each completely filled, the atom has greater stability or is less reactive—as in the case of the inert gas. If the first orbit is filled, the atom is helium (He). If the first and second orbits are filled, the element is neon. This sequence continues through argon (Ar), krypton (Kr), xenon (Xe), and radon (Rn). Radon is both inert and radioactive. Different isotopes of radon are characterized by the numbers of neutrons in the nucleus.

The number of protons in the nucleus determines the element and its atomic number. A given element can have more than one particular number of neutrons. Variation in the number of neutrons does not change the element's chemical properties since the number of electrons is determined by the number of protons, but it does produce considerable change in the stability of the element to radioactively decay. Atoms with the same number of protons but differing numbers of neutrons are called isotopes. For example, if an atom has 86 protons, it is radon. There are three well-known isotopes of radon, containing 133, 134, and 136 neutrons, respectively. The atomic mass number is the sum of the number of protons and the number of neutrons in the nucleus, and this sum is used to label isotopes. The three isotopes of radon have atomic masses of 86 + 133 = 219, 86 + 134 = 220, and 86 + 136 = 222, respectively. Symbolically, these are written as: $^{219}_{86}Rn$, $^{220}_{86}Rn$, and $^{222}_{86}Rn$ or as ^{219}Rn, ^{220}Rn, and ^{222}Rn. The subscript here is redundant with the chemical symbol and is, therefore, often dropped. If the use of superscripts is awkward, then the isotopes can be written as Rn-219, Rn-220, and Rn-222.

There are 27 known isotopes of radon ranging from ^{200}Rn to ^{226}Rn. The half-lives (the time for one-half of the radioactive isotopes to decay) of these isotopes are less than one hour, except for ^{222}Rn (3.8 d), ^{210}Rn (2.5 h), and ^{211}Rn (14.7 h).

The atomic mass numbers are not the exact masses of the atom. They only reflect the total number of neutrons and protons. They are, however, rough approximations of the actual masses. The energy released in radioactive decay comes from the differences in the actual masses of the parent and progeny atoms, determined through Einstein's well-known equation, $E = mc^2$. In this equation, E is the energy, m, the mass, and c, a constant, viz., the speed of light.

It is a general rule of nature that a system will gravitate toward the lowest energy state or the most stable situation possible; e.g., water runs downhill, unlike charges attract each other causing an electron to "fall" into the orbit closest to the nucleus, and snow falls to the ground. In this same sense, if a nucleus can move to a lower energy state (a state in which the protons and neutrons are more tightly bound in the nucleus) by emitting radiation, it will. Such a nucleus is radioactive or unstable to decay. Stable nuclei are unable to lose energy by emitting radiation.

In general, one might expect a nucleus to be able to emit all kinds and combinations of radiation. However, because systems tend to seek stability and because of the nature of the nuclear force, the most likely (or most stable) radiations to be emitted are a helium nucleus (two protons plus two neutrons), called an alpha particle; an electron, called a beta particle; and a type of high-energy X ray called a gamma ray.

The beta particle ejected from a nucleus is created by a neutron decaying into a proton (which remains in the nucleus). As a result of this process, the atom has one more proton in its nucleus and thus has become the atom of a different element with an atomic number one greater than the parent atom. The nucleus can emit energy in the form of a gamma ray without emitting a particle that has rest mass, like the alpha or beta particle. A gamma ray is a form of electromagnetic radiation. Other forms of electromagnetic radiation are light, radio waves, infrared radiation, ultraviolet radiation, and X rays.

The process of alpha and beta radioactive decay leads to the formation of a different element, whereas gamma decay does not. The isotope that decays is called the parent. The resulting isotope, if a different element, is called the progeny. For example, ^{222}Rn decays to the progeny ^{218}Po by emitting an alpha particle. This reaction is written:

$$^{222}\text{Rn} \rightarrow {}^{218}\text{Po} + {}^{4}\text{He}$$

An example of beta decay is

$$^{228}\text{Ra} \rightarrow {}^{228}\text{Ac} + \text{beta particle}$$

Note that in beta decay the atomic mass number remains unchanged while the element changes.

Alpha, beta, and gamma radiations can have many different

energies, and thus they produce different effects as they interact with matter. Each of these radiations is capable of knocking an electron from its orbit around a nucleus and away from an atom. This process is called ionization. If an electron is moved to an orbit further from the nucleus, but still bound to the nucleus, the atom is said to be in an excited state. The atom then will decay when the electron returns to the inner orbit and emits radiation. We see this kind of radiation from a light bulb.

Because the ion is highly reactive, it is by ionization that radiation is detected. Ionization can also be beneficial to humans through its use in therapeutic and diagnostic medicine. Highly reactive ions can, however, lead to deleterious effects in humans such as cancers and leukemias.

Early attempts to accurately measure the output of X rays or emanations from radioactive materials involved measuring the material's heat output with an air thermometer or a bolometer. Other techniques for measuring the changes occurring in a material with radioactive decay include colorometric changes in chemical mixtures, blackening of silver bromide-impregnated strips, and resistance changes in selenium cells.

It was not until 1905 that ionization was first used in measuring radioactive decay. The physical quantity actually measured is the magnitude of the charge created by the interaction of ionizing radiation with matter. Available detection schemes range from that of discharging an electroscope to chambers across which high voltages are applied for collecting the charge.

1.3 RADIOACTIVE SERIES

There are four possible radioactive series or chains of radionuclides. Any one of these can be characterized in terms of the atomic mass numbers (total number of protons and neutrons) of its constituents by the simple expression

$$A = 4n + m$$

where A is the mass number and n is the largest whole integer that can be divided into A and give four with a remainder of m. Thus, the four possible series are $4n$, $4n + 1$, $4n + 2$, and $4n + 3$.

The members of three of these series are found in nature and account for most of the natural radioactivity of terrestial origin. They are the $4n$ series headed by ^{232}Th, which is known as the thorium series; the $4n + 2$ series, headed by ^{238}U, which is known as the uranium series; and the $4n + 3$ series, headed by ^{235}U and known as the actinium series. The radionuclide heading each of these series has a half-life that is long relative to the age of the earth. The fourth series,

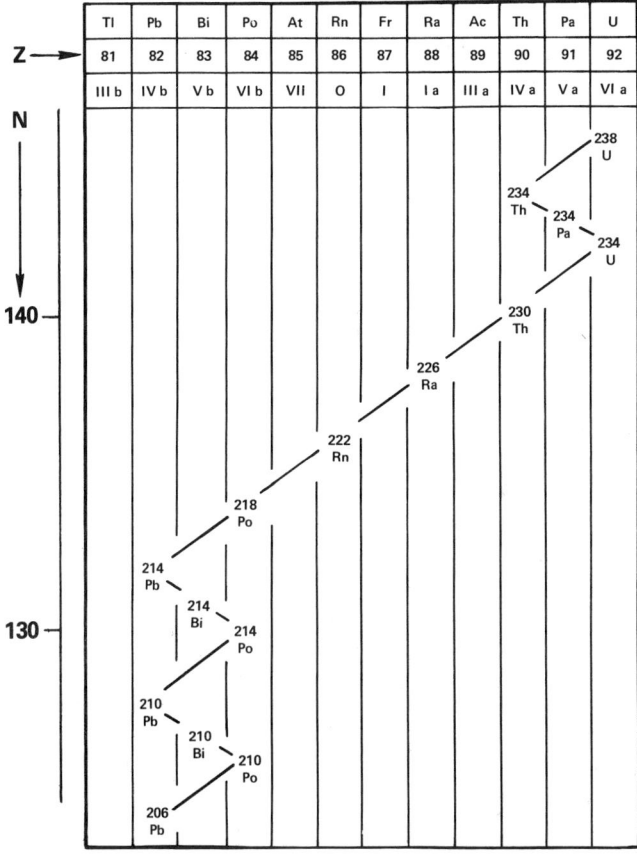

FIG. 1.2. Schematic plot of the major decay transitions in the uranium series starting with ^{238}U. The horizontal scale is Z, the number of protons in the nucleus, and the vertical scale is N, the number of neutrons in the nucleus. Also shown below the proton number are the major categories in the periodic table. Some minor branches are not shown to improve the clarity of the figure. Alpha decays are two steps to the left and beta decays are one step to the right.

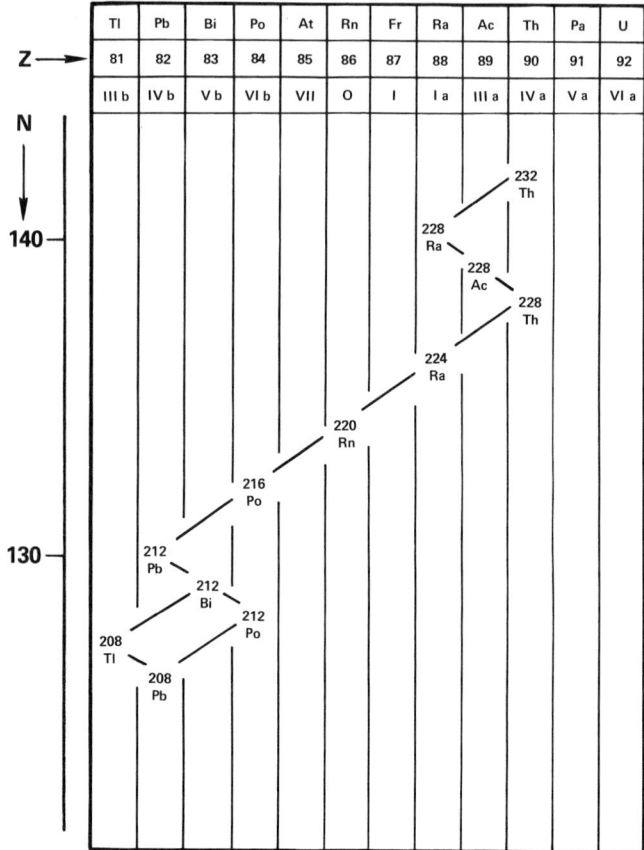

FIG. 1.3. Schematic plot of the major decay transitions in the thorium series starting with ^{232}Th. The horizontal scale is Z, the number of protons in the nucleus, and the vertical scale is N, the number of neutrons in the nucleus. Also shown below the proton number are the major categories in the periodic table. Some minor branches are not shown to improve the clarity of the figure. Alpha decays are two steps to the left and beta decays are one step to the right.

the $4n + 1$ or the neptunium series, has not been found in nature with the exception of its essentially stable end product ^{209}Bi. The major decay steps for the three naturally radioactive series are shown in Figures 1.2, 1.3, and 1.4. These figures are essentially plots of the atomic number, Z, or the number of protons on the horizontal axis versus the number of neutrons on the vertical axis. Also shown across the top of the figure are the abbreviations of the elements and the

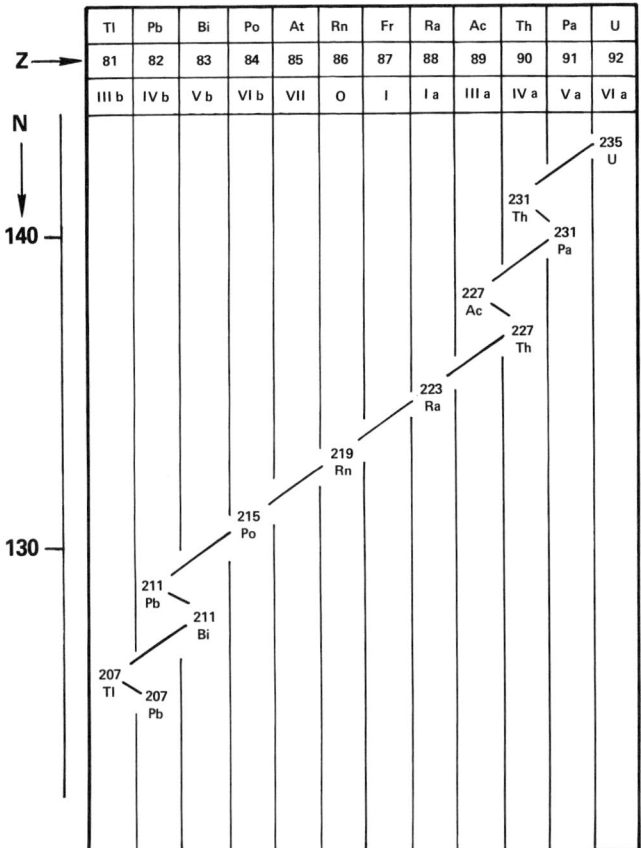

FIG. 1.4. Schematic plot of the major decay transitions in the actinium series starting with ^{235}U. The horizontal scale is Z, the number of protons in the nucleus, and the vertical scale is N, the number of neutrons in the nucleus. Also shown below the proton number are the major categories in the periodic table. Some minor branches are not shown to improve the clarity of the figure. Alpha decays are two steps to the left and beta decays are one step to the right.

corresponding groups in the periodic table. In each series, an isotope of radon is formed about halfway through the sequence.

The details of the three series are partially shown in Tables 1.2, 1.3, and 1.4. The information in these tables includes the radionuclides involved, the decay sequence, historical names, half-lives, and major radiation types and associated energies. Only the properties of the

TABLE 1.2. Uranium Decay Series' Major Characteristics[a]

Radionuclide	Historical name	Half-life	Major radiation energies (MeV)		
			Alpha	Beta	Gamma
^{222}Rn	Emanation, radon (Rn)	3.823 d	5.49		
^{218}Po	Radium A	3.05 min	6.00		
^{214}Pb	Radium B	26.8 min		0.67	0.295
				0.73	0.352
				1.02	
^{214}Bi	Radium C	19.7 min		1.0	0.609
				1.51	1.12
				3.26	1.764
^{214}Po	Radium C'	164 μs	7.69		
^{210}Pb	Radium D	22.3 y		0.015	0.047
				0.061	
^{210}Bi	Radium E	5.01 d		1.161	
^{210}Po	Radium F	138.4 d	5.305		
^{206}Pb	Radium G	Stable			

[a] Some minor branches are not shown.

TABLE 1.3. Thorium Decay Series' Major Characteristics[a]

Radionuclide	Historical name	Half-life	Major radiation energies (MeV)		
			Alpha	Beta	Gamma
^{220}Rn	Emanation, thoron (Tn)	55 s	6.29		
^{216}Po	Thorium A	0.15 s	6.78		
^{212}Pb	Thorium B	10.64 h		0.331	0.239
				0.569	0.300
^{212}Bi	Thorium C	60.6 min	6.05	1.55	0.040
			6.09	2.26	0.727
					1.620
^{212}Po	Thorium C'	304 ns	8.78		
^{208}Tl	Thorium C"	3.05 min		1.28	0.511
				1.52	0.583
				1.80	0.860
					2.614
^{208}Pb	Thorium D	Stable			

[a] Some minor branches are not shown.

PROPERTIES

TABLE 1.4. Actinium Decay Series' Major Characteristics[a]

Radionuclide	Historical name	Half-life	Major radiation energies (MeV)		
			Alpha	Beta	Gamma
^{219}Rn	Emanation, actinon (An)	4.0 s	6.42 6.55 6.82		
^{215}Po	Actinium A	1.78 ms	7.38		
^{211}Pb	Actinium B	36.1 min		0.29 0.56 1.39	0.405 0.427 0.832
^{211}Bi	Actinium C	2.15 min	6.28 6.62		
^{207}Tl	Actinium C"	4.79 min		1.44	
^{207}Pb	Stable				

[a] Some minor branches are not shown.

radon isotopes and their progeny are shown. Details of the higher isotopes in each series can be found elsewhere.[3]

1.4 RADIOACTIVE DECAY PROPERTIES

By 1903 Ernest Rutherford and Frederick Soddy had developed their theory of spontaneous disintegration or the law of radioactive change. They observed that

> in all cases . . . the law of radioactive change . . . may be expressed in one statement—the proportional amount of radioactive matter that changes in unit time is a constant . . . the constant . . . possesses for each type of active matter a fixed characteristic value.[13]

What they had observed was that due to the radiations from a pure sample of radioactive material, the ionization current in their chamber decreased exponentially with time. This can be expressed mathematically as:

$$I = I(0) \exp(-\lambda t) \tag{1.1}$$

where I is the ionization current from the chamber at any time t, $I(0)$ is the initial ionization current, and λ is a constant. Assuming that the quality of the radiation is constant, then the ionization current is proportional to the number of radiations passing through the chamber per unit time.

For more details concerning the differential equations describing radioactive decay for individual radionuclides and radioactive series, see Appendix A.

1.5 UNITS

The study of units of measure is an important exercise. It enables one to develop expertise in an area of science. Because units develop historically, they are often cumbersome, complex, and seemingly illogical. Yet units are essential for quantifying measurements. If units are used improperly or in an imprecise manner, they can obscure the physical phenomena they are intended to describe.

Numerous hardships and even disasters have resulted from failure to standardize units of measure with such diverse things as railroad gauges, screw pitch, fire hose sizes, and boiler manufacture.[5] An excellent discussion of the philosophical approaches to choosing among different systems and the standardization based upon a particular one was written in 1821 by the then Secretary of State, John Quincy Adams.[6] The need for standardization is even more important today because of the increased interdependence between nations and humans in general.

A systematic approach to units in a given field or area of science helps greatly in understanding what is really happening physically. The reader is therefore advised to spend some time digesting the discussion here, as an understanding of the units used will help in understanding the aspects of radon discussed in this book. This is akin to reading the directions for a particular piece of scientific equipment before operating it.

The units described in this section for application to radon and its progeny involve measurement of the rate of radioactive decay, the amount of energy released, and the kinds of particles emitted. Units that relate to the energy absorbed by various human organs and its effect on these organs are also discussed. This latter category will be more fully developed in Chapter 6 in the discussion of dosimetry. It is useful and important to separate mentally the units that describe the rate of decay and the amount of energy emitted from those that relate to body organs and associated human health effects. The former are quantities that are actually measured, and the latter are health effect-related information that is derived from a combination of the former and radiobiological or epidemiological data.

Basically, three systems of units are currently in use; two of these systems are of historical importance, and the third is becoming more universally used. What is here called the conventional system arose historically to describe the amount of radioactivity in an environmental medium and dose imparted by ionizing radiation. Units such as activity were designed to describe the physical properties of sources in the environment, whereas other units such as rad and rem were developed to characterize the risk imposed on a population by some amount of radiation. Some units, such as the working level month, are a hybrid approach, designed to produce a measure of the physical properties of the environment, but they are also approximately valid as indices of risk. Important units within the conventional system are the curie (Ci), the rad, and the rem. During the 1950s, in order to introduce safety standards for radon levels in uranium mines, the additional conventional unit of the working level was created. Finally, during the 1970s the International System of Units (SI) was developed, and it is becoming widely used.

The organization responsible for the international unification and development of the metric system is the General Conference on Weights and Measures (Conférence Générale des Poids et Mesures, CGPM). In 1948 the CGPM instructed its International Committee for Weights and Measures to develop a set of rules for the units of measurements. The International System of Units (le Système International d'Unités, or SI) was developed under this charge. The SI system was adopted by the CGPM in 1960 and was accepted by all signatories to the Meter Convention in 1977. In the United States, the National Council for Radiation Protection and Measurements in its 1984 report on SI units recommended that until 1989 SI units be used with conventional units included in brackets and after 1989 only SI units be used.[7]

The SI system brings with it a systematic and universal approach. It is based on the metric meter-kilogram-second (MKS) general system and includes among others such units as becquerels, grays, and sieverts.

Because of its growing universality, the SI system is used in this book. This approach offers a logical framework in which to develop an understanding of the quantities that will be used. The conventional and uranium miner-based systems of units, however, are still found in the literature and will be described in this section so that the reader can relate them to SI units.

Perhaps the most important feature of the SI system is that no

TABLE 1.5. Greek Prefixes Used in the SI System

Factor	Prefix	Symbol	Factor	Prefix	Symbol
10^{18}	exa	E	10^{-1}	deci	d
10^{15}	peta	P	10^{-2}	centi	c
10^{12}	tera	T	10^{-3}	milli	m
10^{9}	giga	G	10^{-6}	micro	μ^a
10^{6}	mega	M	10^{-9}	nano	n
10^{3}	kilo	k	10^{-12}	pico	p
10^{2}	hecto	h	10^{-15}	femto	f
10^{1}	deca	da	10^{-18}	atto	a

^{a} Because many modern word processing devices do not have the Greek alphabet, the letter u is sometimes used instead of this Greek mu.

cumbersome conversion factors are needed other than unity and powers of ten. This feature is called coherence. When Greek prefixes are used to denote powers of ten, there is a lack of coherence because they are not all the same multiplier. For calculational purposes, it is recommended that the magnitude of a quantity be expressed as the product of a pure number and a unit.

The currently adopted Greek prefixes to denote powers of ten are shown in Table 1.5.

1.5.1 Activity

In general, the activity of a sample is a measurement of the rate of decay of that sample. The SI derived unit of activity is called the becquerel and the unit symbol is Bq. This unit is named for Henri Becquerel, who shared the Nobel prize with Marie and Pierre Curie in 1903 for discovering radioactivity. One becquerel (1 Bq) equals 1 s^{-1}. Becquerel is the special name for s^{-1} when it is used as a measure of activity. Note, however, that hertz is the special name for s^{-1} when it is used as a measure of frequency. In both cases, the unit represents the number of events that occur in one second.

The conventional unit for activity is the curie. It was named after the Curies, a wife-and-husband team who discovered radium and contributed extensively to the development of the field of radioactivity. The magnitude of the curie (unit symbol, Ci) was historically based on the activity of radium and is equal approximately to the rate of decay of one gram of ^{226}Ra. As the accuracy of measurements increased, however, the exact value changed. The value of the curie has been arbitrarily set at 3.7×10^{10} s^{-1} for general use.

Two other activity units are of historic interest. The mache unit was named for Heinrich Mache of the Vienna Radium Institute and was widely used in Europe.[8] The mache unit is the amount of radon in one liter that will produce a saturation current of 0.001 electrostatic units of current and is equivalent to 364 pCi L^{-1}. The eman unit, used in the 1920s and 1930s, is equivalent to 100 pCi L^{-1}.[9]

Because the International Radium Standard Commission held that the curie unit be used only for radium and members of its decay series, a new unit for activity was proposed in 1948. The rutherford was defined as 1.00×10^6 disintegrations s^{-1}.

As an example of how activity is related to the mass of a radioactive substance, consider the activity represented by one gram of ^{226}Ra. Let λ be the radiological decay constant for ^{226}Ra in units of s^{-1} (λ is equal to $0.693/T_{1/2}$, where $T_{1/2}$ is the half-life of ^{226}Ra and is equal to 1602 y). One gram of ^{226}Ra is comprised of a number of atoms equal to $6.02 \times 10^{23}/226$, since one mole of an element contains 6.02×10^{23} atoms and the gram molecular weight of ^{226}Ra is equal to 226 grams. The activity (in disintegrations per second) then is equal to the product of λ and the number of atoms, or

$$A = dN/dt = \lambda N$$

$$= \frac{0.693(6.02 \times 10^{23})/226}{(1602)(365)(24)(60)(60)}$$

$$= 3.7 \times 10^{10} \text{ s}^{-1}$$

In other words, one gram of ^{226}Ra in its pure form produces one curie of activity.

Note that in some publications, particularly those from the U.S. Public Health Service in the 1950s and 1960s, both the activity of ^{222}Rn and the combined activities of both beta particles and alpha particles for ^{222}Rn and its progeny through ^{214}Po are reported. This increased the numbers quoted by a constant amount. Thus, one should be careful in quoting measurements and must be sure what the content of the actual numbers is.

A somewhat simplified overview of the units used in describing radon and its progeny is shown in Figure 1.5. Radon activity is easier and thus less expensive to measure than the activity of the progeny. The activity units are Bq, defined at the beginning of this section, and

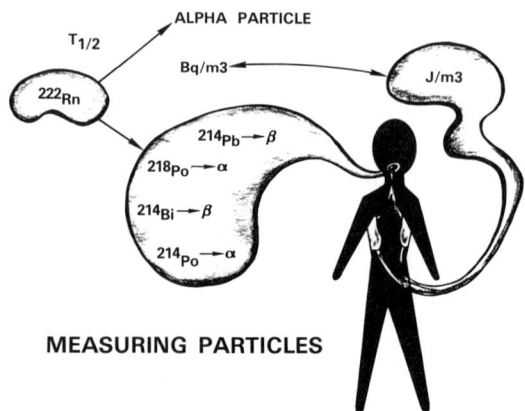

FIG. 1.5. Schematic showing the carrier of the radioactivity, ^{222}Rn, and its activity, which is a function of the half-life. Also shown is a human inhaling the progeny of the decay of the radon, the different progeny, and their emissions in the form of alpha and beta particles. The unit of energy is the central one in determining possible damage to human tissue.

the activity concentration units are Bq m^{-3}. The activity for a fixed number of atoms of an element is a function of the half-life.

The important characteristic to be determined ultimately is the risk assessment for radon involving the effects that result from inhaling the progeny. Thus, it is essential to determine the actual energy, in joules, which will be deposited in susceptible lung tissue by the radon progeny. To relate the activity of the parent radon to the energy deposited by the progeny is the major objective of this section on units.

1.5.2 Potential Alpha Energy

In order to relate atmospheric concentrations of the progeny to the dose delivered to the human lung, several quantities have been developed. The quantity potential alpha energy is the total alpha energy emitted by an atom as it decays through its entire radioactive series. The term potential then refers to the fact that a given atom has a *potential* to produce a certain amount of energy, if it decays through the rest of the decay chain. In other words, potential alpha energy is the energy which could be released in the lung if an atom underwent the entire decay series after depositing in the lung. Only the alpha

PROPERTIES

energy is considered in this definition, since only it is believed to be of radiobiological importance for the ^{222}Rn chain. Because this unit is being developed to describe health effects, one important modification is made. In the uranium series the isotope ^{210}Pb has a 22.3-year half-life, and this is so long that the isotope would probably be eliminated by the movement of mucus before a significant fraction of it decays. Thus, for the uranium series, the important progeny are those with short half-lives and these precede ^{210}Pb. For the thorium and actinium series, the potential alpha energy involves decay to the end of the series, or to ^{208}Pb and ^{207}Pb, respectively. The values for potential alpha energy are shown in Table 1.6. The energies listed in Table 1.6 are in the traditional units of million electron volts (MeV). An electron volt is the energy gained by an electron in falling through a potential difference of one volt.

One electron volt (eV) is given by the product qV where q is the charge on the electron and V is the potential difference. Since the charge on the electron is 1.6×10^{-19} coulombs then, energy of one

TABLE 1.6. Potential Alpha Energy for the Uranium, Thorium, and Actinium Series

Radionuclide	Half-life ($T_{1/2}$)	Potential alpha energy (ϵ_p) (MeV per atom)	Total potential alpha energy per unit of activity (ϵ_p/λ) (10^{-10} J Bq^{-1})
^{222}Rn	3.823 d	19.2	14,600
^{218}Po	3.05 min	13.7	5.79
^{214}Pb	26.8 min	7.69	28.5
^{214}Bi	19.7 min	7.69	21.0
^{214}Po	164 μs	7.69	2.9×10^{-6}
^{220}Rn	55 s	20.9	2.65
^{216}Po	0.15 s	14.6	5.06×10^{-3}
^{212}Pb	10.64 h	7.80	690
^{212}Bi	60.6 min	7.80	65.5
^{212}Po	304 ns	8.78	6.2×10^{-9}
^{219}Rn	4.0 s	20.7[a]	0.191
^{215}Po	1.78 ms	13.94	5.73×10^{-5}
^{211}Pb	36.1 min	6.56[b]	32.8
^{211}Bi	2.15 min	6.56[b]	1.95
^{207}Tl	4.79 min	—	

[a] Includes weighted average of several alpha energies.
[b] Weighted average of 16% of 6.28 MeV and 84% of 6.62 MeV.

electron volt = qV = (1.6 × 10^{-19} coulombs)(1 volt) = 1.6 × 10^{-19} coulomb-volts = 1.6 × 10^{-19} joules.

James Joule was an English physicist who established the principle of the interconvertibility of the various forms of energy commonly called the first law of thermodynamics.

1.5.3 Potential Alpha Energy Concentration

The potential alpha energy concentration is the sum of all the potential alpha energy in a volume of air, divided by the volume of that air. A common unit for this quantity is J m^{-3} and the symbol used here will be c_p. The conventional unit and the one often used for potential alpha energy concentration is MeV L^{-1}. By definition, 1 MeV = 1.6 × 10^{-13} J.

The potential energy per unit of activity and the total potential energy per Bq of activity is the ratio of ϵ_p to λ. The decay constant λ is equal to $0.693/T_{1/2}$ and is the activity per atom, and ϵ_p is the potential alpha energy per atom. Values of ϵ_p/λ are listed in Table 1.6.

If $c_{a,i}$ is the activity concentration of a progeny nuclide in air, then the potential alpha energy concentration is

$$c_p = \sum_i c_{p,i} = \sum_i c_{a,i}(\epsilon_{p,i}/\lambda i)$$

1.5.4 Equilibrium-Equivalent Concentration

In most situations equilibrium between the parent radon isotope and the progeny does not exist, and a single quantity is needed to describe the activity concentration for such a nonequilibrium collection of atoms. The quantity used in this situation is the equilibrium-equivalent concentration, for which the International Commission on Radiological Protection (ICRP Publication 32) uses the symbol EC. The EC of a nonequilibrium mixture of short-lived progeny in air is that activity concentration of the parent gas in radioactive equilibrium with the concentrations of its short-lived progeny that has the same potential alpha energy concentration c_p as the nonequilibrium mixture. Thus the units of EC are Bq m^{-3}, and they represent the activity concentration of the parent that would have to exist in complete equilibrium with the progeny if the short-lived progeny had the same potential alpha energy

PROPERTIES

concentration (J m^{-3}) as in the nonequilibrium mixture. It is important to bear in mind that only the alpha decays of the progeny (and not the radon itself) are considered, as only these are considered to be radiobiologically important.

The EC can best be understood by the example calculation shown below. For the situation starting with 10 Bq m^{-3} of pure ^{222}Rn, a nonequilibrium example for radon progeny after 60 min of ingrowth with no plate-out of progeny is:

Radionuclide	Activity (Bq m^{-3})
^{218}Po	10
^{214}Pb	7.6
^{214}Bi	5.0
^{214}Po	5.0

Then using

$$c_p = \sum_i c_{a,i}(\epsilon_{p,i}/\lambda_i)$$

and the values from Table 1.6,

$$c_p = 10(5.79 \times 10^{-10}) + 7.6(28.5 \times 10^{-10})$$
$$+ 5.0(21.0 \times 10^{-10}) + 5.0(2.9 \times 10^{-16})$$
$$= 380 \times 10^{-10} \text{ J m}^{-3}$$

Then the equivalent-equilibrium concentration is

$$\text{EC}_{\text{Rn}}(\text{Bq m}^{-3}) = \frac{c_p}{\sum_i \epsilon_{p,i}/\lambda_i} \text{ (J m}^{-3}/\text{J Bq}^{-1})$$

$$= \left[\frac{1}{(5.79 + 28.5 + 21.0 + 2.9 \times 10^{-6}) \times 10^{-10}}\right] c_p (\text{J m}^{-3})$$

$$= 6.86 \text{ Bq m}^{-3}$$

The equilibrium factor F is defined as the ratio of the total potential alpha energy for the actual progeny concentrations to the total potential alpha energy of the progeny which would be found if the progeny were in equilibrium with the parent gas. Thus

$$F = EC_{Rn}/c_{Rn}$$

Using the values calculated in the above example,

$$F = 6.86/10 = 0.686$$

1.5.5 Working Level

The unit used widely to describe the potential alpha energy concentration of radon progeny is the working level (WL). Because of the widespread historical use of this unit and because of the complexities involved in defining this unit, it will be discussed in some detail here.

The WL was first suggested in 1955 by representatives of several uranium-mining states who met to determine what a safe exposure level of radon was.[10] They concluded that since the damage to the lungs was due to the progeny involved and since the alpha particles from the progeny with short half-lives were the most significant part of the radiation, any standard should be based on the first four short-lived daughters of radon, viz., RaA, RaB, RaC, and RaC'. It was decided that a standard should be based on a single unit, rather than a specification of the individual concentrations for each progeny. They proposed that a safe level at which to work (a working level) would be 100 pCi L^{-1} of "each of the three daughter products RaA, RaB, and RaC through RaC'." The group felt that this level would be not more than one-tenth of that in the German and Czechoslovakian mines and that "over his working lifetime a worker exposed to this level for a period of forty hours a week would not suffer biologic damage." The representatives observed that this standard was more precisely stated as 1.3×10^5 MeV of potential alpha energy per liter of air. This numerical value is derived from the alpha energy released by the total decay of the short-lived radon progeny products in radioactive equilibrium with 100 pCi of ^{222}Rn per liter of air. The standard appeared formally in a 1957 Public Health Report.[11]

The definition of a WL was formally and more generally stated in 1968 as any level of concentration or burden of radioactivity in a given air environment which produced a prespecified potential alpha energy concentration. Thus, related to environments containing radon and progeny of radon, a WL is represented by any combination of short-lived radon progeny in one liter of air that will result in the emission of 1.3×10^5 MeV of potential alpha energy from the radioactive decay of the radon progeny. This would mean 100 pCi L^{-1} of each of ^{222}Rn, ^{218}Po, ^{214}Pb, ^{214}Bi, and ^{214}Po.[12] Note that only the alpha particles emitted by nuclides down to ^{214}Po in the chain are considered, since nuclides further down the chain have very high chances of being removed from the lung before they decay.

The working level is a measure of exposure rate. It describes how much energy is found in the alpha particles in a liter of air. In order to relate this energy to biological deposition and absorbed dose, many additional factors must be considered. These include the breathing rate, the volume of air per breath, the time the radon progeny remain in the body, and the half-lives of the progeny.

The relationships and notation used here are the same as those developed by Ernest Rutherford and Frederick Soddy at McGill University just after the turn of the century.[13] For a complete discussion of the fundamental principles describing the growth and decay of members of radioactive series, see Reference 14.

The characteristics of ^{222}Rn and its progeny are listed in Table 1.7. The short-lived progeny include ^{218}Po, ^{214}Pb, ^{214}Bi, and ^{214}Po. The total number of atoms per 100 pCi of activity is shown in Table 1.7. Using ^{222}Rn as an example,

$$N_t = (-dN/dt)/\lambda$$

$$= 100 \text{ pCi}/\lambda$$

$$= 100 \text{ pCi} \times T_{1/2}/0.693$$

$$= (100 \text{ pCi})(3.823 \text{ d}/0.693)(3.7 \times 10^{10} \text{ s}^{-1}/\text{Ci})$$

$$\times (\text{Ci}/10^{12} \text{ pCi})(3600 \text{ s/h})(24 \text{ h/d})$$

$$= 1.76 \times 10^6 \text{ atoms}$$

As shown in Table 1.7, for each decay of a ^{218}Po atom there will eventually be two alpha particles emitted, one of 6.00 MeV from ^{218}Po and one of 7.69 MeV emitted by ^{214}Po. Thus, the total alpha particle energy emitted due to the 100 pCi of ^{218}Po radionuclides present is 0.134×10^5 MeV, as shown in Table 1.7. The total energy emitted by

TABLE 1.7. Decay Characteristics Relating to the Determination of the Definition of a Working Level

Radionuclide	Historical name	$T_{1/2}$	Number of atoms per 100 pCi	Alpha particle energy for complete decay per atom (MeV)	Total alpha particle energy for complete decay (MeV)
^{222}Rn	Rn	3.823 d	1.76×10^6		
^{218}Po	RaA	3.05 min	977	6.00 + 7.69	0.134×10^5
^{214}Pb	RaB	26.8 min	8590	7.69	0.661×10^5
^{214}Bi	RaC	19.7 min	6310	7.69	0.485×10^5
^{214}Po	RaC′	164 μs	0.0009	7.69	0.000×10^5
				A total of	1.280×10^5
				or roughly	1.3×10^5
^{220}Rn	Th	55 s	294		
^{216}Po	ThA	0.15 s	0.801	14.6	0.000×10^5
^{212}Pb	ThB	10.64 h	2.05×10^5	7.80	16×10^5
^{212}Bi	ThC	60.6 min	1.94×10^4	7.80	1.51×10^5
^{212}Po	ThC′	304 ns	1.62×10^{-6}	8.78	0.00×10^5
				A total of	17.5×10^5
^{219}Rn	An	4.0 s	21.6		
^{215}Po	AcA	1.78 ms	9.5×10^{-5}	13.94	0.000×10^5
^{211}Pb	AcB	36.1 min	11,600	6.56	0.761×10^5
^{211}Bi	AcC	2.15 min	689	6.56	0.045×10^5
^{207}Tl	AcC″	4.79 min	1530	—	—
				A total of	0.806×10^5

the short-lived progeny of ^{222}Rn is 1.28×10^5 MeV, as also shown in Table 1.7. This is roughly the 1.3×10^5 MeV stated in the definition of a WL.

Because of ^{222}Rn decay characteristics, over half of the contribution to a WL comes from ^{214}Pb and virtually none from ^{214}Po. Thus the definition of a WL is a compromise. It is based on a hypothetical equilibrium atmosphere.

Figure 1.6 gives some idea of the extent of the compromise represented by the WL. The contributions of each of the short-lived progeny, except ^{214}Po which has too short a half-life to contribute, vary with time and are far from equal. At different times the contributions of each of the components are different. For example, in the first few minutes most of the contribution is from ^{218}Po and after an hour most of the contribution is from ^{214}Pb.

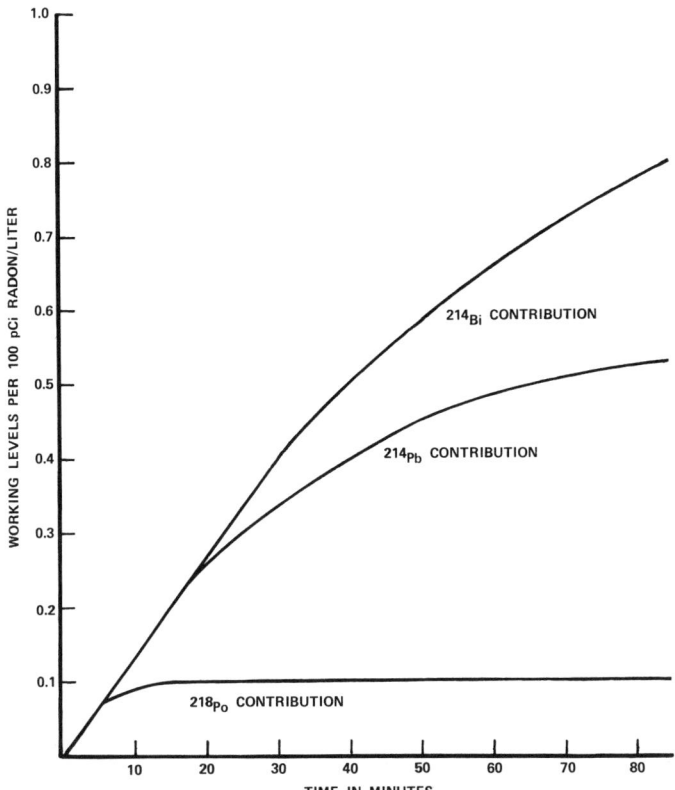

FIG. 1.6. Graph of the working level for initially pure radon, showing the contribution to the working level from ^{218}Po, ^{214}Pb, and ^{214}Bi. The contribution from ^{214}Po is negligible. [Data shown are taken from Reference (16).]

For conversion between systems,

$$1 \text{ WL} = 1.3 \times 10^5 \text{ MeV L}^{-1} = 2.08 \times 10^{-5} \text{ J m}^{-3}$$

and 1 WL corresponds to an activity concentration of 100 pCi L^{-1} = 3700 Bq m^{-3} for ^{222}Rn.

Although the WL was originally only defined for the progeny of ^{222}Rn, it can also be calculated for ^{220}Rn and ^{219}Rn by letting one working level equal 1.3×10^5 MeV of potential alpha energy per liter of air regardless of the source. Thus, 1 WL corresponds to an activity concentration of 7.43 pCi L^{-1} = 275 Bq m^{-3} for ^{220}Rn and of 161 pCi L^{-1} = 5960 Bq m^{-3} for ^{219}Rn.

TABLE 1.8. Potential Alpha Energy Concentration per Bq m^{-3}

Radionuclide	MeV L^{-1}	10^{-10} J m^{-3}	10^{-6} WL
^{218}Po	3.62	5.79	27.8
^{214}Pb	17.8	28.5	137
^{214}Bi	13.1	21.0	101
^{214}Po	1.8×10^{-6}	2.9×10^{-6}	1.4×10^{-5}
^{216}Po	3.1×10^{-3}	5.06×10^{-3}	0.0243
^{212}Pb	431	690	3320
^{212}Bi	40.9	65.5	315
^{212}Po	3.88×10^{-9}	6.2×10^{-9}	3.0×10^{-8}
^{215}Po	3.58×10^{-5}	5.73×10^{-5}	2.75×10^{-4}
^{211}Pb	20.5	32.8	157
^{211}Bi	1.22	1.95	9.36

Table 1.8 shows the necessary conversion factors for relating activity (in Bq m^{-3}) and potential alpha energy concentration for the short-lived progeny of the radioactive series.

As seen in Table 1.7, for the thorium series 17.5×10^5 MeV results from 100 pCi L^{-1} while 1.3×10^5 MeV results from 7.43 pCi L^{-1} of activity in the uranium series.

From the values in Table 1.8, the relationships between equilibrium constant (EC) and the potential alpha energy concentration (c_p) for the different radioactive series can be calculated as follows by dividing the potential alpha energy concentration for each series by the sum of those for the individual radionuclides:

$$EC_{Rn}(\text{Bq m}^{-3}) = \left[\frac{1}{(3.62 + 17.8 + 13.1 + 2.0 \times 10^{-6})}\right] c_p(\text{MeV L}^{-1})$$

$$= 2.90 \times 10^{-2} c_p(\text{MeV L}^{-1})$$

Other similar relationships are

$$EC_{Rn}(\text{Bq m}^{-3}) = 1.81 \times 10^8 c_p \, (\text{J m}^{-3})$$

$$= 3700 c_p(\text{WL})$$

PROPERTIES

by definition, although the calculated value here is 3760. For the thorium series the relationships are

$$EC_{Th} \text{ (Bq m}^{-3}) = 2.30 \times 10^{-3} c_p (\text{MeV L}^{-1})$$

$$= 1.32 \times 10^7 c_p (\text{J m}^{-3})$$

$$= 275 c_p (\text{WL})$$

For the actinium series the relationships are

$$EC_{An} \text{ (Bq m}^{-3}) = 4.6 \times 10^{-2} c_p (\text{MeV L}^{-1})$$

$$= 2.88 \times 10^8 c_p (\text{J m}^{-3})$$

$$= 6011 c_p (\text{WL})$$

1.5.6 Potential Alpha Energy Exposure

The "activity exposure" of an individual to radon is calculated as the time integral of the activity concentration over time, T, to which the individual is exposed:

$$A = \int_0^T c_a(t) \, dt$$

If the activity concentration is a constant, $A = c_a T$, where T is the time involved. A typical SI unit for A is Bq h m^{-3}. It is useful to read this as Bq m^{-3} per hour of breathing rate.

The potential alpha energy exposure (E) is the time integral over the potential alpha energy concentration of the progeny mixture to which the individual is exposed for a time T:

$$E = \int_0^T c_p(t) \, dt$$

When the potential alpha energy concentration c_p is a constant, $E = c_p(t)$ where t is the time involved.

The relationships between units are

$$1 \text{ J h m}^{-3} = 6.24 \times 10^9 \text{ MeV h L}^{-1} = 4.80 \times 10^4 \text{ WL h}$$

or

$$1 \text{ WL h} = 1.3 \times 10^5 \text{ MeV h L}^{-1} = 2.08 \times 10^{-5} \text{ J h m}^{-3}$$

The unit used to describe the potential alpha energy exposure of miners has been the working level month (WLM). One WLM corresponds to an exposure of 1 WL during the reference working period of one month (the assumption is made that there are 2000 working hours per year/12 months, which is approximately 170 h per month). Thus

$$1 \text{ WLM} = 170 \text{ WL h} = 2.2 \times 10^7 \text{ MeV h L}^{-1} = 3.5 \times 10^{-3} \text{ J h m}^{-3}$$

or

$$1 \text{ J h m}^{-3} = 285 \text{ WLM}$$

The relationship between WLM y^{-1} and WL is based on the arbitrary definition of an occupational working month as 170 h. Some earlier definitions used 173 h (see, for example, 30 CFR 57—Code of Federal Regulations). There are 51.6 periods of 170 h in a year. Thus an annual exposure to 1 WL would correspond to 51.6 WLM.

One of the primary limitations to the use of the WLM as an index of risk is that it is a measure of exposure rather than of dose. Its definition does not account for factors such as breathing rates, tidal volumes, or the fraction of progeny unattached to aerosols; each of these modify the relationship between exposure and dose (and hence exposure and risk).[15–17]

1.5.7 Absorbed Dose

The absorbed dose, D, is a measure of the average density of energy absorbed by a mass of cells. The coherent unit of absorbed dose in the SI system is joules per kilogram or J kg^{-1}. The special name given this unit is the gray, for which the unit symbol is Gy. In the conventional system, the unit for absorbed dose is based in the centi-meter-gram-second (CGS) system. It is defined as 100 erg g^{-1} and is called the rad (unit symbol, rad). Thus

$$1 \text{ J kg}^{-1} = 10{,}000 \text{ erg g}^{-1} = 100 \text{ rad}$$

and

$$1 \text{ Gy} = 100 \text{ rad}$$

Louis Harold Gray, for whom the unit is named, received his training at the Cavendish Laboratory in Cambridge during the 1930s. His contribution was a careful study of the absorption of penetrating gamma radiation and the establishment of an accurate method for measuring the energy absorbed in a material by means of the ionization produced in a gas-filled cavity. He applied his findings to the field of medicine and spent his career in the area of medical physics. He was one of the pioneers in radiation biology. His approach to the subject was invariably quantitative and generally simple as he had the vision to see where advances could be made by simple techniques.[18]

1.5.8 Dose Equivalent

The quantity dose equivalent, H, is the product of the absorbed dose, D, the quality factor, Q, and the factor, N, which is the product of all other modifying factors. Dose equivalent is a measure both of the energy absorbed by a mass and the radiobiological effectiveness of the absorbed energy. The special name for this unit is the sievert, which is abbreviated Sv.

The conventional name for this unit is the rem and the unit symbol is rem. Thus,

$$1 \text{ sievert} = 100 \text{ rem} = 1 \text{ J kg}^{-1}$$

Rolf M. Sievert, for whom this symbol is named, laid the foundation for medical and radiation physics in Sweden, and his main contributions to clinical physics were made from 1920 to 1940. He was an active contributor to the science of radiation protection for over 40 years, until his death at the age of 70 in December 1966.[19]

The dose equivalent to an organ may be related to a risk by multiplying H by a risk factor in units of risk per unit dose equivalent.

Unfortunately, the risk factor varies among the organs of the body. In order to make reporting easier, the International Commission on Radiation Protection (ICRP) introduced the idea of the effective dose equivalent.[20,21] In this approach, the dose equivalent to an organ is multiplied by a weighting factor which accounts for the radiosensitivity of that organ. The weighting factor is equal to the risk incurred by an individual and the next generation from irradiation of a given organ to a value of H in that individual divided by the risk which would have been incurred if the whole body of the individual had received a dose equivalent of H. As a result, the actual risk imposed by a specified value of the effective dose equivalent is constant regardless of the organ irradiated. The total risk to an individual (and the next generation of children resulting from that individual) is equal to the sum of the effective dose equivalents for all organs of the body.

Using the weighting factors, the contributions to the total effect on the human body can be calculated by summing the annual dose equivalents to different organs (H_T) after weighting them with the corresponding weighting factors (w_T) as follows:

$$H_E = \text{summation over } T \text{ of } (w_T H_T)$$

where H_E is the annual effective dose equivalent. The term effective

TABLE 1.9. Calculation of the Annual Effective Dose Equivalent to Humans from Natural Background Using Data from the UNSCEAR Report[a]

Organ	Weighting factor	Annual dose equivalent (mSv y^{-1})	Annual effective dose equivalent (mSv y^{-1})
Gonads	0.25	0.97	0.24
Breast	0.15	0.95	0.14
Lung			
Mean dose	0.12	0.96	—
Tracheal/bronchial	0.06	14.0	—
Pulmonary	0.06	1.8	—
Total			1.0
Red bone marrow	0.12	1.1	0.13
Bone surfaces	0.03	1.9	0.057
Thyroid	0.03	0.88	0.026
Other	0.30	0.97	0.29
		Total	1.9 mSv y^{-1} [b]

[a] Reference (3).
[b] Approximately 200 mrem y^{-1}.

dose equivalent is used commonly by health physicists but was not used by the ICRP in its publications 26[20] or 30.[21] The term was formally introduced in another document in 1979.[22] The weighting factor approach has been applied by the United Nations Scientific Committee on the Effects of Atomic Radiation (UNSCEAR) to the estimates of the contributions to human exposure due to natural background.[3] The annual effective dose equivalent for natural background, as is shown in Table 1.9, is approximately 2 mSv y^{-1} (200 mrem y^{-1}). About one-half of the dose equivalent arises from inhalation of ^{222}Rn and its decay products. It is unfortunate that both dose equivalent and effective dose equivalent are expressed with the same unit since this can lead to some confusion.

In the SI approach to units, the major quantities for describing radon progeny include potential alpha energy, potential alpha energy concentration, activity concentration, and potential alpha energy exposure. These quantities are similar to those used in the conventional system. One additional quantity is most useful in the SI system, viz., the equilibrium-equivalent concentration. The symbols used in this volume to identify these quantities are those used by the ICRP in its publication 32.[23]

2

History and Uses

C. RICHARD COTHERN

"He insisted that staff and students should understand orders of magnitude. They must know about how big physical quantitites are . . . Not only did he know instinctively how big things are, he was very good indeed at mental arithmetic. No one else I ever knew could copy a dozen numbers down wrongly, add them up wrongly, and then come up with the right answer. It wasn't really fair. He had furthermore an unrivaled ability to put himself in the place of an alpha-particle in a piece of apparatus and decide just what he would do in the circumstance."

A description of Lord Rutherford by Lord Bowden, a student and later a colleague of Rutherford's at Cambridge, related in *Rutherford, Simple Genius*, by D. Wilson, MIT Press, Cambridge, MA (1983).

2.1 HISTORY

This section on history is organized roughly by subject and is arranged in a somewhat chronological sequence where possible. Because historical events do not always lend themselves to simple chronological presentation, some events will be discussed, in part, by topic.

C. RICHARD COTHERN • Office of the Administrator (A101F), U.S. Environmental Protection Agency, Washington, D.C.

2.1.1 Historical Chronology of Radon

1597	Agricola noted high level of what turned out to be lung cancer among Erz Mountain miners
1896	Becquerel discovers radioactivity of uranium
1898	The Curies and Schmidt discover radioactivity of thorium and also the elements radium and polonium
1898	Rutherford discovers alpha and beta particles
1899	Thomson and Rutherford demonstrate that radioactivity causes ionization
1899	Rutherford discovers thoron and calls it emanation
1900	Dorn discovers the emanation in the ^{238}U series, which is now called radon
1901	Rutherford and Brooks demonstrate that radon is a radioactive gas
1901	Discovery of active deposit of thorium by Rutherford and of radium by the Curies
1902	Rutherford and Soddy discover transmutation
1902	Thomson discovers radon in tap water
1903	Rutherford and Soddy develop equation describing radioactive decay
1904	Geisel and Debierne discover actinon
1913	Arnstein identifies squamous-cell carcinoma in autopsy of miner
1913	Fajans discovers group displacement laws
1914	First medical use of radon
1925	First mention of the name radon in the literature
1940s	Causal link shown between radon and lung cancer
1941	National Bureau of Standards advisory committee[1] adopts an air radon standard of 10 pCi L^{-1} (370 Bq m^{-3})
1955	Concept of a working level (WL) first suggested
1957	Development of the Lucas cell for detection of radon
1957	Rediscovery of radon in tap water (in Maine)
1984	Discovery of high levels of radon in homes on the Reading Prong

2.1.2 Early Miner Experience

Although radon was not discovered until the turn of this century, its effects have been known since the 16th century. At Schneeberg in the Erz Mountains between Saxony and Bohemia, a physician, Agricola, noted in 1597 that a high frequency of fatal lung conditions occurred among the local miners.[2]* Similar effects were seen in miners at Jachymov (the Joachimsthal region of Bohemia) as early as the 17th century. These mines were first worked in the 15th century at Schneeberg and the 16th century at Jachymov. They first produced copper, iron, and silver ores, and in the late 1880s, pitchblende that was used for uranium dyes, and later radium and uranium were produced from the pitcblende.

These mines played a central role in the history of radon. Their output of pitchblende provided the chemist's chief source of radium. Control of the mines was exercised by the Radium Institute of Vienna, which was an imperial organization of the Austro-Hungarian Empire.[3]

In 1879, two German physicians, Hartung and Hesse, pointed out that most of the Schneeberg mine deaths were lung cancers.[4] The Schneeberg miners who had worked in the mines for more than ten years developed the Erz Mountain disease, called bergkrankheit, which was variously diagnosed over the centuries as lymphosarcoma, tuberculosis, malignant tumor of the lungs, bronchogenic carcinoma, squamous-cell carcinoma, and small-cell carcinoma. Although the diagnosis of bergkrankheit was revised in 1922 to bronchogenic carcinoma, in 1913 Arnstein first identified squamous-cell carcinoma in tissues sampled as autopsy material from miners.[5] In 1921 Margaret Uhlig suggested that radium emanation might well be the cause of lung cancers.[4]

2.1.3 Discovery of Radioactivity

The history of the discovery of radon and the determination of its properties and characteristics are deeply rooted in the history of the discovery of radioactivity and the experiments involved in its study during the first two decades of this century. The laboratories involved in these developments were primarily located in England, France, Germany, Canada, and, to a lesser extent, the United States. Perhaps the best-known researchers involved in the study of radon are Ernest Rutherford, Frederick Soddy, and Marie and Pierre Curie, but others include Ernst Dorn, William Crookes, André Debierne, and J. J. Thomson. To establish a perspective concerning the history of radon,

---

* The translation of this original work was done by former president and mining engineer Herbert Hoover and his wife.

some of the major studies in the area of radioactivity are presented here along with those for radon.*

In 1896 only a few physicists believed in atoms, and the idea of transmutation was definitely unacceptable. Chemists believed that gold, lead, and all the other elements in Mendeleev's table were unchangeable and were the basic substances out of which the material world was built.

By the end of the first week in January 1896, the world knew of Wilhelm Roentgen's discovery of X rays and their power to reveal the bones in a human hand. He and others at that time suspected that this power was related to fluorescence. This possible relationship motivated Henri Becquerel of Paris to conduct studies of fluorescence for different materials. Acting on a suggestion by Poincaré, Becquerel wrapped photographic plates and exposed them to uranium that had been exposed to the sun. For several overcast days during that period, the uranium lay on the wrapped photographic plates. Becquerel subsequently developed the plates and to his amazement they were exposed by the uranium even though the uranium had not been exposed to the sun. He studied fluorescence and phosphorescence using the hypothesis that the emission of X rays was involved in each. Fluorescence refers to prompt emission of light following excitation, whereas the emission of light after excitation has ceased is referred to as phosphorescence. An atomic decay time of approximately 10^{-10} s is often used as the demarcation between fluorescence and phosphorescence.

Becquerel used several materials and reported in February 1896 on his studies with the double sulfates of uranium and potassium, both being phosphorescent substances.[7] In his experiments he laid the substances on top of a photographic plate well wrapped in black paper and exposed them to the sun. When the photographic plate was later developed, it revealed the silhouette of the phosphorescent substance. In March, he reported that the exposure to sunlight was unnecessary. By May, he was able to report that after the materials were shielded from light in his darkroom for two months, the photographic plate was still exposed without perceptible weakening of the image.[8]

There is a general lack of prehistory in the field of radioactivity as there are no recorded accounts of anyone coming close to discovering the phenomenon of radioactivity or of discovering closely related phenomena. The history developed in spite of the fact that the mysterious penetrating rays from uranium did not give pictures of bones as X rays did and thus were not as interesting to scientists for

---

* General references for more reading about the history of radioactivity are given in Reference (6).

practical applications. Perhaps it was the inexplicability of the rays from uranium that attracted the interest of Marie Curie and others.

2.1.4 Early Studies

Gerhard C. Schmidt of Erlangen and Marie Sklodowska Curie of Paris (see Figure 2.1) independently discovered in 1898 that thorium also exhibited radioactive powers and verified the existence of the phenomenon. Also in 1898, Marie and her husband Pierre went on to discover two other substances with the same properties and called them polonium and radium. Their paper was the first to use the word radioactivity. Polonium was named after Marie's native Poland. Together with Gustave Bémont, they discovered radium and named it from the Latin word "radius," meaning ray or emitting rays.[9] Because of their similar properties, they first referred to radium as radioactive barium or eka-barium.

In 1898, shortly before he went to McGill University, Ernest Rutherford (see Figure 2.2) showed that uranium rays could not be reflected, refracted, or polarized and thus were not electromagnetic in nature. He further showed that there were two components of these uranium rays. While one component could scarcely penetrate matter, the other had considerable penetrating power. For convenience he named these components alpha and beta. The intensely penetrating gamma rays also emitted were found the following year by Paul Villard of Paris. By 1900 Henri Becquerel had shown that the beta component was an electron. Ernest Rutherford and Frederick Soddy, in 1903, demonstrated that the alpha component was a helium ion (we now know it is the nucleus of a helium atom).

It was discovered in 1899 by J. J. Thomson and Ernest Rutherford at Cambridge University that uranium rays could split molecules in a gas to form ions. If an electric field is applied to a chamber that contains ions created by radioactivity, a current is produced whose saturation level is proportional to the intensity of the incident radiation. In 1899 the Curies coined the word radioactivity to describe the spontaneous emission of penetrating and ionizing rays. It was Rutherford who showed that uranium produced rays which ionize in the same way that X rays do.

In general, the scientist around the turn of the century had two basic instruments with which to detect the ions created by the newly discovered rays from uranium and other radioactive materials. The two

FIG. 2.1. Marie Curie in her laboratory. Photograph provided by the Niels Bohr Library of the American Institute of Physics, 335 East 45th Street, New York, NY 10017.

HISTORY AND USES

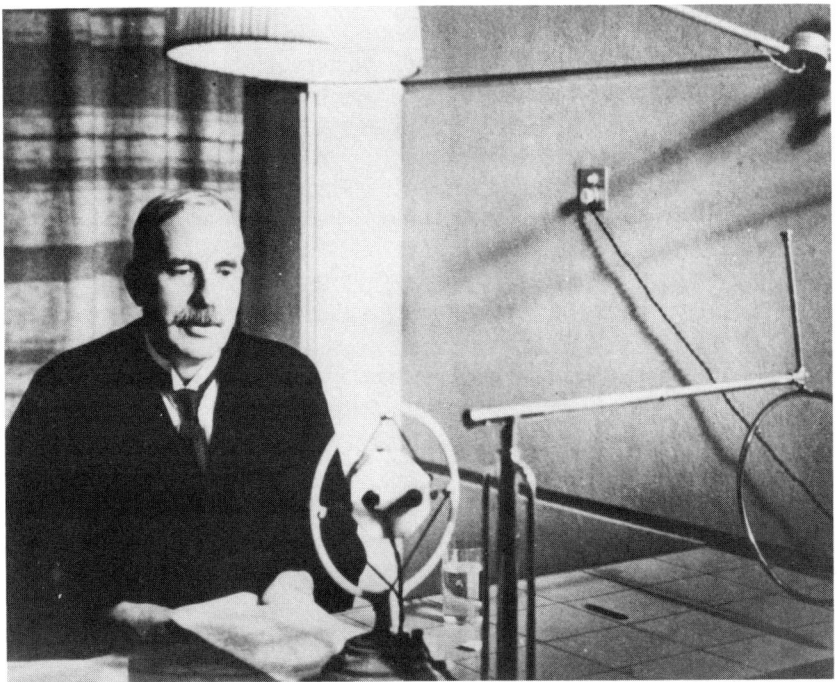

FIG. 2.2. Ernest Rutherford. Photograph provided by the Niels Bohr Library of the American Institute of Physics, 335 East 45th Street, New York, NY 10017.

classes of instruments were the quadrant electrometer and the electroscope; the latter often had a gold leaf. When radioactive substances were brought close to the charged plate of a gold-leaf electroscope, they caused the rapid collapse of the gold leaves. Up to about 1914 both instruments were widely used, but the operational difficulties of the quadrant electrometer caused it to fall from favor. Since spectroscopic methods are about 5000 times less sensitive than electrometric ones, it might have taken much longer to discover radium were it not for the electrometer. Because of the sensitivity of the electrometer, Rutherford threatened the severest penalties for anyone who allowed emanation to escape into the laboratory, since this would lead to contamination from the progeny.[10]

Figure 2.3 shows a quadrant electrometer as advertised in 1911. These electrometers generally had eight quarter-circle metal plates arranged on insulating rods that provided an upper and lower set of four plates. A thin aluminum needle was suspended between these sets

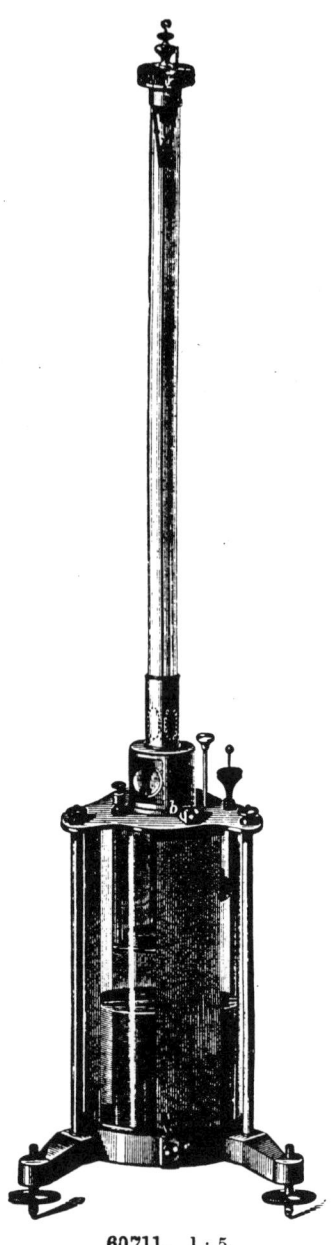

60.711. 1 : 5.

FIG. 2.3. Quadrant electrometer as shown in a catalogue circa 1911 with the following description. Thompson-Weinhold's with replenisher: Maxwell vane, all parts heavily gilt; with mirror, and arranged for subjective as well as objective reading, with amber insulation. The catalogue was from Max Kohl A. G. Chemnitz (Germany) and the copy was provided by the Niels Bohr Library of the American Institute of Physics, 335 East 45th Street, New York, NY, 10017.

HISTORY AND USES

of plates on a fine wire and the entire apparatus was housed in a glass chimney. Above the needle was a small mirror which would deflect light rays to indicate how much the needle had rotated. Ions from the radioactive decay were applied as an electric current to the electrometer plates, changing their potential and so causing a deflection of the needle, which was charged from an external battery. Hence the deflection was proportional to the radioactivity present.[11]

Ernest Rutherford, a master of the experimental technique, had at his disposal instruments like the gold-leaf electroscope and electrometers. An electrometer detects an amount of electricity corresponding to about 10^{-18} g of hydrogen. Since there are 6×10^{23} molecules of hydrogen in a gram, at standard temperature and pressure, this is an extremely small amount of mass and shows how sensitive the measurement methods were. Thus, although crude by today's standards, the electrometer was a highly sensitive instrument and the primary instrument available for studying radioactivity in the early days.

Marie Curie tested all of the known elements for radioactivity and found that only uranium and thorium were radioactive. She found that the activity of the uranium mineral, pitchblende, is four or five times as great as one might expect it to be from its uranium content. Indeed, after the uranium was extracted, powerful radioactivity remained. As she had tested all the elements then known, it could be concluded that this residual activity must have been due to an unknown element or elements. Four years of work was needed to extract a decigram of the unknown material from tons of pitchblende residues.[12]

2.1.5 Thorium Series and Thoron

The properties of thorium seemed to be more difficult to investigate than those of uranium and radium. When Rutherford took a post at McGill University in Montreal in the fall of 1898, he passed this problem on to Robert Owens, a Professor of Electrical Engineering at McGill. Owens had an arrangement to spend time at Cambridge working under J. J. Thomson, and his work with Rutherford was preparation for his studies at Cambridge. Owens conducted several careful experiments and, in the process, encountered situations where the activity seemed to be varying erratically as the levels of ionization shifted unexpectedly from one level to another. He traced this problem to air currents in his apparatus, and when he closed it up in a tight box, he was able to stabilize the ionizing effect of the thorium. He noted that if he blew air through his apparatus, the ionization dropped sharply and did not recover until his sample had been left for several minutes. It is likely that Rutherford decided in May or June of 1899 that emanation (now we call it thoron or ^{220}Rn) was the answer. He took the summer to prove this and, as was his style, intuition was followed by complete and thorough proof. He announced the discovery of emanation on September 13, 1899.[3]

Ernest Rutherford's two papers concerning work in 1898 and 1899, published in early 1900, are classics in scientific experimentation.[13,14] He and his colleagues became renowned in the scientific world because of their enthusiasm, miraculous intuition, and visionary interpretation of the meager and conflicting information gathered on what we would today view as very limited equipment.[3,15] These early publications show how carefully and meticulously Rutherford and his colleagues approached their studies. At one point Rutherford noted that thorium compounds gave out a

> type of radiation similar in its photographic and electrical actions to uranium and Roentgen radiation. In addition to this ordinary radiation I have found that thorium compounds continuously emit radioactive particles of some kind, which retain their radioactive powers for several minutes.

He called this new source of radiation "emanation" as he was unsure of its character and properties. The definition found in Webster's unabridged dictionary of emanate is "to come out from a source" or "to give out." An emanation means either "the action of emanating" or "the origination of the world by a series of hierarchically descending radiations from the Godhead through intermediate stages to matter." It is not clear which of these definitions Rutherford intended. It seems most likely that he was just unsure of the nature of this new radioactivity and wanted to select a description that was vague enough to fit the character that was eventually determined.

In measuring the properties of emanation, Rutherford got a surprise—its radioactivity decreased in geometrical progression with time. The radionuclides observed up to that time had had such long half-lives that the phenomenon of decay had not been discovered.

Marie Curie had, by 1901, discovered that radiation was emitted from the samples of radium and thorium. She felt that the gas produced when radioactive materials were heated was radioactive and that even without heating, it was escaping into the air from the samples of radium and thorium.[16] She attempted to see the spectrum of a gas in a Geissler tube but was not able to, probably because too few atoms were present. She reported in her D.Sc. thesis that a rare-gas spectrum was seen later by Eugene Demarcay [see reference for thesis in Reference (6)]. It is generally thought that the first correct spectrum of radon emanation was seen by Rutherford and Royds in 1908.

2.1.6 Other Properties of Radioactive Series

Ernest Rutherford compared the emanation "property" of different thorium compounds and found thorium oxide to have the greatest. This physical property is the ability of the compound to emit the emanation and is not related, as we know now, to the compound's radioactive characteristics. Rutherford investigated several properties of this newly discovered emanation. He showed that it could pass through several thicknesses of paper and diffuse through the gas in its neighborhood. He was able to measure the half-life of the emanation and found it to be about one minute. In showing that the emanation passes through a plug of cotton wool without any loss of its radioactive powers and that it is unaffected by bubbling through hot or cold water or through weak or strong sulfuric acid, he had demonstrated that it behaves like an ordinary but inert gas.

After the discovery of the emanation from thorium, Ernst Dorn in Germany discovered that radium produced an emanation similar to that given out by thoria (which is what Rutherford called the thorium compound that he was studying and that is now known to be thorium oxide). Dorn read his paper at Halle, Germany, on January 20, 1900.[17] However, Dorn noted that the emanation from radium retained its radioactive power much longer than that from thorium and that the presence of moisture in air increased the emanation capacity. He also incorrectly reported an emanation from polonium.

In 1901 Rutherford and Harriet Brooks performed an experiment to determine if the emanation from radium was a vapor of the substance, a radioactive gas, or particles of matter that contained a large number of molecules.[18] It was clear by that time that the emanation caused the glass tube which contained it to phosphoresce. If left for some time, the emanation blackened the glass tube. The investigators further noted that the emanation caused every substance in the vicinity of the radioactive source to become radioactive for several days. To determine what the emanation of radium was, Rutherford and Brooks measured its diffusion into air and compared the coefficient of diffusion with that for other gases and vapors. They concluded that the emanation was a radioactive gas with molecular weight in the range of 40 to 100.

They had to guess at the weight, because it was higher than any of the others with which they had to compare.

In the midst of these experiments, Rutherford found another temporary radioactivity on the nearby bodies, and again downwind from the thorium source. He called this newly discovered phenomenon "excited radioactivity" and he determined that its activity dropped to one-half of its initial value in 11 hours.[14] By 1905, Rutherford was calling the progeny of the excited radioactivities of radium, thorium, and actinium the active deposits.[20] The "excited radioactivity" of radium had been discovered by the Curies a few months earlier. Rutherford showed that this excited radioactivity was strongly attracted to a loop of negatively charged wire, thus showing the activity was in the form of positive ions.

Rutherford showed that the emanation passed through metals if they were sufficiently thin. In 1901 Rutherford and Soddy further examined thorium oxide and found that by heating it, they could enhance its emanating power and with intense heat they could cut off the release of the emanation.[19]

Soddy tried a series of carefully planned chemical reactions and, being unable to get emanation to react with anything, he thought that it could be a member of the family of inert gases which William Ramsay had discovered during the previous seven years. Soddy and Rutherford concluded from their studies that the emanation behaved like a rare gas of the argon family and that it was not thorium but a progeny, thorium X (what we now know is ^{224}Ra), that was releasing the gas.

Henri Becquerel and Pierre Curie in 1901 got radiation burns, but radium was also used for therapeutic purposes such as curing deep-seated and skin cancers. For stomach cancer one could drink "liquid sunshine" to bathe the affected parts of the body.[21]

Besides using labels like radium A, B, C, and X, thorium A, B, C, and X, etc. (see Tables 1.2, 1.3, and 1.4) in the early days of research on radioactivity, other names were coined. For example, radium emanation was sometimes called niton by Ramsay to go with the rest of the noble gases, emanation was called emanium, and the parent of ^{226}Ra was at one time called ionium. The *nit* in niton comes from the Latin meaning shining or the property of emitting light. The suffix, -on, was used to place it in the noble gas series of the periodic table. It was not until about 1925 that the name radon began appearing in the literature. Historically the decay products of radon have been referred to as "radon daughters" and this label still persists in some of the literature. However, most modern literature uses the label "radon progeny" to describe the decay products of radon.

2.1.7 Transmutation

The chemical separation of thorium X was formally reported in 1902 and was a result of chemical separations by Soddy and Rutherford.[22] Soddy was recruited by Rutherford from McGill's Chemistry Department where he was a junior member. It was in 1902 in a paper that they reported the explanation of results that had been puzzling in the previous investigations. They showed that the residual liquid from a thorium solution from which the thorium had been precipitated as the hydroxide possessed about as much emanating power as the original solution, whereas the thorium hydroxide was more or less free from emanation. In a section entitled "General Theoretical Considerations" at the end of the paper, they pondered the possibility of atoms "breaking up" and the "production of new matter." This was the first realization that they were dealing with transmutation of the elements.

Earlier the Curies had criticized Becquerel for suggesting the idea of transmutation or spontaneous atomic disintegration but now Soddy and Rutherford could reach no other conclusion from the results of their studies. Soddy at this time stated that "alchemy must be regarded as the true beginning of the science of chemistry. Transmutation has always been the real goal of the chemist."[22] In 1902 this was a revolutionary hypothesis, but by 1904 it was well established and used as an intellectual tool by Rutherford to explain the growth and decay of radioactivity. Rutherford was awarded the Nobel prize for chemistry in 1908 and was kidded about his transformation from physicist to chemist.

2.1.8 More on Characteristics

Ernest Rutherford and Frederick Soddy's paper on radioactive change was published in 1903 and states that

> in all cases ... the law of radioactive change ... may be expressed in one statement—the proportional amount of radioactive matter that changes in unit time is a constant ... the constant ... possesses for each type of active matter a fixed and characteristic value.[23]

The analysis and notation of this paper are used today (see Section 1.4).

The actinium emanation, called actinon, was discovered independently by Friedrich Giesel and André Debierne in 1904.[24] Debierne measured its half-life at 3.9 s. Giesel called it emanium and Debierne called it actinon. A year later, the two were shown to be the same substance.

In 1903, Rutherford and Soddy measured the half-life of the radium emanation as 3.7 d (now known to be 3.823 d) and for the active deposit found a half-life of less than one hour. We now know it to contain several progeny with half-lives in the range of microseconds to years.

When it arrived in 1903, Rutherford and Soddy put McGill's new liquid-air machine to use immediately, condensing the emanations and further showing their properties as a gas. They also separated the concept of the rate at which emanation decayed from the speed at which the emanation could escape, which they had been referring to as "emanating power."[25] The first determination of the emission spectrum for the radium emanation was made in 1904 by Ramsay and Collie.[26] They obtained the visual observations of the wavelengths of 11 lines and stated that the spectrum was similar to that observed for the other rare gases.

Soddy suggested as early as 1903 the use of radium emanation for treatment of tuberculosis patients. It was felt then that the four-day half-life was short enough that the substance would soon be harmless. Emanation soon became a fad used for several complaints (see Section 2.2). Spray devices and masks were used and "emanatoria" and "inhalatoria" were built. Emanation baths were popular for respiratory diseases. Water impregnated with emanation was given by mouth and injection. One of the reasons behind the public enthusiasm for emanation may have been the perception that uranium mine workers never suffered from rheumatism or gout. In the years preceding World War I the emanation craze reached its peak—just when evidence was mounting that the miners' major malady was cancer of the lung.

In a series of experiments on induced activity in air and on the electrical conductivity produced in gases when they pass through water, J. J. Thomson in 1902 discovered radon in water.[27] Led by this report, H. F. R. Von Traubenberg in 1904 demonstrated that the radium emanation was contained in the tap water of his native Freiburg, Germany.[28] Independently and on a suggestion from J. J. Thomson, Henry Bumstead and Lyndie P. Wheeler at Yale found the emanation of radium in the surface and ground waters near New Haven, Connecticut. They showed that the gas had a half-life of about four days.[29] Other measurements at that time showed abnormally high concentrations in spring water at Hot Springs National Park, Arkansas. The researchers of that period were unable to find the emanation in rainwater, although we now know that rainwater contains measurable amounts of radon and the active deposits of radium and thorium.

It was not until 1908 that Rutherford and Hans Geiger were able to prove that Rutherford's intuition was correct and the alpha particle was a helium nucleus. The electrical method Geiger developed for this experimental work in 1908 is what we call the "Geiger counter."

Soddy estimated the atomic weight of the radium emanation by liquefying it at low temperatures, putting the weight at between 40 and 100 based upon work of Rutherford and Brooks. Bumstead and Wheeler at Yale put it at 180; we now know it to be 222. It was not until 1913 that Soddy realized that the emanations of radium, thorium, and actinium could have different atomic weights. Rutherford and his team had already developed the concept of atomic number in 1910.

Charles E. S. Phillips was a founding member of the Roentgen Society of London and was credited with being the first radiological physicist. In 1906 he visited Rutherford and attempted to interest him in a committee he was trying to get together to look into the possibility of setting a standard for radioactivity.[3]

Rutherford went as Britain's official representative to the international congress, which met in September 1910 to set up a radium standard. The congress was poorly organized and developed into a public shambles, although the participants did agree to call the unit of radioactivity the curie. They did appoint an international committee to develop the official international standard for radium. The committee consisted of Bertram Boltwood, Marie Curie, Debierne, Eve, Geitel, Hahn, Meyer, Rutherford, Schweider, and Soddy. Their three decisions were (1) that Madame Curie would prepare a radium standard containing about 20 milligrams of the substance, (2) that this standard would be kept in Paris and would be only used for comparisons with national secondary standards while it remained under control of the committee, and (3) that the international unit of radioactivity would be based on an equilibrium of radium and its emanation. This standard was formally accepted by the international committee in the spring of 1912 and installed at the International Bureau of Weights and Measures.

2.1.9 Group Displacement Laws

By 1907 the study of radioactivity was considered mysterious and wondrous enough to be described as the "chemistry of phantoms."[30] Because of the small quantities involved in the research and the transitory nature of radioactivity, the traditional chemists were reluctant to place these newly discovered "elements" into the periodic table. An additional problem was that by this time, over thirty radioelements were known. No more than twelve boxes existed in the periodic table in which to house them. Rutherford and Soddy's theory of transmutation was already well established, and in 1908 the measured atomic weight of radon put it near radium in the periodic table. In 1911 Soddy proposed that when a radionuclide decayed, it moved two boxes away from the parent and generally in the direction of lower atomic weight, but possibly in the opposite direction. The characteristics of the radionuclides that emitted beta particles were too uncertain at this time to determine how they fit into the periodic table. Kasimir Fajans, one of Rutherford's team at Manchester, finally described how the radionuclides fit together by the group displacement laws that he developed and described in a series of papers in 1913.[31] He deduced the correct relationships by organizing all the evidence from electrolysis and from dipping metals in solutions that he and his colleagues had generated. Upon analysis of this data he deduced that alpha emission left the progeny more electropositive; that is, it went into solution as a positive ion. Beta emission produced progeny that were more electronegative or nobler and went into solution as negative ions. Thus Fajans developed the rules that the electropositive transitions of alpha decay involved a move two boxes to the left in the periodic table and that beta decay involved a move one box to the right. With these rules he was able to place the known radionuclides into the periodic table. Although others were close, the final, correct interpretation of the group displacement laws belongs to Kasimir Fajans alone.

By 1913 Rutherford, in his book,[6] was able to describe in detail much of what we now know about the properties of the emanations, including that they only emit alpha particles, in some cases of different energies, and that they condense around $-150°C$, are absorbed by water according to Henry's law, and are strongly absorbed by coconut charcoal.

2.1.10 The 1920s, 1930s, and 1940s

By the end of World War I, the chemistry of natural radioactive elements (radiochemistry) had been worked out and was virtually complete. The physical side of this area of endeavor evolved into what we now call atomic and nuclear physics. Thus, in the first two decades of this century the science of radioactivity was born, grew, and had established a place in the history of science.

In 1932 Robley Evans demonstrated, in his 1.5-L ionization chamber, that the lower level of detection of radon was at a background statistical variation value of 2×10^{-12} Ci (0.074 Bq). The device he developed for this determination combined a double ion chamber with a string electrometer. By 1935 this approach could determine radon concentrations down to 0.1 pCi L^{-1} (3.7 mBq m^{-3}).[32,33] By 1940 a simple collection apparatus had been developed for the determination of thorium in air by collecting the active deposit.[1]

Analytical measurement of radon improved considerably in 1943 with the substitution of a low-grid-current vacuum tube for the electrometer.[34] By 1953 the lower limit of detection was pushed to 5×10^{-15} Ci L^{-1} (0.18 Bq m^{-3}).[38] In 1957 the "Lucas cell" was developed using the alpha-scintillation properties of silver-activated zinc sulfide on the inside of a steel shell with a quartz window.[35] This is the cell commonly used today.

Studies were conducted exposing young rats continuously in a emanatorium containing 2000 mache units (over 7000 WL at equilibrium). Not surprisingly, the rats became distinctly ill after four weeks and died in the following one or two weeks. The researchers concluded that the animals succumbed because of external radiation.[36,37]

Even by 1932 it was not yet clear that radon was causing lung cancer in miners. In a review of the Jachymov miners dying in that period, it was concluded that radon was "the most probable cause of the tumors," but "this question, however, requires further investigation."[39] It was not until the 1940s that a cause–effect relationship was firmly suggested between lung cancer and radon.[40] The toxicity of radon waters was first suggested in 1932.[41]

The experience of the radium watch dial painters in the 1920s led to serious endeavors to decrease the danger involved in handling radioactive materials such as radium and thorium. Efforts included monitoring radon and thoron in expired air with an electroscope.[42] Also methods were developed that reduced the time of exposure in the purification process such as used in hospitals.[43] By 1940 there was enough experience with radon to recommend 10 pCi L^{-1} (370 Bq m^{-3}) as a safe working concentration in plants, laboratories, and offices.[44] This value was incorporated in Report 5 of the National Council for Radiation Protection and Measurements (NCRP) in 1941 (then called the U.S. Advisory Committee of X-Ray and Radium Protection) and adopted by the International Commission on Radiation Protection (ICRP) in its Committee II report in 1959.

Radon was used as a source of alpha radiation in the study of the effect of radiation on chemical reactions.[45] It was used because, as a gas, it could be introduced in small volumes in the interior of many systems and its loss was not a big problem because of its short half-life. The chemical reactions studied included oxidation, hydrogenation, decomposition, polymerization, and condensation in gaseous systems.

In his Nobel prize-winning research, Enrico Fermi selected for his source of neutrons the reaction of alpha particles from radon with beryllium powder. He used radium as the source of radon. He and his team at the University of Rome in 1934 bombarded the elements in the order of their appearance in the periodic table. When they reached uranium, they found what happened to be a different kind of radioactivity. However, they failed to realize that they had fissioned the uranium atom—that remained for Otto Hahn and F. Strassman five years later. Later these same radon-beryllium sources were used to initiate fission in nuclear piles.[46]

2.1.11 Recent Events

A renewed interest in the environmental consequences of radon in indoor air started in the New England states due to its discovery in the water used for drinking and other purposes in houses and other buildings.

The rediscovery of the radioactivity in well water occurred by accident in Raymond, Maine, shortly after Christmas, 1957.[47] A 13-year-old boy was testing a portable survey meter used for prospecting and observed a high count rate when holding the beta-gamma probe close to a pressure tank in the water system at his father's business. His father immediately recognized this as unusual and after reproducing the high activity reading, reported this condition to the State of Maine, who requested the U.S. Public Health Service to investigate.

Starting in April 1958, over 45 wells in the Raymond, East Raymond, and North Windham, Maine, areas were analyzed by the Public Health Service and the State of Maine. These wells varied in depth from 30 to about 600 feet. The activities found varied from 2500 to 583,000 pCi L^{-1} (93 to 21,600 kBq m^{-3}) total radon activity plus that of the progeny and the long-lived alpha particles varied from 0 to 666 pCi L^{-1} (25,000 Bq m^{-3}).

A nationwide interest in indoor levels of radon was created by the discovery of high levels of radon in homes in the Reading Prong area

HISTORY AND USES

of Pennsylvania, New Jersey, and New York starting in 1985.[48] The discovery of radon in homes in this area began with an employee named Stanley Watras at the Limerick nuclear-power plant in eastern Pennsylvania.* He had been setting off the radiation portal monitor upon leaving the building. Since the reactor had not yet gone critical, there was only minimal concern about these warnings. One day, on an impulse, Watras turned upon entering the building and went through the monitor. The alarm went off, showing that he was contaminated just as he had been when he left the plant on other days. Watras asked the utility to investigate the source of this contamination, and they hired a contractor to trace the source of the radiation. When they monitored the Watras home they found the source. The levels of radon were in the range of 16 to 20 WL.

Once the utility realized that the problem was in the Watras home, they called the State Department of Environmental Resources in Harrisburg, Pennsylvania, on December 19, 1984. Also notified was the U.S. Environmental Protection Agency's Office of Radiation Programs. Scientists from the state gathered up monitoring equipment and went to the Watras home in Boyertown (in Colebrookdale township) on December 26. The family left the house for the Christmas holidays and the measurements began. The background level outside the house was approximately 300 mrem y^{-1} (3 mSv y^{-1}). At the rafters of the basement the radiation level was about 900 mrem y^{-1} (9 mSv y^{-1}). Alpha track detectors revealed levels of 2700 pCi L^{-1} (100 kBq m^{-3}) in the basement, roughly 1000 pCi L^{-1} (37 kBq m^{-3}) in the air of the upper floors of the house, and about 100 pCi L^{-1} (3.7 kBq m^{-3}) in the attached garage. The radioactivity was from radon, and no amount of thoron was found. Measurements using the Kusnetz method showed levels in the range of 12 to 22 WL.[49]

On December 28 the state contacted the township to report that this was not an individual problem but a general public health problem. On January 5 the Secretary of the Pennsylvania Department of Environmental Resources wrote a letter strongly urging the Watrases to leave their home.

The soil samples in the area averaged approximately 1 pCi g^{-1} (0.037 Bq g^{-1}) although some were as high as 20 pCi g^{-1}), but this is not out of the ordinary. The septic system showed the same levels as the basement, and this ruled out building materials as the source. A

---

* The discussion concerning the Watras home is based on information obtained by the author in an interview with Margaret A. Reilly of the Pennsylvania State Department of Environmental Resources, Harrisburg, PA.

survey of other homes in the area showed a few as high as 1 WL. The house next to the Watrases was about 0.006 WL.

The source of the radon is a granite that is enriched with uranium and has high permeability. This granite formation is called the Reading Prong and it extends under eastern Pennsylvania, northern New Jersey, and southern New York.

Pennsylvania, where the radon levels appear to be the highest in the Reading Prong, has provided free radon monitoring to the residents of Berks, Bucks, Lehigh, and Northamptom counties. Of the 11,590 homes in this survey approximately 60% exceeded 0.02 WL, about 13% exceeded 0.1 WL, and 0.7% exceeded 1 WL. These data include measurements made with alpha track detectors and those using the Kusnetz survey method.[49]

2.2 USES

2.2.1 Medical Applications

Shortly after radium was discovered, its medical applications began. These applications were both external and internal to the body. Since radium decays to radon and then to its progeny, medical applications of radium are clearly related to those of radon. The discussion here will focus primarily on those applications of radon and its progeny. [For more information, see Reference (50)].

In medical applications, radon has some advantages over radium. A single radium source can generate numerous radon sources, and the radon sources have a higher specific activity as a gas can occupy a smaller volume. The half-life of radon is relatively short and this allows permanent implantation. There is less chance of loss of the more expensive radium source because it is handled less.

Radon was first used medically in 1914, with the collaboration of John Joly, Professor of Geology and Mineralogy at the University of Dublin and Walter C. Stevenson, a Dublin surgeon and radiologist.[51] They treated several malignant and nonmalignant conditions. By 1920 radon had largely replaced radium for use in most larger hospitals. It was used in the treatment of numerous diseases through the 1930s and into the 1940s.

Radon encapsulated in gold seeds has been used for medical purposes primarily for the treatment of malignant tumors. In general,

HISTORY AND USES

it has been used to treat patients with two different types of localized cancers by permanent interstitial implantation. One category consists of patients with relatively small accessible cancers that can be treated on an outpatient basis, inserting the radon seed under local anesthesia. The other major category includes intrathoracic or intraabdominal tumors that cannot be removed because of local extension or because removal would cause severe functional disability. Radon seeds have also been used in treating dermatological, bladder, oral, and other types of cancer.

In the 1930s and 1940s radon seeds were used for dermatological disorders such as vascular and neoplastic tumors and acne. Because of the short half-life of radon, its radioactivity is considerably diminished after a month. However, one seed removed after 33 years was estimated to have given the patient approximately 11 rads for the ensuing 32 years after implantation, primarily from progeny.[52]

For some tumors many radon seeds are implanted. For example, in treating an infiltrating carcinoma of the urinary bladder, as many as 20 to 30 seeds are needed. Experience with 146 patients with this affliction revealed that after excision and electrocoagulation of the tumor, the insertion of radon seeds led to a five-year survival rate of 21% and a cure for 8%.[53] Other examples of implanting multiple seeds were for patients having cancer of the prostate or lung or metastatic disease of lymph nodes of the neck. Radon seeds were also used as an adjunct to surgery of certain oral and nasopharyngeal tumors.[54,55] The decision to use radon seeds depends on the location, size, origin, and sensitivity of the tumor.

Radon seeds can be left *in situ* because of their short half-lives, whereas needles of longer-lived radionuclides such as radium must be removed. Since radon emits no gamma rays, it is not directly the source of radiation. Its progeny ^{214}Pb and ^{214}Bi are the radionuclides that supply most of the gamma radiation. However, for other reasons such as cost, availability, and practicality, other radionuclides are preferred for implants.[54] During the 1950s and 1960s a large share of the market that had been the domain of radium and radon products was taken over by radionuclides made in accelerators and nuclear reactors. Among those widely used are cobalt-60, cesium-137, tantalum-182, iridium-192, and gold-198.[56]

2.2.2 Earthquake Prediction

Of the many geochemical parameters that provide predictive information about movement in the earth's mantle, radon is the most useful and the most studied. Earthquake prediction is a relatively new field and has been the subject of three recent international meetings and two journals.[57] Studies of examining radon movement as a possible predictor of earthquakes have occurred primarily in Russia, the Republic of China, and the United States.

Radon emanating from soil and measured in groundwater would seem to be a good indicator of crustal activity such as earthquakes. Such movement would be expected to change the radon levels. However, the current literature describing the possible correlation between radon levels and earthquake activity uses such qualifying and cautious words as possible, apparent, limited, could sometimes, may be, and suggestive. It is clear that in some cases there are precursor changes in radon levels, but that the causal relationship or mechanism relating these to earthquake activity is not understood.

The measurements of potential earthquake precursors involve radon from groundwater in deep wells of a few thousand meters or from soil within one meter of the surface. The measurements span time periods of months and sometimes years and are usually done in the vicinity of a fault line or area of high seismic activity. The earthquakes studies are generally of a magnitude greater than 4 on the Richter scale, although some possible correlations have been reported for smaller magnitudes. Figure 2.4 shows a radon concentration spike preceding a Chinese earthquake.

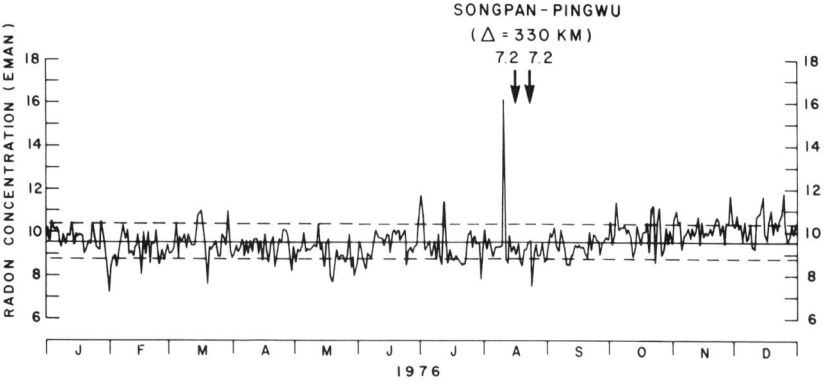

FIG. 2.4. Time series radon concentration at Guza station in China, showing an elevated value before earthquakes of 7.2 on the Richter scale. Reproduced by permission from C. Y. King, Radon studies in China, *Earthquake Predict. Res. 3*, 47–68 (1985).

HISTORY AND USES

An earthquake is the failure of a rock under the stress of an applied force and results in a shock. The corresponding changes in radon levels can be long-term or can occur just before the earthquake. The long-term or bay-form is a gradual change whereas the short-term or peaked change is sharper.[58] A long-term or several years buildup of radon levels has been observed such as before the 1966 Tashkent earthquake. Some short-term anomalies have been seen in Japan and China. In some cases, an anomaly in radon concentration is seen within a day or so of the event.

Radon has a mean diffusion length (the average distance it moves by diffusion) in dry porous soil of about one meter and a few centimeters in groundwater, and thus a limited ability to migrate due to its short half-life. An earthquake may therefore either change the rate of release of radon and thus change the convection gradient at shallow depths or the extra pressure on the pore volume may cause an increased emanation. On the atomic level the theory is based on the recoil energy of the radon progeny of about 100 keV. These atoms travel through hundreds of crystalline lattice sites and often get lodged in a microcapillary in the mineral structure either in the aqueous or air medium.

Several complications arise in trying to relate radon levels with earthquake activity. Rain contains radon washed out of the atmosphere and can change levels at the surface and in groundwater. Pressure and temperature affect the emanation levels of radon and thus introduce a seasonal variation. The local geological structure also is a factor as is the conductivity through it. Additional complications include tidal motions and other general seismic events. In spite of all the problems, it has been claimed that anomalies in radon levels have been detected tens to hundreds of kilometers from the epicenter of a subsequent earthquake. It is predicted that they could be measured as far away as 4000 km for an $M = 8$ (Richter scale) earthquake.

• Earthquakes clearly need to be predicted both in location and magnitude. The measurements in the past have brought some successes and some failures for radon changes as an indicator. The equipment is costly and measuring is tedious, but the rewards could be significant if many lives can be saved by predicting an earthquake.

2.2.3 Atmospheric Transport

Radon and its progeny have been used to study atmospheric vertical advection, residence and transit times of atmospheric molecules over land and oceans, and the boundaries of continental air masses. It is an ideal tracer for atmospheric transport because of its 3.8-day half-life, which is its only significant sink. The sources of radon are primarily over land masses, the oceans being a very minor source. These lifetime and source characteristics are of the same order as those of many air pollutants such as NO_x, CO, methane, and many of the moderately reactive hydrocarbons. Thus radon is a good surrogate for studying the transport of atmospheric pollutants.

Radon has been used to study the vertical transport of pollutants between the regions near urban centers and the free troposphere. It has been shown that the concentration of radon decreases less with height in the troposphere in the summer and that upward mixing in the mid-troposphere is efficient. This is reflected in eddy diffusion coefficients that show low values at ground level, larger values in the mid-troposphere, and a decrease as the tropopause is approached.[59]

The relative amounts of ^{222}Rn and its progeny ^{210}Pb and ^{210}Po are indicative of the residence and transit times over land and oceans[60,61] and the exchange between stratosphere and the troposphere.[62] Radon has been used to show the general flows of air in North America[63] and India.[64] In the latter study, the data suggested that continental air was introduced into the lower current from above.

Other isotopes of the gas have been used to study atmospheric transport. Because of its short half-life of 55 s, thoron has been used to study mixing of air at the earth's surface. It has also been used to study the atmospheric electrical characterisitcs within the first meter of the earth's surface.[65]

2.2.4 Mines, Spas, and Questionable Medical Applications

In the early days of the study of radiation, it was immediately clear to all that X rays were useful in diagnosis. However, other applications were proposed that bordered on the bizarre, including the use of X rays as a cure for criminal behavior and for inducing a dog to salivate by projecting an image of a bone onto the animal's brain.[66] Similar applications were suggested for the use of radioactivity, including placing radon emanators near livestock water tanks, although the purpose is not clear.

The early days also produced its share of quacks and charlatans. Included among those verified cases of quackery were many which claimed to cure cancer, including Rupert Wells' Radol, the Chamlee

HISTORY AND USES

Cancer Cure, the Radio-Sulfo Cancer Cure, a radium healing balm, and radium oil.[67] In the early 1920s a toothpaste with radium that emanated radon was marketed by Allgemeine Radiogen AG of Berlin. It was called Radiogen and was supposed to prevent the formation of dental plaque.[68] Luckily, most of the quacks of this period were simply crooks and use of their product involved little exposure to radiation.

Radon-charging devices, often called emanators, were produced in the 1920s and 1930s. These were often crocks into which water was introduced and allowed to sit for a few hours or days. These crocks were made from rock containing uranium or radium, and the water would thus contain uranium, radium, and radon. The radon levels in these devices have been shown to be as high as 180,000 pCi L^{-1}.[69] The Denver Radium Service (DRS) offered a Radiumactive Vitalizer as late as the 1940s to provide "radiumized, gas-charged health water." Figure 2.5 shows a Vitalizer similar to that developed by DRS, who used a photograph showing the response of narcissus bulbs to sprouting water from their vitalizer jar.

Radon, too, has had its share of questionable medical applications. For example, the Denver Radium Service[70] in 1914 claimed that radium emanation was a diuretic, a laxative, a stimulant, and a "goad" to the vital processes. They also claimed that it cured rheumatism, neuritis, toothache, hypertension, myocardial affections, insomnia, hay fever, lumbago, malnutrition, and many other afflictions. As late as the 1940s they were marketing a number of products for internal use that contained radium and radon, including tablets, suppositories, and ointments.[71] Also in the 1940s radon was proposed as a cure for deafness, and when introduced into petroleum jelly as an ointment, it was proposed for use in the treatment of radiation injuries and necrosis.[72] It was not until 1949 that the American Medical Association reported that a radon salve, Alphatron, was unacceptable.[73]

By the late 1920s and 1930s several groups (including the Federal Trade Commission, the Better Business Bureau, and the American Medical Association) had joined the campaign against fraudulent and hazardous radioactive remedies and devices. These activities, the bone cancer cases appearing among the radium watch dial painters, and the Depression all probably contributed to declining sales of products using radium and radon in the 1930s.[69]

In 1938, the Federal Food, Drug and Cosmetic Act served to prohibit many of the hazardous uses of radium and radon. However, as late as 1953, a Denver firm advertised a contraceptive jelly containing radium and in 1965 a Florida company was found manufacturing devices that used radon for charging water.[71]

FIG. 2.5. On the left of the photograph is a Radium Vitalizer circa 1926–1927. On the right of the photograph is a Zimmer Radium Emanator manufactured by a Los Angeles company. This device has a strength of 50,000 m. u. indicated on it and was distributed by Viva Radon Ltd. and reportedly sold for $300.

One of the more controversial and questionable applications of radon was its use in bathing, drinking, and extended exposure in specially closed rooms called emanatoria or inhalatoria or radiotoria established at health spas and old uranium mines. Muds from these places were applied to the body. Almost all the information made available about these activites was in the form of personal testimonials, and the reactions of physicians ranged from charges of quackery regarding the claims to tacit blessings from the patients who found no other relief.[74,75] Perhaps the best example of such activities in the United States occurred at the Free Enterprise uranium mine near Boulder, Montana. In the early 1950s newspapers and magazines began carrying testimonials of people who had been to this mine. The most common claim was the reduction or elimination of pain. The owners of the mine claim no curative effect, but many people who were there claimed to have been cured of arthritis, sinusitis, asthma, and bursitis. People came from all over the world on crutches and in wheelchairs to this mine. Other mines in the area with a similar attraction are the Merry Widow, Earth Angel, and the Radon Tunnel. Some have called these "health mines." The radon level in the Merry Widow mine is such that a uranium miner could work eight hours a day, five days a week for about three and a half months before receiving the maximum allowable yearly exposure. The tunnels are large enough to be walked in upright and are often furnished with benches and tables. Visitors may sit in the mine and wash in and drink the water from the mine. Some spas have advertised the radon content of the water. As early as 1904 it was recognized that mineral waters were radioactive due to dissolved radon.[76]

Radon spas are also found in Europe. The treatment used in these spas is called balneological treatment and has been known for its curative effect since Roman times.[77] Radon baths can be found in the USSR, Austria, and Poland.[78] In the USSR about 25,000 radon baths are prescribed daily by the National Health System.[78] Some workers in these spas are exposed to levels above the annual limit of 3.5 WLM.

2.2.5 Other Uses

The distribution of concentrations of radon in groundwater reflects the vertical and horizontal hydrogeological structure.[79] Buried faults or dislocations show anomalously high surface radon levels. Fluctuations as great as 100-fold have been seen in radon levels from soil due to uranium deposits. Most detectable deposits are within a few meters of the surface. Radon in groundwater has also been used as a measure of the influence of ground motion in subsidence, the pressure on groundwater, the mixing of surface water and groundwater, landslides, and land subsidence. It can also be used as an indicator of natural gas sources. The studies of these varied physical occurrences may show an effect on the radon levels.[80] However, it is not clear that there is a definite correlation. Some have concluded that no relationship exists between these physical controls and radon levels in groundwater.

Large discontinuities have been found in radon concentrations at the air–sea and sea–sediment interfaces. These measurements have led to a better understanding of ocean circulation.[81]

Radon has also been shown to be useful in uranium exploration.[82,83] When radon from soil is monitored, it can show variations within a few meters of a uranium deposit. If the soil is basically gravel, the anomaly may be distinguishable for a few kilometers. However, the presence of any moisture can reduce the distance over which significant levels of radon are measured.

Radon has been used in the exploration for petroleum. It has been suggested by field measurements that radon in an oil field will migrate to the perimeter and thus be at a minimum in the center of the field.[84] This was suspected of being due to the deflection of upward-moving solutions of radium by the brine near the petroleum deposit. There is continuous controversy and rebuttal about this technique.

> # 3

Measurement

DOUGLAS J. CRAWFORD-BROWN AND
JACQUELINE MICHEL

3.1 MEASUREMENT OF RADON AND PROGENY IN AIR

The discussion in other chapters indicates that the prediction of risk from the ^{222}Rn decay chain in air requires a knowledge of the average concentration of each of the progeny throughout the period of exposure. The most accurate calculations of the dose to the lung use the separate estimates of the concentration of each of the progeny, requiring that the contribution of each to the total exposure be measured separately. An alternative approach is simply to measure the working level concentration directly, which implies that only the total potential alpha energy is measured in the sample of air. While this approach avoids several problems with instrumentation needs, it suffers from the fact that there is not a one-to-one correspondence between working level months (WLM) and dose to the lung. Under many environmental conditions, however, the correspondence is close enough (accurate to within about 20%) to justify use of the WLM as the index of exposure. In addition, the concentration of the progeny can also be estimated by measuring only the ^{222}Rn concentration and then applying standard equlibrium ratios to estimate the concentration of the progeny. Since these ratios can depend on atmospheric conditions and ventilation rates, care must be taken to ensure that correct ratios for the particular

---

DOUGLAS J. CRAWFORD-BROWN • Department of Environmental Science and Engineering, School of Public Health, University of North Carolina, Chapel Hill, North Carolina.   JACQUELINE MICHEL • Research Planning Institute, Inc., Columbia, South Carolina.

structure are employed. The reader should refer to the chapter on exposures (Chapter 5) for a discussion of typical equilibrium ratios.

Regardless of the method used to estimate the concentration of radon or progeny, there are a few features of radon and progeny measurements common to designing any measurement program. These features are essential in ensuring that the measurements reflect the actual conditions of exposure throughout the period of interest. The most accurate and precise measurements will yield unrealistic risk estimates if the measurements are not performed at times representative of exposures.

One of the first considerations is that the radon concentration, and hence the concentration of the progeny, can vary by more than an order of magnitude over the course of the year.[1] These variations arise from changes in the water content of soil and fluctuations in atmospheric pressure, each of which influence the emanation of radon from soil. In addition, ventilation rates in buildings depend on the season, as doors and windows are opened and closed in response to the changes in temperature. Finally, the radon concentration also can vary widely within a given day,[1] usually displaying a distinct pattern of rises and falls throughout the cycle of day and night.

Because of the daily and seasonal fluctuations, measurements should be performed throughout the year and at varying times during each day. For example, the Grand Junction Remedial Action Criteria,[2] which was designed for the decommissioning of the inactive uranium mill sites, required that measurements be taken in six separate periods, each of 100 h in duration and with the measurement periods spaced at least four weeks apart. Regardless of the sampling schedule adopted, the measurement times should accurately reflect the distribution of ventilation conditions. Failure to do so can result in large errors (as much as an order of magnitude) in estimates of the average annual exposure.

The measurement scheme also must account for potential differences in concentration throughout a structure. Radon and progeny concentrations can vary at different points within a single building, since rooms may differ in distance from the source (such as soil) and in local ventilation conditions. It usually will be necessary to perform separate measurements on the different floors of a building, and any final estimates of the radiologically significant exposure in a building then would account for the distribution of exposure time on these floors.

With these factors in mind, a choice then must be made of the method to be used in collecting measurement data. Conceptually, measurement techniques can be divided into three broad categories[3]: (1) grab sampling, (2) continuous and active sampling, and (3) integrative sampling. The choice between these categories will depend upon the costs involved, the time over which an instrument can be devoted to measurements at a single location, the kind of information required, and the desired accuracy with which measurements can be related to an estimate of risk.

Grab samples consist of essentially instantaneous measurements of the radon or radon progeny concentration in air over time intervals that are short (on the order of minutes) compared to the time scale of fluctuations in concentration. Continuous sampling involves the automatic taking of measurements at closely spaced time intervals over a long period of time. The result is a series of measurements which can give information on the pattern with which the concentration varied throughout the measurement interval. Finally, integrating devices collect information on the total number of radiation events (such as alpha decays) which occur throughout some fairly long period of time, usually on the order of several days to months. The result from integrating devices is an estimate of the approximate *average* concentration through the measurement interval. Unfortunately, information about the temporal pattern of concentration within the measurement interval is lost in the integration. The following two sections will discuss various approaches to measurement utilizing each of these general approaches, although the discussion is organized around the issue of whether radon or the progeny are being measured.

Before getting into a detailed discussion of particular measurement procedures, a few general comments are in order concerning the manner in which the concentration of a radionuclide is determined. All of the methods discussed detect (and quantitate) the radionuclides through a measurement of the various radiations emitted. Reference to the radon decay chain indicates that these radiations include gamma rays, beta particles, and alpha particles, each of which may be measured and related to activity. For many of the measurement methods, only the alpha particles are counted, since the background for alpha counting usually is much less than for gamma rays and beta particles. In addition, many of the progeny decay primarily by alpha-particle emission, and alpha-particle spectroscopy offers a ready tool for separating the different decays. While the reduction in background is a large advantage of any system employing measurements of alpha particles, however, the very short range of the alpha particles (a few centimeters in air) requires that the detector be placed very close to the source. In addition, alpha-particle spectroscopy systems often are not available. A few of the methods described in these next sections, therefore, make use of the gamma-ray and beta-particle emissions.

Radon is, of course, only part of a decay chain. As a result, it is possible for the composition of a sample to change dramatically with time after sampling. This feature is particularly important in two regards. First, any sample of radon will drop in activity by a factor of two every 3.83 d. Samples of the progeny decay much more rapidly, with most of the initial activity gone after 4 h. Care must be taken, therefore, to ensure that air samples are counted quickly after they are taken, since the sensitivity of a measurement will generally decrease with time. Knowledge also must be obtained of the change in the relative activities of the progeny with time if a fixed time of sampling and counting is not used.

Second, many measurement techniques for radon rely on the growth of the progeny into the sample. The progeny are allowed to grow into equilibrium with the radon and then all of the emissions from the chain are counted. This method increases the sensitivity of the measurement, but it is essential that sufficient time be allowed for the growth of the progeny to a known degree of equilibrium. A period of 4 h is a typical length of time for allowing equilibrium to be reached.

Some of the measurement techniques for the radon or the progeny concentration require that measurements of activity be performed while a sample is changing in composition. The reader may find it necessary, therefore, to develop equations relating the initial activity of each radionuclide at the time of sampling to the activity at any time, t, during counting. Since measurement schemes differ in how the sampling and counting times are chosen, it is not possible here to present a discussion of the mathematics involved in each method. Detailed discussions may be found in Appendix A. It is possible, however, to give a general discussion of the differential equations of radioactive growth which underlie all techniques.

In general, let $N_i(t)$ be the number of atoms of the ith radionuclide in a decay chain present in a sample at time t. The differential equation describing the rate of change of $N_i(t)$ with respect to time is:

$$\frac{dN_i(t)}{dt} = \lambda_{i-1} N_{i-1}(t) - \lambda_i N_i(t) \qquad (3.1)$$

where λ_j is the radiological decay constant for the jth radionuclide in

MEASUREMENT

the chain of decay. Equation (3.1) is simply a differential equation of the Bernoulli form, and the generalized solution is:

$$N_i(t) = e^{-\lambda_i t} \int_0^t e^{\lambda_i t} \lambda_{i-1} N_{i-1}(t) \, dt + C \, e^{-\lambda_i t} \tag{3.2}$$

In this relation, C is a constant given by the boundary conditions and usually will be equal to the number of atoms of the ith radionuclide present at the start of counting or the end of sampling (depending upon how the time $t = 0$ is defined). For the first radionuclide in a decay series, the intergral in Equation (3.2) will be zero, and the familiar exponential behavior is observed. If branching fractions are involved, the decay constants must be multiplied by the appropriate fractions.

3.1.1 Measurement of the Radon Alone

One of the earliest methods for measuring the concentration of radon in air is the scintillation cell,[4-6] which usually is utilized as a grab sample. For various reasons, this cell has also been known historically as a Lucas cell. The scintillation cell itself is available in many sizes (usually 0.1 to 2 L) and shapes, but it in general consists of a chamber whose walls are coated with a scintillating material such as silver-activated zinc sulfide phosphor. The chamber usually is constructed using available materials such as plastics, and the phosphor can be obtained as crystals which are mixed with glue and either spread on the interior walls or applied as thin sheets. The chamber is capable of being evacuated to low pressure.

Once the chamber has been evacuated and sealed, the cell is brought to the environment to be sampled and is opened. When the chamber air has equilibrated with the external air, the cell is closed and placed on the photomultiplier tube which views the cell through a window built into the cell walls. Counting is performed with a single-channel analyzer after allowing for the ingrowth of the progeny. The progeny in the original air of the sample can be excluded from entering the cell by placing a millipore filter over the entrance to the cell. An alternative method of removing the contribution from the original progeny is to allow no filtration at the entrance but to wait for 4 to 8 h after sampling to begin counting, which allows the original progeny to decay away so that only those produced by the radon in the cell are left. Table 3.1 displays the features of the scintillation cell, as well as the characteristics of the other systems discussed in this section.

The scintillation cell has been used both as a continuous and as an integration device for the measurement of radon.[7-9] The system used as an integration device is the same as described above, but the sampled air is drawn at a low rate into the chamber over a longer period of time. The cell then either is continuously coupled to a photomultiplier and scaler, or the cell is closed after a long collection time and returned to the lab for counting. In either case, the result is a measurement, or series of measurements, reflecting longer-term radon exposures than are measured with grab samples. An alternative approach has been to collect radon in another structure (such as an impervious bag) and then transfer it to the scintillation cell back in the lab.

In the early 1970s, the U.S. Atomic Energy Commission (USAEC) Health and Safety Laboratory (HASL) developed a method for measuring radon which relied on introducing the air directly into an ionization chamber.[10] The current in the chamber then is proportional to the concentration of radon. The system could be used either as a continuous or grab sample monitor, since the air could be introduced immediately in a batch or allowed to flow continuously through the chamber at a low rate.

Several methods for measuring radon rely on its ability to diffuse through passive barriers, such as foam rubber, which prevent passage of the progeny.[11,12] The radon diffuses through the cover and then

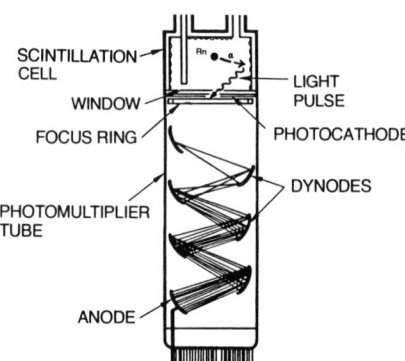

FIG. 3.1 This figure shows a typical counting system which uses a scintillation cell. The cell is detachable and lined internally with a scintillating material. A vacuum is drawn in the cell and it is opened in the field. After closure, the cell is placed on top of a photomultiplier tube which converts the light pulses into an electrical signal. (Photo courtesy of Belanger.)

TABLE 3.1. A Summary of Currently Available Radon Measurement Devices

Instrument type	Application	Sensitivity[a]	Reference(s) in text	Figure number
Scintillation cell	Grab or continuous	<3.7 Bq m^{-3}	4–9	3.1, 3.5
Ionization chamber	Grab or continuous	<3.7 Bq m^{-3}	10	
Passive barrier with collection of progeny on scintillator	Continuous	<3.7 Bq m^{-3}	11, 12	3.2
Passive barrier with collection of progeny on TLD	Integrating	0.8–8.1 Bq m^{-3}	14	
Two-filter method	Grab or continuous	<3.7 Bq m^{-3}	15	
Activated charcoal	Integrating	7.4 Bq m^{-3} for 100-h exposure	4, 17, 18	3.3
Alpha track	Integrating	<18.5 Bq m^{-3}-month	19–21	3.4

[a] Defined as the lowest quantity which can be distinguished from background with a 95% confidence level.

FIG. 3.2. Radon monitors have been constructed based on a design known as the Wrenn chamber. At the center of the hemisphere is a ZnS scintillator covered by aluminized Mylar (to provide an electric field). The outer wire mesh is covered with foam, which prevents the progeny from entering. Once inside, the radon atoms decay and the progeny are swept to the scintillator surface by the electric field. The pulses then are counted with the photomultiplier tube attached to the scintillator. Apparatus to measure radon concentration in air as a function of time and weather parameters. Equipment from left to right: Rainwise Weather Station, interface, Ludlum Instrument Company radon in air detector, Ortec power supply, Ludlum Instrument Co. 2200 scaler, Hewlett Packard 85 computer. (Photo courtesy of Tom Hess.)

decays in the space between the cover and detector, which may be a scintillator. The newly formed charged progeny then could be attracted to the scintillator surface by placement of an electric field, established by charging a surface near the scintillator. All of the rest of the electronics are identical to the scintillation cell method, and the system can be used as a continuous monitor for radon. Modified versions of this approach have been reported involving replacing the scintillator with a thermoluminescent dosimeter (TLD) chip to yield an integrating device or drawing away the progeny with an electric field so that only the radon decays are counted.[13,14]

MEASUREMENT

FIG. 3.3. Charcoal canisters can be manufactured simply with a small can covered by a screen. The charcoal is contained in the space below the screen, and the screen is held in place by a ring. A top is fitted over this arrangement until exposure, at which time it is removed. The top is replaced and sealed at the end of exposure, and the entire can placed onto a NaI gamma detection system for counting.

Care must be taken in interpreting the results of these passive barrier systems since the diffusion barrier introduces a time lag between changes in radon concentration in the ambient air and changes in the equilibrium activity on the surface of the scintillator or TLD. In addition, progeny can build up inside the system over time, requiring that this buildup be accounted for in calculations of activity or concentration.

Another historically early device for the measurement of radon in air is the two-filter method.[15] Air enters a small tube of fixed length (usually 0.3 to 1 m) and is filtered at the entrance to remove the progeny. As the radon travels the length of the tube, it decays to ^{218}Po, which is trapped on an exit filter. After a suitable length of time for sampling, which depends on the radon concentration and desired

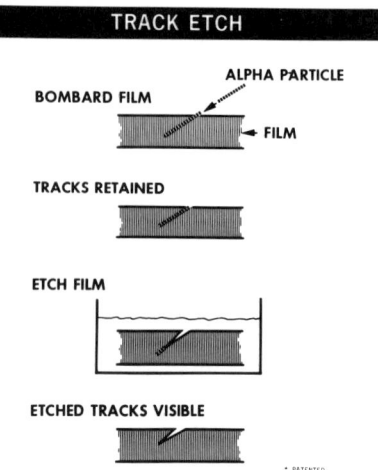

FIG. 3.4. Alpha track devices consist of a material, such as a film, which sustains damage along the track of an alpha. The material then is placed into etching fluid, which enlarges the track by extending the region of damage. Once the tracks have been sufficiently enlarged to become visible, their density at the surface of the material is determined. (Photo courtesy of Terradex.)

FIG. 3.5. There is a wide range of devices suitable for continuous measurements of radon. One such device, manufactured by Pylon, is shown in this figure. The system contains a pump which draws air at intervals into a scintillation cell. The results then are stored or printed at intervals, yielding a picture of the temporal pattern of radon concentration. (Photo courtesy of Pylon.)

sensitivity, the filter is removed and counted by measurement of the alpha decays. The method also is useful as a continuous monitor if the second filter is left in place and counted continuously through either the beta-particle or gamma-ray decays (since it is difficult, but not impossible, to place an alpha-particle detector sufficiently close to the filter during continuous operation). Continuous and integrative monitoring has been accomplished through the use of a TLD chip at the exit filter.[14,16]

The amount of ^{218}Po growing into the stream of air flowing through the tube can be obtained by solving the decay equations for the first progeny in a sample of initially pure radon at a time equal to the transit time of the tube. Longer tubes clearly permit greater ingrowth of the ^{218}Po, although increasing the length also increases deposition within the tube which is not available for collection at the exit filter. In any event, this wall deposition must be accounted for in calibrating the system.

In recent years, the activated-charcoal canister and alpha track methods for measuring radon have become quite popular due to the low cost and small size of the devices. The activated-charcoal device utilizes the fact that radon is partially adsorbed onto the surface of activated charcoal and will remain loosely attached until heating.[17,18] The method is very simple to use and requires a minimum of equipment, most of which is available in most counting labs. The activated charcoal can be purchased in bulk and is loaded into small containers using almost any available materials (although the containers must be impervious to radon diffusion once they are sealed). Approximately 50 to 100 g of the charcoal is placed into each container and the container is sealed. It then is opened in the area to be monitored and left exposed for 2 to 5 d to allow equilibration with the ambient air. After this time, the canister is sealed and returned to the lab for counting.

Counting typically is performed using a NaI system coupled to either a multichannel analyzer or single-channel analyzer centered about the pertinent peaks. The gammas from the ^{214}Bi decays alone can be used, or increased accuracy can be obtained if the gammas from the bismuth and the other gamma-emitting progeny are combined into a single region of interest. Other approaches to determining the activity of the progeny on the canister charcoal have relied on dissolving the charcoal in liquid scintillation fluid. This is followed by counting of the fluid in a liquid scintillation detector or by de-emanation of the radon from the charcoal into a scintillation cell followed by alpha-particle counting.

The equilibrium activity of radon progeny on the charcoal is a function of the radon concentration in the ambient air, the exposure time, and the relative humidity.[18] Since the progeny are used in measurements, sufficient time must be allowed following closure of the canister to allow for ingrowth of the progeny prior to counting. Ten-minute counts usually are sufficient for a fairly accurate (within 20% at the 95% confidence interval) determination of the radon concentration. The trapped radon then is released from the charcoal by heating the opened canister for several hours at approximately 200°C, and the charcoal then may be reused. When calibrating the system, it is important that any canisters used for calibration be heated to the same degree as those to be employed in the field. One drawback of the charcoal method is that it is useful only for an integration measurement, and it appears to give undue weight to the final day or two of exposure. In addition, separate lots of the charcoal can differ dramatically in performance, so calibration should be performed whenever a new shipment is obtained.

The final method for measurement of radon to be discussed here, and one also suited ideally to large sampling programs, involves direct exposure of alpha track material to ambient radon.[19-21] In a typical system, a small slab of alpha track material is placed at the bottom of a small cup (any material for the cup will suffice so long as the progeny cannot diffuse through the walls). Often, a filter is added across the front of the cup to remove the progeny. The radon alone then is free to diffuse into the space of the cup, where it will decay to the progeny. The alpha emissions from the radon and any progeny produced in the cup then irradiate the alpha track material, leaving microscopic damage along the track of each alpha. After exposure for the desired period of time (extending up to a year), the material is removed from the cup and etched in an acidic or basic solution (usually KOH or NaOH) operated upon by an alternating electric field. This etching extends the region of damage around each track, and the separate tracks become visible as small holes on the surface of the material. These tracks then are counted visually using a wide-screen microscope, with the number of tracks being proportional to the number of alphas striking the material surface during exposure. The accuracy of the method can be modified partially by varying the fraction of the surface examined for tracks. A wide range of materials are available for use as the alpha track material, and several commercial firms now market services which include rental of the device and counting of the tracks. For periods of time of up to a year, the system provides a true integrated measurement of the average exposure conditions (at least within the variability of the method itself).

3.1.2 Measurement of the Radon Progeny

Most of the techniques for measuring radon progeny (see Table 3.2) are similar in principle to those discussed for radon. As with the case of radon, the devices can be separated into the categories of those representing grab samples, continuous measurements, or integrations. In addition, it is useful to distinguish between active devices, which involve the pumping of air, and passive devices, which rely purely on diffusion. Whenever measurements are performed in support of risk calculations requiring the concentration of each progeny, alpha-particle spectroscopy is used to separate the decays on the basis of the distinct alpha energies. When performing spectroscopy, it is important to bear in mind that the presence of air spaces between the source and detector can smear the spectra, and that similar smearing can result from the self-absorption of the sample. The introduction of helium to replace the air has proven effective in reducing this source of degradation of the spectra.

Many of the early methods for measuring radon progeny are simple modifications of a method originally developed by Kusnetz[22] and Tsviglou et al.[23] Air is drawn through a filter at a fairly low volumetric rate and the progeny attach to the filter. Since the progeny are attached to aerosols to a large extent, the filter should be capable of retaining particles down to at least 0.1 μm in diameter. The radon does not attach to the filter due to the fact that radon is an inert gas.

When the air has been sampled for the desired length of time (usually on the order of 10 minutes to an hour), the filter is removed for counting. Alpha-particle spectroscopy then can be used to determine the activity of each progeny on the filter, although other authors have made use of the gamma-ray[24] and beta-particle[25] emissions since this lowers the cost. It is usual to count the filter during two or three separate time intervals to aid in resolving the original concentration of each of the progeny in the sampled air. If only the working level concentration is desired, the total number of alpha-particle decays can be measured and related to the potential alpha energy. Equations relating the number of decays of each of the alpha-emitting progeny on the filter to the concentration in air are conceptually simple to generate but are too long to present here. The reader should examine the original paper by Martz et al.[26] for a complete presentation. The equations make use of the decay chain solutions presented in this chapter in the form of Equation (3.2). In general, the sensitivity of the

TABLE 3.2. A Summary of Currently Available Radon Progeny Measurement Devices

Instrument type	Application	Sensitivity[a]	Reference(s) in text	Figure number
Kusnetz–Tsviglou and modifications	Grab or continuous	0.0005 WL	22, 23	
Modified Kusnetz with gamma or beta counting	Grab or continuous	Not published	24, 25	
Passive alpha track	Integrating	Several WL h	29, 30	3.4
Alpha track with pump	Integrating	1 WL h	28	
TLD with pump	Integrating	0.0005 WL	31–34	
Filter and pump with surface barrier	Integrating	0.005 WL or less	36	
In vivo	Integrating	Accurate to within a factor of 4, slightly less accurate for urinalysis	37, 38	

[a] Defined as the lowest quantity which can be distinguished from background with a 95% confidence level.

system is affected by the manner in which the sampling and counting times are arranged.

Several notes of caution are in order concerning use of the Kusnetz–Tsviglou approach. First, the flow of air may cause particles to dislodge from the filter during sampling. It is necessary, therefore, to test the filters for collection efficiency under the conditions of application in the field. Second, most filters will not retain the unattached fraction of the progeny. If this proves to be the case, a simple method for measuring the unattached fraction is to place a finely meshed wire screen downstream from the filter. The surface charge on a metal screen collects the unattached ions, and the screen then is treated and counted exactly as the filter. For accurate determinations of dose, an estimate of the unattached fraction usually is desired.

Short collection times—on the order of 2 to 10 min—usually are considered to constitute a radon progeny grab sample. Very low flow rates, however, can extend the sampling times up to an hour or longer. If the filter is to be removed and counted, there is little advantage to sampling longer than an hour, since the rapid decay of the progeny gives primary weight to those collected in the last hour. A continuous device, however, can be constructed by automating the collection and counting.[27] The filter is built into a strip which is held in place during sampling and then moved under the detector for counting. The detector measures the alpha particles and/or beta particles and the filter strip is moved to the next position for a new sample. In this manner, a stream of results can be generated over a long period of time. Such systems usually are used only to determine the working level concentration unless an alpha spectroscopy system can be devoted.

As in the case of radon measurements, passive alpha track devices have proven useful in measuring radon progeny.[28–30] The devices usually yield, however, only an estimate of the working level concentration. While it was possible to exclude the progeny from a measurement of radon using the devices with a diffusion barrier, it usually is not possible to exclude the radon from a measurement of the progeny. As a result, it is necessary to obtain a separate measure of the radon and subtract this contribution from the radon-plus-progeny measurement. This can be accomplished by using two alpha track chips, one of which is shielded from the progeny by a diffusion barrier. An alternative is to use a single chip and assume a standard equilibrium ratio. In either case, the alpha track system acts as an integration device, yielding an estimate of working level month (WLM) or average working level (WL) concentration throughout the exposure interval.

Working level monitors have been developed[31–34] which use a low-volume air pump in conjunction with a TLD chip (LiF or CaF_2). The air is drawn at a very low rate (on the order of 1 L min^{-1}) through a filter which is positioned close to the TLD in a common holder. Following the sampling period, the TLD is removed and counted. The device can be used as a personal monitor and yields an estimate of the WLM or average WL concentration to which the worker was exposed. Similar systems replace the TLD by an alpha track device.[28,35] Measurement capabilities are similar to those of the devices using the TLD.

The scintillation cell used for measuring radon also is ideal for obtaining a grab sample of the progeny. As in the case of the passive alpha track system, however, it is necessary to obtain a separate measurement of the radon concentration. Fortunately, this measurement can be accomplished simply by the scintillation approach. The sample is counted very quickly (within an hour or two) to obtain an estimate of the total WL concentration (progeny) plus radon concentration. By then waiting an additional 4 to 10 h, it is possible to allow the original radon progeny to decay, leaving only the contribution from the progeny which have grown into equilibrium with the original radon. This radon contribution then is counted and subtracted from the first measurement. The result is an estimate of the WL concentration at the time the sample was obtained. Since the radon progeny decay quickly and the relative concentration of each progeny changes with time, the system must be calibrated and utilized under a constant counting schedule.

Finally, the most recent method for obtaining an integrated measurement of exposure to radon progeny uses a pump to draw air slowly through a filter.[36] The filter then is counted by a surface barrier detector similar to the diode used in normal alpha spectroscopy. A distinct advantage of the device is that it can be operated in a semicontinuous mode if the output from the detector is printed at regular intervals. Results are generated more quickly than is the case with systems based on TLDs, but is is important to realize that surface barrier detectors are fairly fragile and rather expensive.

3.1.3 Measuring Progeny in the Body

Measurements of radon progeny in air are performed ultimately to provide an estimate of the dose delivered to lung tissue (see Chapter 6 on dosimetry). One of the major problems in relating air concentration to dose arises in the necessity to specify the fraction of a radionuclide deposited in the lung and the rate at which it is cleared from the lung.

MEASUREMENT

This problem of modeling deposition and clearance could be overcome if the radon progeny could be measured directly in the body, and such measurements could also be used to estimate the average concentration of progeny in air during past exposures. This approach is especially useful in historical studies of workers (such as the uranium miners) who were exposed in the past to high levels of radon progeny and for whom detailed information on air concentration is not available.

Two primary methods are available for estimating past exposures to radon progeny based on direct measurements of the amount present in the body. These involve estimating the concentration of activity in the urine and external measurements of the gamma rays emerging from the body as a result of decays in internal organs. In each case, it is either the long-lived lead or polonium which is used as an indicator of exposure.

Direct measurements of emerging gamma rays typically use the gamma rays from ^{210}Pb and rely on decays occurring in lung or bone tissues.[37,38] The method is known as in-vivo counting and utilizes a system of either NaI or germanium detectors placed over the body. Problems with background counts are lessened by placing the detection system in a shielded room. Care is taken to filter the air, since an influx of radon progeny can greatly affect background. Average exposures to the radon progeny in the past then are determined from the body burden measurements by reversing the application of metabolic models which detail the movement of the radionuclides within the organs and tissues of the body.

These models also can be used to determine exposures by measuring activity in urine samples.[39] Both ^{210}Pb and ^{214}Po have proven useful in quantifying past exposures. These radionuclide concentrations in urine can be determined either by counting of decays on a NaI system or by use of liquid scintillation. As with the in-vivo counting method, the major problem with the use of urinalysis results to estimate past exposures lies in the often large uncertainties surrounding metabolic models, uncertainties which are particularly acute for decay chains. These uncertainties make it unlikely that these two approaches can yield estimates of exposure to within better than a factor of four to five, particularly when values specific to individuals (rather than population averages) are required.

3.1.4 Calibration

No discussion of measurement techniques would be complete without at least a brief discussion of how to calibrate a system. Measurement systems yield only the number of radiation interactions occurring within the detector. The process of calibration is intended to provide a relationship between the number of such interactions and the amount of radioactivity in the original sample of air. The calibrations also are performed to ensure that a system has not degraded in accuracy or precision. Depending upon the system, the relationship between radiation interactions and air concentration can vary from multiplication by a constant through to a complicated function involving volumetric flow rate, filter efficiencies, humidities, etc.

One of the central needs of a calibration system designed for use with measurements of radon and progeny is an exposure chamber containing a known, and controlled, concentration of the radon and/or progeny. Ideally, the concentrations should be capable of being maintained at a constant level throughout any exposures. This requirement often means that the chamber must be fairly large, since opening the chamber to insert the measurement devices can result in a large change in the concentration if the air which is exchanged represents a significant fraction of the chamber volume. In addition, the temperature and humidity of the chamber air should be controllable if the calibration factor for the device is a function of these parameters. Controlling the humidity is especially important for the charcoal canister method, since the uptake of water increases attenuation of the gammas and changes the equilibrium activity of radon on the charcoal surface.

An exposure chamber can be used for any of the passive methods of measurement described previously. The device to be calibrated simply is placed in the chamber for the desired period of time, removed, and counted. Counting geometry and instrument settings must be identical to those employed in routine measurements. The measurement result (counts per minute, number of tracks, etc.) then is determined and divided into the known concentration of the chamber. The chamber concentration usually is measured through use of a scintillation cell calibrated with a National Bureau of Standards source of radon and progeny. The determined ratio then is assumed to hold true for any future measurement involving a similar device. Some devices (such as charcoal canisters) are affected in their calibration factor by length of exposure and humidity, so it may prove necessary to obtain separate calibration factors for each set of exposure conditions.

Active devices also involve use of a pump and usually a filter. The pump must be calibrated by measuring the flow rate against a calibrated flow rate or using a calibrated flow rate meter. Filter efficiencies can be determined by pulling air at a measured flow rate and predetermined radon progeny (or particle) concentration through the filter. It is important that the flow rate used in the calibration of the filter be as close as possible to the rate used in the field measurements, since this factor can influence filter efficiency. An alternative to developing separate calibration factors for components is to simply use active devices in conjunction with exposure chambers for purposes of calibration. Air from the exposure chamber at a known concentration then is used as input into the entire active device and filter efficiencies, flow rates, counting efficiencies, etc. are already included in the calibration factor. While this approach is easiest, it requires that new calibrations be performed whenever components are altered or suspected of having degraded.

During the past seven years, facilities in the United States have built exposure chambers for radon and/or progeny and offer the chambers for national use as central calibration facilities. They also conduct national certification tests for programs attempting radon or progeny measurements. Two of the major facilities are operated separately by the U.S. Environmental Protection Agency and the Environmental Measurements Laboratory (formerly the Health and Safety Laboratory). These services are particularly important for national measurement programs involving many subcontractors, since it is possible to calibrate devices for a large number of subcontractors with a single source of exposure. The reader should contact these calibration facilities before setting out on the construction of a new exposure chamber.

3.2 MEASUREMENT OF RADON IN WATER

3.2.1 Introduction

The determination of radon in water is relatively simple compared to air measurements. The rate and magnitude of variations are much lower, and there are fewer sampling problems. Nevertheless, the measurement of radon in water has its own set of sampling issues and analytical difficulties, and these will be discussed in detail. All methods require correction for decay of radon during the delay between sampling and analysis.

3.2.2 Sampling Methods

Probably the most difficult aspect of the measurement of radon in water is the collection and handling of samples. The first sampling issue is where the sample should be collected. During most surveys for exposure data, the sampling point is at the tap in a home. A sample is collected after the water is run for a few minutes. For a public water supply, such a sample would be representative if either there was only one source or well for the water or all the water was blended prior to distribution. However, oftentimes individual wells or well clusters provide water to different parts of the distribution system so one sample would not be representative of the entire system. In addition, there may be temporal variations in the relative source contributions; that is, depending upon time of day or demand, different water sources may be used. In these cases, the radon content can vary widely depending on the concentration and percentage of contribution of each source to the system. Multiple-well systems are particularly difficult to characterize because of these variations. One solution is to know the distribution system and to design the sampling plan to collect samples representing each source. Alternatively, the radon content of each well can be determined from samples collected at the wellhead prior to distribution.

A second serious sampling problem is associated with the actual transfer of a water sample into a container. Because radon is a gas, special care is required to collect a sample without significant outgassing of radon in the process. The standard sampling protocol is to turn on the tap for several minutes to allow the water in the pipes to be replaced with fresh water. If sampling a domestic well source, the water is run until the pump starts and runs for several minutes. The actual transfer of water from the tap to the sample container depends on the analytical method to be used. In all cases, every effort is made to prevent loss of gas. The water can be allowed to overflow a container for several container volumes, and a sample then is collected from the bottom. The sample can be transferred by sucking water into an evacuated bubbler flask, usually 500 to 1000 mL in volume. Smaller volumes can be collected by syringe and transferred to vials. No samples should be collected in the field for later transfer because of the likelihood of radon loss.

The greatest contribution to the analytical uncertainty in determination of radon in water is due to sampling. In one experiment, a series of samples were collected sequentially and without stopping from a single well.[40] The variation among these samples was about ±25% and was larger than the variation in the well measured over a three-year period. As a result of this uncertainty due to sampling errors, many researchers collect duplicate samples and reject results when the standard deviation of the two values exceeds that normally expected.

The temporal variability of radon in groundwater has been reported as both low and high. Studies of several wells, both sand and rock wells, in South Carolina showed a maximum variation in samples collected at the wellhead of 15% over three years.[40] Studies in Houston showed temporal variations of less than a factor of two.[41] In contrast, results from rock wells in Finland showed temporal reductions in the radon content of up to 80%.[42] The reason for the large temporal changes in Finland was postulated to be dilution by relatively low-activity surface water, although no assessment of the contribution of sampling errors was made. Short-term temporal variations in the radon content of groundwater collected from wells can also occur when the well casing is perforated at several depths and draws water from multiple aquifers. The radon content will vary in proportion to the amount of water drawn from each aquifer. Usually, a quasi-equilibrium will be reached in several hours, although there can be seasonal variations due to changes in water level and hydraulic head.

It is likely that temporal variations in the radon content of groundwater, not related to changes in water sources, are very low. In areas where there is a significant seasonal or long-term change in the water table, the flow path of water may be different enough to affect its radon content seasonally. Barometric pressure changes and pressure changes due to the tidal cycle can have relatively small effects. However, the most common source of differences in the radon content of groundwater from a well probably is due to the actual process of collecting a sample.

3.2.3 Analytical Methods of Analysis

There are two primary methods for the measurement of radon in an aqueous sample, although there are several variations for each method. The standard technique for radon measurements for many years was the radon bubbler and alpha scintillation cell method.[5,43] In this method, a carrier gas is passed through the sample in a bubbler flask to purge out the dissolved radon. The purging can be accomplished in either a once-through or a recirculating system. The released radon is then transferred either directly or after a concentrating step into an evacuated scintillation cell. The scintillation cell, referred to as a Lucas cell in honor of its inventor, Henry Lucas, is coated with a ZnS:Ag phosphor and has a quartz window. After transfer of the radon to the Lucas cell, the sample is stored for three hours to allow equilibrium to be reached between radon and its short-lived progeny. The cell is then placed directly in a photomultiplier tube for detection of the pulses generated by alpha particles striking the phosphor. This method has very low background. In fact, ^{226}Ra is determined in water samples by purging a sample first, then allowing radon ingrowth followed by measurement as described above, a procedure known as the radon-emanation technique.

The detection limit for the bubbler method is very low, about 50 Bq m^{-3}, even without concentration. The analytical precision of this technique has not been determined by collaborative testing but is reported in the literature to be about ±10%. The biggest limitation to this method is that the sample should be collected in the field in the same bubbler flask used during measurement. Only glass can be used because most plastics are permeable to radon. Handling and transport of glass flasks that are traditionally round-bottomed is at best awkward, and loss of sample integrity is a frequent problem. The preparation time takes about 15 min per sample, in addition to the time needed for sample changing in the counting system and measurement of the sample volume. Counting time is usually less than 10 min at the concentrations frequently found in groundwater.

These problems led to the development of a faster analytical technique for radon in water, which has become the most common method in use today: liquid scintillation counting.[44,45] In this method, 10 mL of sample are injected into a glass vial containing 5–10 mL of liquid scintillation solution. The vial is tightly capped and shaken vigorously. The samples are then returned to the laboratory and counted with a liquid scintillation counter, usually with an automatic sample changer. The detection limit is about 370 Bq m^{-3} for a 40-min counting period.

The only intercomparison results available for radon in water are for an informal study conducted by the University of Houston in 1979. An analysis of variances was run on a small sample of data from six different laboratories using four variations of methods for two different levels of activity.[46] The interlaboratory precision for the high levels of activity (188,000 Bq m^{-3}) was 4.4%. The interlaboratory precision for the low-level samples (30,000 Bq m^{-3}) was 6.7%. Thus, the precision in the analytical determination of radon in water is low (about 5%) when compared to other sources of error.

4

Sources

JACQUELINE MICHEL

4.1 INTRODUCTION

Indoor air radon concentrations can vary over four orders of magnitude. *Why?* These ranges occur because of the wide variations in the rate at which radon is generated from its sources (primarily soils, groundwater, and building materials), in the modes of radon's transport through various materials, and in the means of its entry into structures. In this chapter, the sources and mechanisms of transport of radon emanating from natural and man-made environments will be discussed, focusing on those factors which result in elevated indoor radon concentrations.

Because radon is a radioactive, noble gas, with no chemical reactivity under normal conditions, its concentration at any point of measurement is a function of three primary factors:

1. Concentration and distribution of its parent in the source material.
2. Efficiency of transport processes which bring it into the biosphere.
3. Its half-life.

Each of these factors will be discussed below, emphasizing its role in producing elevated radon concentrations.

Because radon is a short-lived member of the ^{238}U decay series and a progeny of ^{226}Ra, its concentration is a function of the levels of

---

JACQUELINE MICHEL • Research Planning Institute, Columbia, South Carolina.

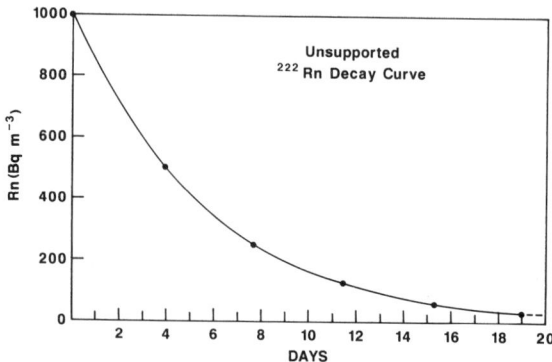

FIG. 4.1. Decay curve for unsupported radon, showing that every 3.83 d the concentration of radon decreases by half.

^{238}U and ^{226}Ra in the source material. However, there is no strict correlation between radon in water or soil and the radium or uranium concentrations. In fact, good correlations exist for only the two extremes: if the radium content of the source material (rock or sediments) is very low or very high, then the radon is likely to be very low or very high, respectively, in the adjacent water and soil gas. However, if the radium content is not very low or very high, the radon content of the adjacent water and/or soil gas can range widely. Other factors are then dominant in controlling radon concentrations. To be able to predict where radon will be high (or low), the processes which transport radon from the solid matrix to the air or water in pore spaces and then to the point of exposure need to be understood. Extensive research is currently being conducted worldwide to understand these processes.

The third primary factor which controls radon concentrations is its half-life. Figure 4.1 shows the decay curve of radon after it has been removed from its parent, a condition known as unsupported activity. The radon activity decreases exponentially, so that in five half-lives or about 20 d, the activity is about 1% of the original level. Transport processes which occur over a period greater than 25–30 d do not affect radon concentrations appreciably. Therefore, unsupported radon usually does not accumulate in groundwater over distances greater than 50–100 m. Radon transport in soils in normal conditions by diffusion is limited to several meters due to radon's short half-life. Thoron, with a 55-s half-life, would be limited to even smaller distances.

4.2 THE SCALE OF RADON TRANSPORT PROCESSES

Study of the sources and transport processes of radon is here approached on three different scales—macroscopic, mesoscopic, and microscopic—all of which are very important contributors to our understanding of radon.

First, there are parameters or processes of *macroscopic scale*, which can be defined or quantified regionally, for areas of size on the order of square kilometers. The best example is characterization of different rock types as to their average uranium content. Figure 4.2 shows the distribution and mean of uranium in four different types of igneous rocks. Mafic igneous rocks (like basalt) generally have low uranium content, with a mean value of 0.9 parts per million (ppm). In contrast, the more acidic igneous rocks tend to have higher uranium content, with the highest values found in granites, which have a mean of 4.7 ppm. Granitic rocks with 20–40 ppm uranium are not uncommon.

Another trend shown in Figure 4.2 is that the distribution of uranium in rocks can be described as log normal, meaning that there will always be a significant number of low values and few high values for any parameter. Therefore, most basalts will have very low uranium content, but there will be scattered instances of high values. Conversely, many granites can be characterized as having low uranium content. There will be significant overlap in the distribution of uranium in different rock types.

Macroscopic factors are those which are intrinsic to the rock type or geologic setting. Those factors which affect uranium content and distribution in rocks are fairly well understood. Uranium exploration studies conducted over the last 30 years have provided detailed geochemical models for uranium migration and enrichment. One of the most basic properties of uranium and thorium is the tendency to be enriched in the more volatile phases as molten or partially melted rocks cool, processes associated with the formation of igneous and metamorphic rocks. Therefore, rock types which have a low melting point, such as granites, have higher uranium content than those with high melting points, such as diorites and basalt. Uranium can be enriched or depleted by this same tendency to concentrate in the fluid and volatile phases during the heating which frequently occurs with metamorphism. As the rocks are heated and partially melted, the uranium can be mobilized and depleted from the rock being metamorphosed and concentrated in the volatile or liquid phases. By this process, uranium, thorium, and other volatile elements are enriched in features such as pegmatite veins.

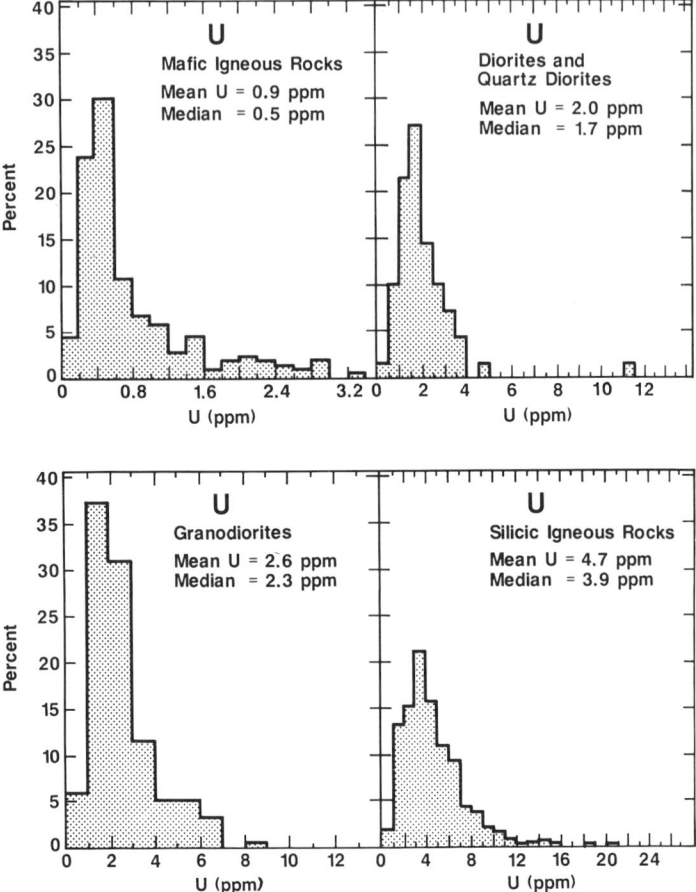

FIG. 4.2. Histogram showing the distribution of uranium for different igneous rock types.[1]

Uranium can be transported from the earth's mantle to its crust by this fractionation on a global scale. Metamorphic rock terrains would be expected to have a wide range of uranium distribution, with some areas greatly enriched and others depleted, depending upon the mobilization of uranium.

A second mechanism for uranium transport and enrichment that is macroscopic in scale is the ease with which uranium can be oxidized and mobilized by groundwater flow. Uranium in the +4 oxidation state forms compounds that are very insoluble at normal temperatures. However, uranium in the +6 oxidation state can readily form soluble compounds. Thus, under oxidizing conditions, uranium can be leached from rocks and soil and transported in solution by groundwater. Once reducing conditions are encountered, the uranium is rapidly precipitated from solution. This leaching and fixation process is not stable, but it can be quite dynamic because of changing oxidation/reduction conditions in groundwater systems. In fact, there can be multiple leaching and fixation cycles over periods of time spanning millions of years. Many of the common "sandstone-type" uranium deposits in Colorado, New Mexico, and other western states were formed by this process of leaching of uranium from one area and transport by groundwater flow until reducing conditions caused uranium to precipitate out. Hydrothermal fluids also can leach uranium from its source rock and enrich it in adjacent deposits.

In the study of radon, these secondary transport and enrichment processes are important in that they can greatly alter, via a type of large-scale weathering, the uranium content of a rock or sediment. These mechanisms of uranium redistribution are very important to the processes of *mesoscopic scale* which affect the radon content and emanation potential of rocks and sediments. Locally, uranium and radium can be enriched on a scale in the range of square meters, such as redistribution within a soil profile or along shallow faults and fractures. Thick soil overburdens can attenuate radon exhalation, such as where glacial deposits overlie bedrock. Fractures in rocks can increase the surface area in contact with groundwater, enhancing the amount of radon in solution. Fractures and faults can serve also as conduits for greater diffusion of radon to the surface. These processes affect the local distribution of radon within a region or unit.

On a *microscopic scale*, there are processes which enhance radon release from a solid into the pore space (water or air). For example, the location of the parent uranium or radium in a mineral grain greatly affects the probability that radon will be able to escape from the solid. It is estimated that 30–70% of the uranium in granites is locked up in very resistant minerals such as monazite or zircon. However, up to one-third occurs as loosely bound, interstitial oxides which readily release radon to the intergranular pore spaces. The distribution of uranium in the rock will greatly affect its radon emanation. Metamorphic rock processes again can play an important role at this scale. At medium-to-high metamorphic grades, uranium which had been distributed as interstitial oxides or sorbed onto mineral grains can be recrystallized into more resistant minerals, decreasing the radon emanation of the rock. In the Piedmont region of Virginia, North Carolina, South Carolina, and Georgia, there is a broad zone of metamorphic rocks known as the Monazite Belt,[2] because of the presence of abundant monazite, a thorium-rich, heavy mineral, in the high-grade metamorphic rocks. In some cases, the formation of monazite may have reduced the potential for elevated radon. However, there are such wide ranges in the original rock type and degree of metamorphism that the radon distribution will also vary widely.

4.3 RADON TRANSPORT MECHANISMS

There are three types of radon transport that are important to understanding and predicting radon migration:
1. Transport from the solid to either the gas or liquid in the pore spaces.
2. Transport of radon relative to the gas or liquid (molecular diffusion).
3. Transport of radon with the gas or liquid (convection or groundwater flow).

The transport of radon from the solid is a process known as emanation, and the emanating power is defined as the ratio of the number of radon atoms which escape from the solid to the number of radon atoms formed by decay of radium in the solid. This quantity is known as the emanation coefficient, escape ratio, and percent emanation. It is always a measured value for a specific type of material in both composition and form and is strongly a function of the test conditions. The emanation coefficient of a material is a very important factor in estimating high radon concentrations.

4.3.1 Mechanisms of Transfer from Solids

Because radon is part of a radioactive decay series, there is a unique process by which it is transferred from solids into pore spaces called alpha recoil. During decay of a radionuclide by alpha emission, the alpha particle is ejected from the nucleus, carrying off most of the excess energy, much like a bullet being shot from a gun. The newly created progeny radionuclide actually recoils in the opposite direction (much like the gun), with energies of 10^4–10^5 times the typical chemical bond energies. The recoiled atom can move within the mineral grain's structure; the recoiling radon atom has a typical range of 20–70 nm.[3,4] Alpha recoil is important because it can break chemical bonds, physically move the recoiled atom to a different position in the solid, and damage the crystal structure. The progeny ion also has different chemical properties from the parent, so that chemical bonds may not necessarily be reformed.

There are several different mechanisms by which alpha recoil enhances radon emanation, as shown in Figure 4.3. Atom "A" is located at a depth greater than the recoil range for the grain; therefore, it is retained in the grain even after recoil. Atom "B" escapes from the original grain but is embedded into the adjacent grain. Atom "C" also escapes from the original grain, but in this case, it is recoiled into the water in the pore space, which absorbs its remaining recoil energy and stops the atom. It is now free to diffuse through the pores. Atom "D" recoils into air in the pore space where little energy is absorbed, so it travels across the pore space and is embedded into the adjacent grain.

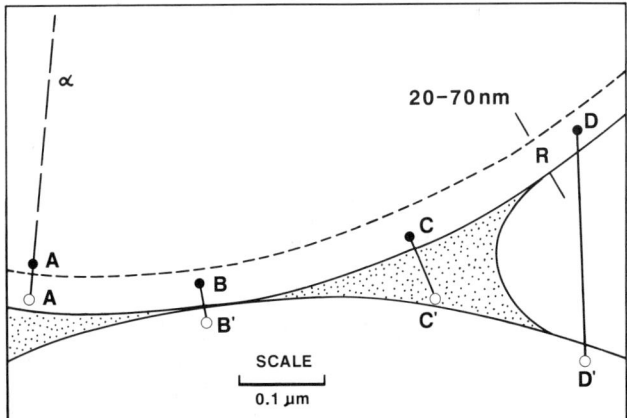

FIG. 4.3. Schematic diagram showing the radon emanation process for different types of alpha recoil. The spheres represent two grains. The stippled area represents pore water, whereas the open zone represents an air-field pore space. See text for discussion.[3]

The radon atoms that end their recoil in the pore space are called the *direct-recoil fraction* of the emanation coefficient. It is important to note that the presence of liquid in the pore spaces increases the direct-recoil fraction because water slows the atom's travel better than air. It is important also to note that, in this model (Figure 4.3), the recoil range, R, is a critical factor in how much radon escapes from the solid. It would appear that for grains which are much larger than the recoil range, few recoil atoms will be able to escape by direct recoil. The recoil range, which is 20–70 nm, is usually only a very small percentage of the total grain dimensions and its radium content, assuming that the radium is homogeneously distributed. The recoil fraction would thus be related to grain size. For example, if the grain size was 1 μm in diameter, about 5% of the atoms may recoil out of the grain, whereas for millimeter-size grains, 0.005% may leave the grain by direct recoil. The radon emanation measured from various materials is much greater than that calculated assuming direct recoil only. So there must be other processes of radon loss from mineral grains.

When the recoil atom is buried in the adjacent grain, such as atoms "B" and "D" in Figure 4.3, the energy absorbed is sufficient to melt or volatilize the solid material along the recoil path. Therefore, the recoil atom could escape back into the pore before the path could cool, a process called *indirect recoil*.[4] There are no estimates of what percentage of the radon emanation is the indirect-recoil fraction, though it could not be significantly greater than the direct-recoil fraction.

A third process of radon transfer is by diffusion, which is migration of the gas relative to the medium, be it solid, liquid, or gas. Radon obeys the standard laws of diffusion, so the flux is proportional to the concentration gradient. Radon diffusion coefficients for various media are shown in Table 4.1. Diffusion coefficients for radon in crystalline solids are extremely small, and the radon will decay before moving any

TABLE 4.1. Radon Diffusion Coefficients for Various Media

Medium	Diffusion coefficient ($cm^2 \ s^{-1}$)	Diffusion length (m)
Air	10^{-2}	2.4
Water	10^{-5}	
Sand	3×10^{-2}	1.5
Argillite	8×10^{-5}	
Concrete	2×10^{-5}	0.04–0.26
Mineral crystals	10^{-9}–10^{-20}	

distance, regardless of crystal size. The presence of water in the pore space decreases radon migration, because the diffusion coefficient for radon in water is about three orders of magnitude smaller than in air. Therefore, there must be other paths along which radon can diffuse more efficiently. In minerals which contain high concentrations of uranium, there are areas of radiation damage to the crystal due to frequent alpha recoils. It has been hypothesized that radon diffusion would be enhanced by radiation damage, but several experiments have not detected any increase in emanation due to radiation damage.[5,6] Therefore, a range of characteristics of the material account for the high radon emanations measured, including grain size which affects the outer surface area, distribution of radium in the solid (especially where radium may be concentrated in the surface layer and at grain boundaries), porosity (both of the material and individual grains), and the amount of interstitial fluids.[7] The relative contribution of each of the factors is difficult to estimate or measure, but experiments have been conducted to show when certain factors predominate.

Transport by molecular diffusion of radon in soils is limited because of its short half-life. Migration by diffusion ranges from about 5 m in gravel to about 2 cm in saturated mud or clay, and distances greater than 1 m are probably unusual.[8]

The importance of porosity of individual grains (internal voids) in contributing to the release of large amounts of radon has been demonstrated by recent studies.[9,10] Emanation coefficients have been measured for dry grains several millimeters in size and found to be similar to those for powdered samples. Therefore, a major part of the radon comes by alpha recoil from *within the grains*, not only from the outer surface, because the grains have many openings with large surface areas. The radon recoils out of the walls of these openings into the internal pores and then diffuses out to the intergranular pores. These internal pores are called "nanopores" because the openings were estimated to be about 10–20 nm wide. One of the most compelling arguments in support of these conclusions was that both ^{220}Rn and ^{222}Rn emanated in proportion to their production rates from their parent isotopes for dry samples, indicating both radon isotopes are generated by recoil into the nanopores. But in wet samples, only ^{222}Rn emanation was detected. If diffusion was a factor governing the escape of radon from a solid, the emanation ratio of ^{220}Rn to ^{222}Rn would be approximately 1:80, the ratio of their half-lives. Instead, recoil is the dominant mechanism.

Radon diffuses out of the nanopores into the intergranular pore space. In wet wamples, ^{220}Rn decays before it can escape the nanopores because of the slower diffusion rates for radon in water. In dry samples, both radon isotopes diffuse out equally, even though ^{220}Rn has a 55-s half-life.

Subsequent experiments were conducted on various sizes of single crystals with high uranium and thorium content.[10] Emanation coefficients for ^{222}Rn for single crystals of apatite, monazite, and uraninite from 0.0002 to 0.2 kg in weight were 0.5–25.0%. The radium content of these crystals is homogeneous, so there is little likelihood that there would be any surface layer still richer in radium on the surface. The emanation coefficient from surface area recoil would be much less than 1%; therefore, it was concluded that these crystals contained an extensive network of nanopores which were connected to the surface.

The only crystals that had low emanation coefficients were zircons (0.001–0.01%), in line with other studies that have shown that zircons, in general, have low emanation coefficients.[11] Most zircons contain water in their mineral structure which, while increasing the direct-recoil fraction, would slow the diffusion of radon out of nanopores relative to other minerals during "dry" experiments.

The presence of nanopores that greatly increase the surface area which generates direct-recoil atoms explains why there is so much unsupported ^{222}Rn in groundwater. As a gas, radon can migrate out of the nanopores, whereas the nongaseous isotopes, which are introduced in approximately equal quantities to the nanopores, are readily adsorbed onto the surface. Because of its short half-life, ^{220}Rn does not diffuse out of the nanopores. Therefore, large amounts of ^{222}Rn can enter intragranular pores, resulting in high radon concentrations, particularly when the intergranular pore volume is relatively low, such as in crystalline rock aquifers.

Radon emanation coefficients can be higher than expected, when the radium content of the solid is not homogeneous. This situation is important in two very different settings. In fresh rocks, particularly granites, a significant amount (up to 30%) of the uranium (and radium) content of the rock can occur as intergranular films, which would generate a large direct-recoil fraction of radon. In weathered rocks and soils, uranium and thorium released during disintegration of the various minerals are readily adsorbed onto the surface of clay particles that are formed during the weathering process. Coprecipitation with iron oxide and formation of secondary minerals in pore spaces or fractures commonly scavenge uranium and radium from solution. Thus, the surface of grains can be relatively enriched in radium and enhance radon emanation. This effect, caused by the distribution or siting of radium, can be very important, especially for soils. A series of experi-

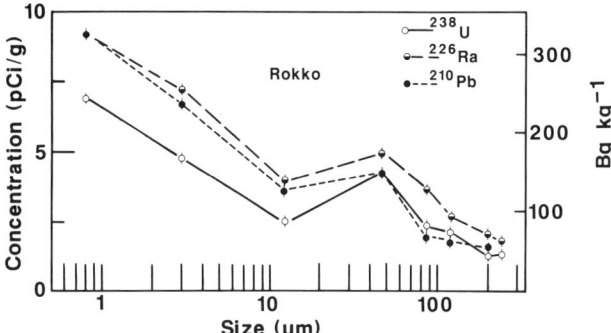

FIG. 4.4. Distribution of uranium-series isotopes in a granitic soil versus soil particle size.[13]

ments demonstrated this effect of distribution on radium siting and radon exhalation, which is a measure of the rate of radon release from a material.[12,13] Using two different soils derived from weathering of granites, the distribution of the ^{238}U series nuclides on the soil particles in relation to their size was determined, as shown in Figure 4.4. The concentrations of ^{238}U, ^{230}Th, ^{226}Ra, and ^{210}Pb all increase with decreasing grain size below 100 μm. Surface sorption plays an important role in this enrichment, and the surface area per gram of soil particle increases with decrease in particle size. For larger soil particles, the amount of a nuclide adsorbed onto the surface is negligibly small in comparison to that contained in the soil originally because the surface area per gram is relatively small. The calculated radon exhalation rates also increased with decreasing grain size, from 0.04 to 0.29 Bq m^{-2} s^{-1}.[12] Thus, two soils with similar radium content can generate significantly different radon concentrations in the pore spaces, depending on the geochemical distribution of the radium and the grain-size distribution of the soils.

As discussed earlier, moisture can both positively and negatively affect the radon emanation rates from solids in several ways. Moisture in pore spaces between grains can increase the direct-recoil fraction relative to air-filled pores because of the shorter recoil distance in water. A water film of about 0.1 μm thickness is sufficient to stop the recoiled atom in the pore. If there is a moisture content gradient in the soil, active transport of radon on water molecules may take place. However, if the internal pores within individual grains are filled with water, the rates of radon diffusion from within the grains are reduced. In fact,

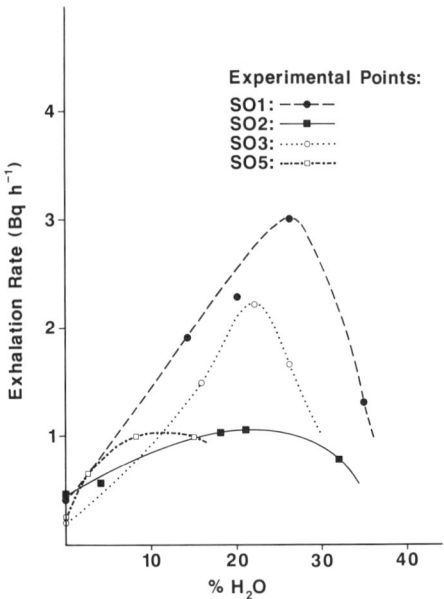

FIG. 4.5. Radon exhalation rate versus moisture content for different soil samples.[14]

Rama and Moore[9] used the observation of decreased radon emanation from wet versus dry samples to support the hypothesis of the presence and role of nanopores in radon transfer from solids. Radon diffusion in water is about 10^{-3} times that in air (Table 4.1). Water in intragranular pores as well slows radon diffusion. The effects of increasing moisture content on radon exhalation rates for various soils can be seen in Figure 4.5. Up to a certain level of moisture in a soil, the positive effects of enhancement of the alpha-recoil fraction can increase the exhalation rate by 50–250%. Above some moisture content, the reduction in diffusion predominates. The same trends were observed in experiments for shale, soil, and concrete.[14]

4.3.2 Transport by Groundwater

Because of the chemical inertness of radon, its transport in groundwater systems is controlled only by molecular diffusion and the flow of groundwater itself. With a diffusion coefficient of 10^{-5} cm^2 s^{-1} in water, diffusion is an important transport mechanism at scales on the

SOURCES

order of 10^{-2} m. As discussed in the previous section, diffusion is an important process by which radon from the microcrystalline fractures in a grain is transported into the pore spaces of the aquifer.

Groundwater flow is the dominant mechanism by which radon is transported in aquifers. The limiting factor in the transport length of radon in groundwater is its half-life. In 30 d, the radon content of groundwater will be less than 1% of its original activity. The only other process that has any other significant effect on radon, once it is in solution in groundwater, is outgassing either by exposure to large air spaces, such as in caves and cavities in limestone, or by elevated temperatures. Radon is most soluble in organic liquid phases (when compared to gas and water phases), so there may be unique settings where radon transport is affected by organic pollutants.

4.3.3 Radon in the Atmosphere

The concentration of radon in the atmosphere is governed by the source strength and dilution factors, both of which are strongly affected by meterological conditions. The largest source of atmospheric radon is from the soil surface, although secondary contributors include the oceans, natural gas, geothermal fluids, volcanic gases, ventilation from caves and mines, and combustion of coal. The primary mechanism of radon removal from the atmosphere is by decay, whereas radon progeny can be removed by other processes such as wash-out and sorption onto particles which settle out. Consequently, radon progeny are generally depleted relative to radon in the lower atmosphere.

There have been many studies of the sources and transport of atmospheric radon, primarily concerned with the use of radon as a natural tracer of air masses.[15-17] A comprehensive review was conducted by Gesell,[18] which summarizes background atmospheric radon concentrations. Radon levels in the atmosphere have been found to vary with height, season, time of day, and location.

Because the source of radon is primarily the soil/air interface, there is a pronounced vertical gradient in radon concentrations. Figure 4.6 shows typical radon profiles with height above ground, with each curve having a different scale factor. One observation from these data of human health concern is that a child breathing outdoors at a height of 0.5 m is exposed on the average to about 16% higher radon levels than an adult breathing at 1.5 m.

Seasonal variations in radon levels have been monitored through-

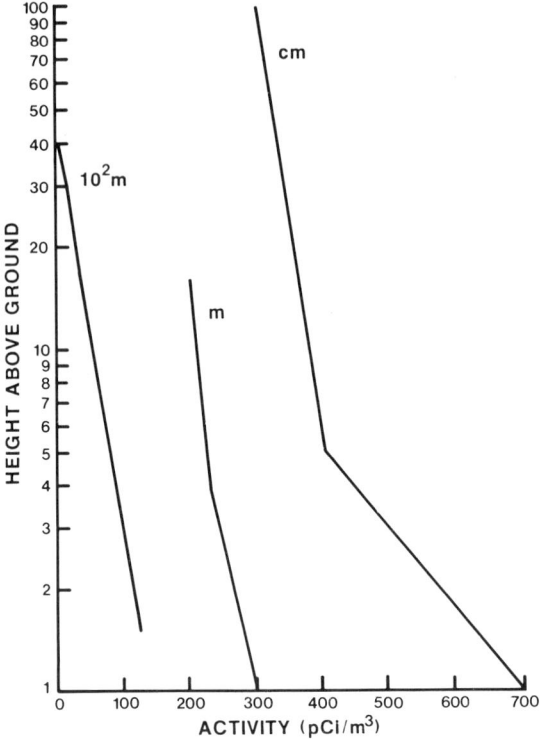

FIG. 4.6. Variation of atmospheric radon or radon progeny concentration with height above the ground surface. Note the changes in scale for each curve.

out the world by both continuous and intermittent measurements. The results are varied, depending on the location (Figure 4.7). In New Jersey, a pronounced maximum was observed in midsummer (June/July). In Cincinnati and Washington, D.C., a spring minimum and a fall maximum were reported. These results and data from other countries have been used to generalize that a maximum value of radon occurs in the autumn and early winter corresponding to a minimum in turbulent transfer. The minimum in radon occurs in spring due to the opposite condition.[19] These studies also showed that high mountain regions have little seasonal variation; areas with little soil exposed have very low, constant values; and seasonal variations generally range by a factor of 2–4 between the maximum and minimum values.

Radon varies significantly with time of day, which is a function of

SOURCES

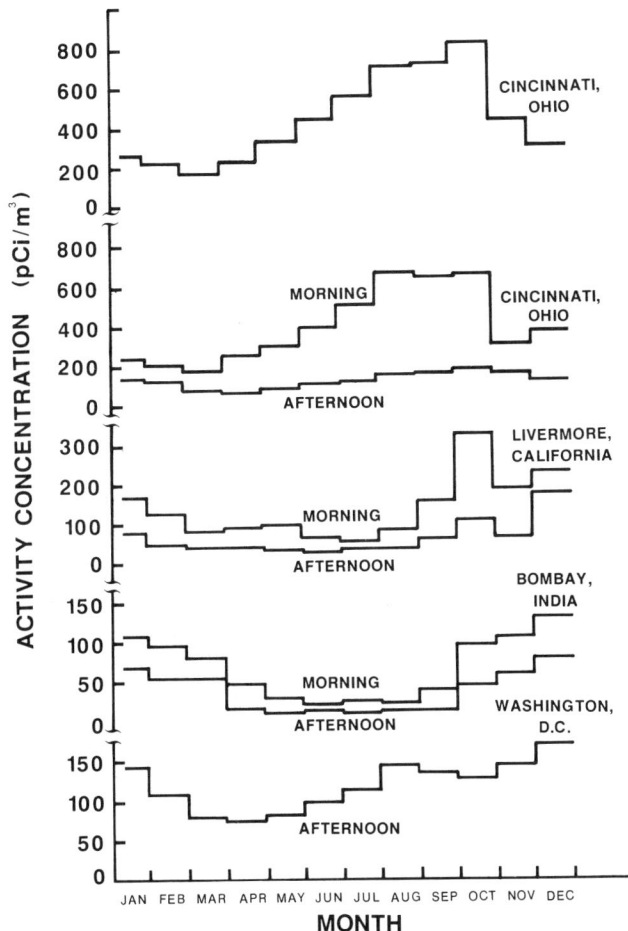

FIG. 4.7. Annual variation of atmospheric radon concentration obtained from measurements made once or twice per day. Radon is inferred from radon progeny measurements.[18]

atmospheric stability. Figure 4.8 shows diurnal variations in atmospheric radon or radon progeny at selected locations, plotted as the average of numerous daily measurements. All locations show the same pattern. As summarized by Gesell,[18]

> Early morning atmospheric temperature inversions lead to an extremely stable atmosphere which restricts vertical turbulent mixing and leads to relatively higher near-ground radon concentrations.

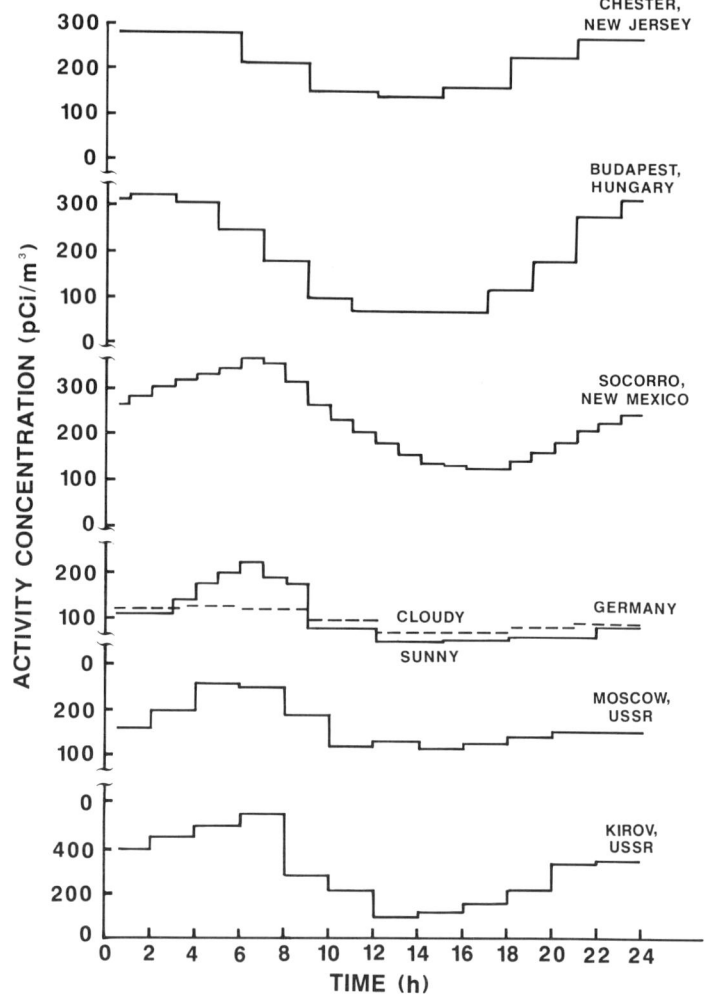

FIG. 4.8. Diurnal variation of atmospheric radioactivity (radon or radon progeny).[18]

Shortly after sunrise, solar radiation warms the ground which in turn warms the lower atmosphere, breaking up the inversion and leading to a substantial decline in concentration. Concentrations remain low until late afternoon when radiant cooling of the surface leads to increase in atmospheric stability and a corresponding increase in radon concentration.

Diurnal variations range by a factor of 2–5, with early-morning highs and midday lows.

TABLE 4.2. Average Radon/Radon Progeny Concentrations in the United States[a]

Location	Period of measurement	Average value $(Bq\ m^{-3})$
Grants Mineral Belt, NM	November	22.8
Grand Junction, CO	Annual	27.8
Laguna, NM	June	18.5
Cincinnati, OH	Annual (morning)	16.3
Cincinnati, OH	Annual (afternoon)	5.1
Argonne, IL	Late spring/summer	11.1
Socorro, NM	Annual	8.9
Chester, NJ	Annual	8.1
Lloyd, NY	Summer only	7.4
Lloyd, NY	March/April	3.0
Washington, DC	Annual (afternoon)	4.5
Hawaii	May/June	1.0
Wales, AK	Annual (afternoon)	0.7
Kodiak, AK	Annual (afternoon)	0.4

[a] Modified from Reference (18).

The largest variations in atmospheric radon occur spatially, as would be expected. Even though there have been numerous studies of radon and its progeny in air throughout the globe, comparison of these data is difficult because of large variations in sampling frequency, measurement techniques, and units of reporting. The best compilation of data was conducted by Gesell,[18] whose results are summarized in Table 4.2. The highest concentrations occurred in the Colorado Plateau, a region of extensive uranium mineralization with levels of 18.5–27.8 $Bq\ m^{-3}$. More "average" values of 4.5–11.1 $Bq\ m^{-3}$ were found throughout the continental United States. Lowest values occurred in Alaska and Hawaii, probably due to ice cover and small land mass. Based on these and other data, the mean value for atmospheric radon in normal areas of the contiguous United States lies in the range of 4–15 $Bq\ m^{-3}$ and is probably about 9 $Bq\ m^{-3}$.[18]

4.4 SOURCES OF RADON IN INDOOR AIR

4.4.1 Introduction

Study of indoor radon has evolved and changed directions many times over the last 20 years. Early work with radon was on its use as a tracer of transient processes such as atmospheric mixing. It was used extensively as a uranium and petroleum exploration tool and to predict earthquakes. The environmental concentrations of radon began to be more intensely studied when high levels were found in homes in Sweden constructed with concrete containing alum shale, a high-uranium-content material commonly occurring in Sweden. As a result, much of the research on radon related to indoor air up until the early 1980s was focused on the radium content and radon exhalation of building materials. Areas with enhanced radium concentrations, such as sites where uranium and phosphate tailings were used as fill, were also studied in the 1970s. The U.S. Environmental Protection Agency (USEPA) started to evaluate radon in public drinking water supplies in the early 1980s as part of its program to develop regulations for contaminants in drinking water. When researchers started to report elevated indoor radon levels in areas outside of known uranium mineralization, it became apparent that elevated indoor radon levels were occurring more frequently than expected. Yet it was not until the discovery of extremely high radon in homes in the Reading Prong region of New Jersey and Pennsylvania in 1985 that the potential magnitude of the problem was realized by state and federal agencies in the United States. This realization has sparked a new effort into the study of the sources of radon and its transport from soils into structures. Most of this work is in progress and promises to add greatly to our present understanding of radon.

In the next section, the primary sources of indoor radon—soils, groundwater, and building materials—will be discussed in detail. Data on minor sources of radon—uranium and phosphate mill tailings and fuels—also will be presented. There are several comprehensive reviews on the sources of indoor radon[7,18,20] that are recommended for additional background reading.

4.4.2 Radon in Soils

Studies of soils as sources of indoor radon have focused on two primary areas:

1. Factors which affect radon concentration in soil gas.
2. Factors which enhance the transfer of radon from soils into the overlying air or adjacent structures.

High values in either of these two general parameters, soil gas radon

SOURCES

and transport efficiency, can result in elevated radon values in structures. High values in both can generate dangerously high levels.

Factors that affect the radon concentrations in soil gas include the radium content and distribution, the porosity (both intergranular and intragranular), moisture, density, and, in some cases, the underlying bedrock. Although the radon content of soil gas is not directly a function of the radium concentration of the soil, radium concentration is an important indicator of the potential radon source in soils and bedrock. The data base on radium and uranium concentrations in rock is much more extensive than the one on soils; however, the radium content of soils should reflect the relative levels in the different types of parent bedrock from which the soils were derived. Table 4.3 lists the uranium content of major rock types, and Table 4.4 lists the uranium and radium content of soils. The range in the average concentrations is relatively small, although individual samples can vary by a factor of 100 or greater, even for nonmineralized rocks and within samples separated only by a few meters. Rock types with high uranium concentrations are granites, metamorphosed igneous rocks, and shales, with average concentrations of about 50 Bq kg^{-1}. For rock types with uranium mineralization, concentrations of uranium-bearing sandstones can be as high as 500,000 Bq kg^{-1} and phosphates generally have up to 1500 Bq kg^{-1}.

Radium in unmineralized soils is typically in the range of 10–100 Bq kg^{-1}. Even though there has been no synthesis of measurements by soil type, the relationships should be similar to those for rocks with the radium content generally a little lower than the uranium content. In one survey, 330 soil samples were collected along major highways in 33 states, resulting in a range of ^{226}Ra of 8.5–160.0 Bq kg^{-1} and an arithmetic mean of 41 Bq kg^{-1}.[24] Other studies using aerial radiometric data for seven western states showed a range of mean radium concentrations, averaged over different geologic units, of 4–130 Bq kg^{-1}.[25] But average radium content cannot be used to estimate soil-gas levels of radon. Studies being conducted by the U.S. Geological Survey have shown that soil-gas radon levels vary widely in small areas (within a housing lot) and are not well correlated with radium content of the soil.[23] Table 4.4 shows the available data on the radon soil-gas concentrations in different soils. Current research by several groups has indicated that soil porosity is a critical factor causing high soil-gas radon levels, even in areas which have "normal" radium concentrations.[8,21,7] In Spokane, Washington, in soils formed on coarse glacial outwash deposits which have 2–3 ppm uranium, the soil-gas radon is 7000–37,000 Bq m^{-3}.[23] In Sweden, the coarse gravel soils associated with

TABLE 4.3. Uranium Content in Rocks, Minerals, and Sedimentary Deposits[a]

Rock type	No. of samples	Mean (ppm)	Range (ppm)
Igneous rocks			
Mafic igneous rocks	236	0.9	0.0–3.4
Diorites	69	2.0	0.0–11.0
Granodiorites	156	2.6	0.5–9.0
Granitic rocks	535	4.7	0.5–21.0
Minerals of igneous rocks			
Quartz		0.7	
Biotite			1.0–40.0
Potassium feldspar			0.2–3.0
Plagioclase			0.2–5.0
Zircon			300–3,000
Apatite			5.0–150.0
Monazite			1.0–30.0
Metamorphic rocks			
Gneiss		2.0	
Schist		2.5	
Slate		2.7	
Sedimentary deposits			
Black shales		8.0	3.0–250.0
Gray and green shales, North America	52	3.2	1.2–12.0
Mancos Shale, western United States	102	3.7	0.9–12.0
North American carbonates	516	2.2	0.6–8.8
Florida limestones	10	2.0	0.5–6.0
California limestones	27	1.3	0.03–4.9
Orthoquartzites	18	0.4	0.2–0.6
Phosphates			50.0–300.0
Beach sands (United States)	83	3.0	

[a] Modified from Reference (1).

glacial eskers which have 4–10 ppm uranium can have 10,000–20,000 Bq m^{-3} of radon.[22]

High permeability of soils can have two effects which increase indoor radon levels in adjacent structures. In highly permeable soils, especially sand and gravel, radon migration by diffusion is great, up to a depth of several meters. Thus, radon in the soil gas can accumulate from a much larger area. Also, the convective flow of radon-bearing soil gas is proportional to permeability, and permeable soils allow rapid transport of radon in soil gas into the structure under very small pressure differentials. Conversely, relatively impermeable soils, such as silt and clay, do not have sufficient porosity to allow transfer of significant amounts of soil gas into a house. Clays usually have high

TABLE 4.4. Uranium and ^{226}Ra Content of Soils and Radon in Soil Gas

Soil type	U content (ppm)	^{226}Ra (Bq kg^{-1})	Soil-Gas radon (Bq m^{-3})	Reference(s)
Sweden				
Till, normal		15–62	5000–30,000	21
Till, with granitic material	5–30	30–125	10,000–60,000	22, 21
Till, with U-rich granite		125–360	10,000–200,000	21
Till, alum shale	10–100	175–2500	100,000–>1,000,000	22, 21
Esker gravel	4–10	30–75	10,000–200,000	22, 21
Sand, silt		6–70	2000–30,000	21
Clay		25–100	10,000–80,000	21
United States				
Glacial outwash	2–3		2000–40,000	23
Reading Prong, NJ			40,000–>1,000,000	23

water content, which tends to further decrease soil-gas levels. If the ground fails to pass the percolation test that is used to measure the suitability of a site for a septic drain field, it will probably have low permeability and low radon potential.[8] The only exception to this rule is for dry, desiccated clays that can have extensive cracks which may allow radon transport.

Permeability is also important in enhanced radon transport in rocks. Sheared and fractured rock, resulting from certain types of metamorphic and deformation processes associated with mountain building, have a higher potential for emanating high radon levels. Fractures allow transport of radon from greater depths than unfractured rock, and uranium can be preferentially located along fractures. Fractured bedrock may be an important contributor to the very high radon levels found in the Reading Prong area.

Soil moisture and depth to water table can have temporal effects on soil-gas concentrations by slowing diffusion. In soils wetted by the capillary fringe, the emanation coefficient is much higher, in the range of 15–55% for granular soils due to enhanced direct recoil effects of intergranular moisture. In comparison, desiccated soils have emanation coefficients of only a few percent. In Sweden, lower values of soil-gas radon were obtained when the water table in a till with high amounts of alum shale was very near the surface.[26] The high water table hindered the radon gas from the alum shale from reaching ground level.

Factors which affect the transfer of radon in soil gas into the air and structures are porosity and permeability of the soil, barometric pressure, pressure-driven flow of air through the soil, moisture, wind, thickness of the soil over bedrock, and, to a lesser degree, temperature. Each of these factors is discussed below.

A factor which has been found to be a major mechanism of transport of radon from soils into structures is pressure-driven flow. Much of this work has been conducted by researchers at the Lawrence Berkeley Laboratory, University of California. Flow-inducing mechanisms are stack effect, wind, precipitation, and barometric pressure.[27] The "stack effect" is due to the pressure differential that occurs across walls separating air masses of different temperatures. It is a major contributor to air infiltration, especially in winter. In buildings 5 m in height, a temperature difference of 20°C can cause a pressure differential of 4 Pa from floor to ceiling, pressures which can result in infiltration rates of the order of 100 m^{-3} h^{-1}.[27] Some of the infiltration can occur through the floor, drawing radon-rich air from the soil. Wind also generates pressure differentials that induce air infiltration into a

SOURCES

structure on a scale similar to temperature effects, but with a different spatial distribution. Winds can enhance the exchange of air between the soil and overlying structure, by forcing air flow in from the soil on the windward side and out into the soil on the leeward side of the structure.

The influence of depressurization of a structure on soil-gas migration was investigated using a tracer gas (SF_6) and indoor radon measurements.[28] The tracer gas was injected into the soil at two different residences and its concentration monitored over time at various points in the soil and in the houses. Different experiments were designed to monitor the movement of air in the soil itself (shallow and deep) and the entry of soil air into the basements as well as to map the pressure field in the soil. Depressurization of 25–50 Pa was shown to have a substantial effect on the transport rates of the tracer gas within the soil, with effective transport velocities exceeding 1 m h^{-1}. In addition, radon levels in the basement of one residence were monitored during the six-day experiment, the results for which are shown in Figure 4.9. Depressurization caused a marked increase in the indoor radon concen-

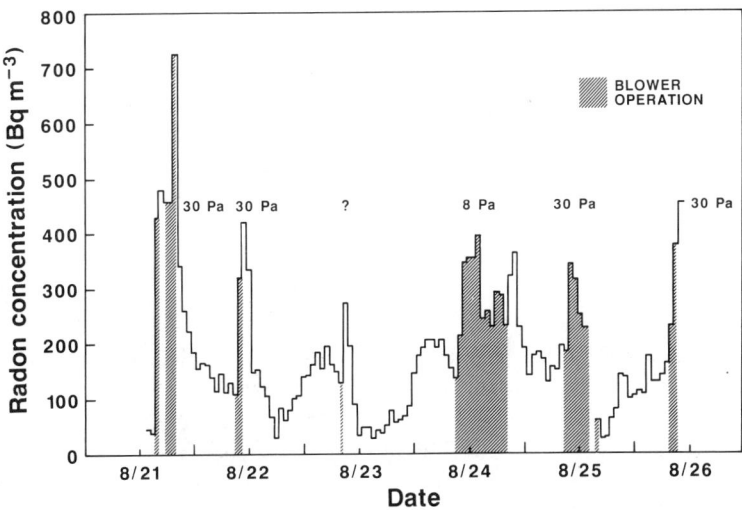

FIG. 4.9. Radon concentration versus time in a house in which a blower fan was operated to induce depressurization. Radon concentrations increased due to pressure-driven flow of soil gas into the house.[28]

trations, further supporting the conclusion that pressure-driven flow can cause substantial transport of soil gas into a structure.

Precipitation effects on flow of radon into structures are not well understood, but some studies have shown increased indoor radon during heavy rainfall. It has been postulated that the increased moisture content of the soil and/or water percolation through the soil forces soil air to flow into the crawl space.[27] Changes in barometric pressure usually occur during rainfall so it is difficult to separate out the two effects. Atmospheric pressure changes of 1–2% associated with the passage of frontal systems has been shown to produce inversely proportional changes in the radon flux from soils of 20–60%.[29] Increasing pressure forces low-radon air into the ground and decreasing pressure draws radon-rich air from the ground.

All of these processes discussed above affect the physical flow of air through the soils (and into the structure), not the rates at which radon is generated or diffuses through the soil. Increases in temperature, on the other hand, have been shown to increase exhalation of radon from soils in laboratory experiments, probably due to decreases in the amount of radon adsorption onto the soil materials. Radon exhalation rates approximately doubled from 5° to 22°C and again from 22° to 50°C.[14] The effects of soil thickness on radon transfer have been studied only indirectly. In Sweden, the alum shale bedrock is a major source of radon in soils and is overlain by various types of glacial till, which has a shielding effect and can reduce soil-gas levels significantly.[26] There have not been any detailed surveys measuring the radon flux in soils of varying thickness formed over a uniform bedrock.

Extensive studies are being conducted to identify those conditions most likely to generate high radon soil-gas levels. Several approaches have been used, but all efforts focus on the production of maps delineating areas with potentially high radon levels in soil gas. As examples, approaches used in Sweden and the United States will be discussed.

In 1979, when the problem of high radon in Swedish homes became widely publicized, building codes were passed requiring that radon progeny in new buildings could not exceed 70 Bq m^{-3} and in existing buildings could not exceed 400 Bq m^{-3}. Infiltration of radon from the ground was the main contributor in Sweden to elevated indoor radon levels.[30] Specific building techniques had to be used when developing areas with high soil-gas radon levels. The Geological Survey of Sweden was commissioned to produce maps of all areas and rock types known to be radioactive so that the high-risk areas could be identified. Two types of maps have been produced: the first were the GEO-radiation maps which delineated high radon areas. The areas with high risk for elevated soil-gas radon levels were considered to be those with a bedrock of uranium-rich alum shale, uranium-rich granites or pegmatites, and areas of glacial drift cover comprised largely of these rock types.[31] Most of the bedrock in Sweden is covered by 4–10 m of glacial till, which frequently contains various amounts of fragments derived from these uranium-rich rock types.

The GEO-radiation maps have been shown to adequately define the areas at risk of high soil-gas radon levels due to elevated uranium content in the bedrock or soil cover. However, measurements of soil gas in areas considered to have normal uranium levels (2–10 ppm) have detected radon concentrations from 1000 to 200,000 Bq m^{-3}, with the higher levels observed in soils formed on glacial eskers which have extremely high porosity. Thus, the GEO-radiation maps cannot be used to identify all high-radon areas. Subsequently, radon-risk maps have been developed for determining the need for building requirements.[32]

The radon-risk classification on the maps is based on geological criteria, as follows:

High-risk areas (radon concentration in soil gas: >50,000 Bq m^{-3})

- Uranium-rich granites.
- Pegmatites.
- Alum shale.
- Highly permeable soils.
- Building requirements: "radon-safe" construction, i.e., thicker reinforced foundation, ventilation below foundation.

Normal-risk areas (radon concentration in soil gas: 10,000–50,000 Bq m^{-3})

- Rocks and soils with low or normal uranium content.
- Soils with average permeability.
- Building requirements: "radon-protective" construction, i.e., no open holes in the foundation.

Low-risk areas (radon concentration in soil gas: <10,000 Bq m^{-3})

- Rocks and soils with very low uranium content.
- Limestone.
- Sandstone.
- Basic igneous and volcanic rocks.

- Soils with very low permeability (clay and silt).
- Building requirements: none.

An example radon-risk map is shown in Figure 4.10.

In the United States, several approaches and levels of detail have been used to identify high-risk areas. As a first nationwide effort, in 1986 the USEPA produced a map of the United States showing areas with potentially high radon levels in soils (Figure 4.11). This approach was admittedly simple, based primarily on geological data synthesized from the National Uranium Resource Evaluation (NURE) program. Five rock types or geological settings were identified as having the potential for high radon:

1. Near-surface occurrence of potential uranium resources.
2. Granitic rocks not differentiated by uranium content.
3. Near-surface occurrence of uraniferous phosphate deposits.
4. Uraniferous black shales.
5. Major thorium and rare-earth mineral deposits (as areas and localized occurrences).

Thus, this map is based solely on the uranium content of near-surface rocks and does not address soil characteristics. The shaded areas in Figure 4.11 represent those areas which may have a higher percentage of houses with elevated radon levels as compared to the nonshaded areas. The USEPA estimated that 12% of homes nationwide may have annual radon levels greater than 150 Bq m^{-3}, the level recommended by the USEPA as of 1986 as a corrective action target.

Another approach that has been studied in the United States and Canada is the delineation of areas that have high radium in the near-surface soil and rock using data acquired by the National Airborne Radiometric Reconnaissance (NARR) Program. The coverage of the NARR is extensive and would provide a systematic, cost-effective approach to estimate radon source strengths. A detailed analysis of the feasibility of this approach of using NARR data only was conducted for California and the Pacific Northwest.[25] The results showed that significant regional variations in radium concentrations could be detected, but only readily at the 1° × 2° quadrangles scale, about 1:250,000. This approach could be used to rank areas by potential radon source strength and to identify areas which should have higher priority for detailed study.

There have been other efforts to use NARR data in conjunction with geologic mapping, ground measurements of gamma-ray activity, radium content of soils, soil-gas radon, and indoor radon levels. In

SOURCES

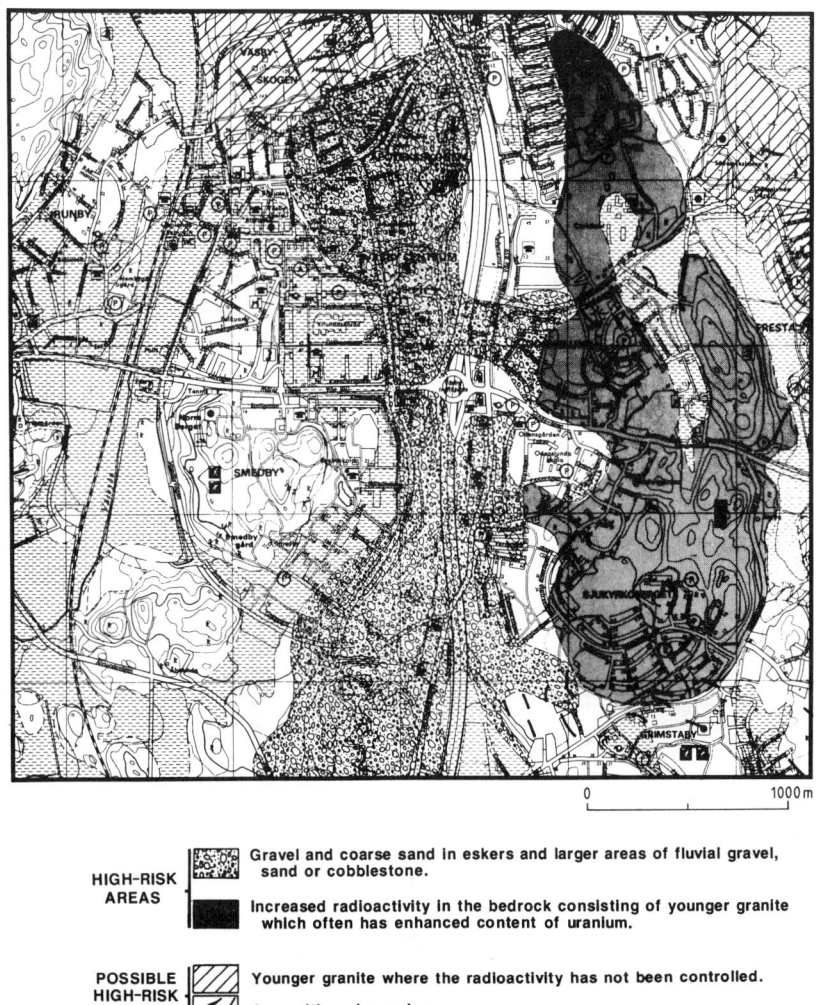

| HIGH-RISK AREAS | Gravel and coarse sand in eskers and larger areas of fluvial gravel, sand or cobblestone. |
| | Increased radioactivity in the bedrock consisting of younger granite which often has enhanced content of uranium. |

| POSSIBLE HIGH-RISK AREAS | Younger granite where the radioactivity has not been controlled. |
| | Area with end moraine. |

| NORMAL-RISK AREAS | Till or bedrock with normal radioactivity. |

| LOW-RISK AREAS | Sand, silt or clay. |

FIG. 4.10. Example of a radon risk map produced in Sweden identifying high-, normal-, and low-risk areas. These maps are used to identify areas requiring specific building codes to prevent elevated indoor radon levels.[32]

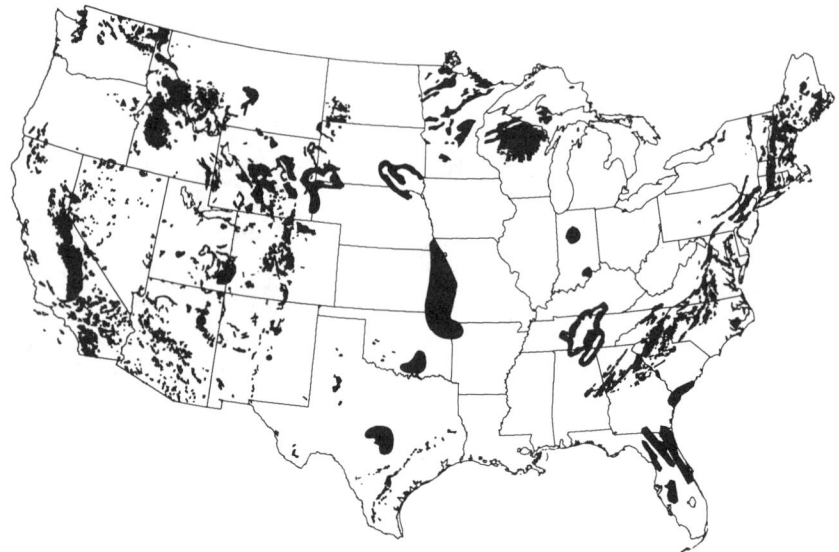

FIG. 4.11. Map of the United States showing areas with potentially high radon levels in soil gas, based primarily on geological reports and modification of NURE data.[33]

Canada, indoor radon levels were compared with airborne radiometric data, showing a positive correlation.[34] GEO-radiation maps in Sweden were based on this combination of data.[31] From the results of these studies, it can be concluded that airborne radiometric data are useful in identifying areas with potentially high radon but need to be used in conjunction with other data for making predictions on radon levels.

4.4.3 Radon in Surface Waters

Radon concentrations in surface waters are generally very low, particularly in water bodies used as drinking water sources. There are three sources of radon in surface water: (1) radon supported by ^{226}Ra in solution and suspension; (2) discharge of radon-laden groundwater; and (3) diffusion from bottom sediments.

Figure 4.12 shows the typical radon profile in a water column. Because of its short half-life, radon is generally in equilibrium with ^{226}Ra in oceanic water. However, radon is lost rapidly from the water surface to the atmosphere by diffusion. The magnitude of radon loss

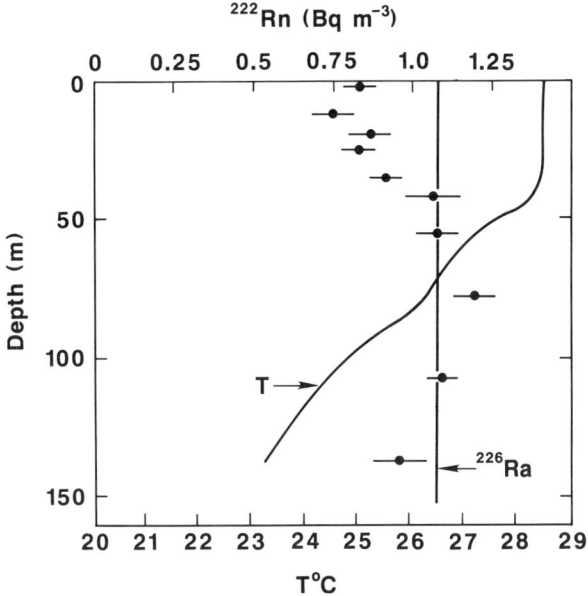

FIG. 4.12. Dissolved radon concentration as a function of depth in the ocean surface. The radon deficiency relative to ^{226}Ra in the upper 50 m is caused by loss to the atmosphere.[35]

depends on gas-exchange rates between the air and water and the rate of vertical mixing of the water column. The gas-exchange rate is controlled by molecular diffusion through a stagnant film tens of micrometers thick at the air–water interface. The thickness of this layer is a function of the rate of turbulent mixing caused by winds, waves, and currents. Mixing below this layer is by eddy diffusion. Radon has proved to be an excellent tracer in the study of air–water gas-exchange rates in lakes, rivers, estuaries, and oceans.[36–38]

The radon profile at the sediment–water interface is opposite that discussed above for the air–water interface (Figure 4.12). Radon occurs in excess relative to ^{226}Ra near the sea floor, due to the diffusion of radon out of bottom sediments. In fact, the ^{226}Ra profiles show increases with depth, because ^{226}Ra also diffuses out of bottom sediments, but the flux of radon, as an inert gas, is about 1–2 orders of magnitude greater than that of ^{226}Ra. Thus, analysis of the ^{222}Rn:^{226}Ra activity ratio allows study of vertical mixing rates of deep bottom water.

Concentrations of ^{226}Ra in the ocean away from effects from river discharges range between 1 and 7 Bq m^{-3}.[35] Although ^{226}Ra increases with depth in the upper water column, the bottom profile changes very slowly because of its 1622-y half-life. The radon profile can change very rapidly, as shown in Figure 4.12, in a quasi-exponential manner. This profile matches that predicted by equations for eddy diffusion mixing, such as

$$K_2 \frac{\partial^2 C}{\partial z^2} - \lambda_{222} C = 0$$

with K_2 the vertical eddy diffusion coefficient, z the depth above the bottom, and C the excess radon concentration. Profiles such as that shown in Figure 4.12 have been used to determine the value of K_2 under different conditions. Deviations from the predicted radon profiles for steady-state conditions have proven to be very important in detecting transient effects, such as lateral mixing off of adjacent topographic highs and turbidity currents.

4.4.4 Radon in Groundwater

Although the occurrence of radon in groundwater was first reported in 1902 by J. J. Thompson at Cambridge University in England, it was "rediscovered" in Maine in the 1950s (see Chapter 2). Since that time, there have been numerous studies of radon in groundwater in specific areas (usually for an area as large as a state). A nationwide survey of radon in public water supplies was conducted from 1980 to 1982 for the USEPA.[39] In this survey, samples were collected at the tap from more than 2500 public water supplies from 35 states, representing 45% of the water consumed by groundwater users in the United States. However, only 5% of the total number of groundwater suppliers in the 48 contiguous states were sampled because mostly suppliers serving at least 1000 people were sampled. Therefore, the results of this survey represent the radon levels in larger systems, a fact that is important in review and extrapolation of the data.

In contrast, most of the detailed studies conducted by university researchers or state agencies have focused on smaller public supply systems or domestic wells. Sampling strategies among these studies have varied significantly. Samples were frequently collected as close to the source well as possible in some studies, allowing the water to run for several minutes so that samples representative of groundwater were collected. Other studies collected samples at the point of use, at the tap.

These inherent differences in the objectives, design, and sampling strategies need to be considered in comparing the results among all the studies.

A common conclusion from many studies is that the small public groundwater supplies and private wells tend to have the highest radon-in-water levels.[40] This relationship is not due, however, to any functional relationship between system size and radon concentration, such as losses by aeration or decay through extensive distribution systems. The correlation reflects the relationship between system size and aquifer composition or rock type. Larger systems obtain groundwater from aquifer types which tend to have lower radon than those which are used by smaller systems and domestic supplies. For example, crystalline rock aquifers (e.g., granites) do not produce enough water for large users so surface-water sources are used. Smaller users could rely on groundwater from a granitic aquifer. A statistical test of the relationship between system size and radon concentration was conducted for 272 public systems in North Carolina.[41] The results showed that system size does not systematically affect radon concentration independent of aquifer composition or rock type.

Most detailed studies have shown that radon concentration is directly related to the composition and rock type of the aquifer. The occurrence of radon by aquifer type is discussed below. Data on radon concentrations in groundwater from different aquifer types are given in Table 4.5.

4.4.4.1 Igneous and Metamorphic Rock Aquifers

Aquifers composed of igneous and metamorphic rocks generally produce limited amounts of water and so are used by small suppliers. Yields are seldom greater than $0.3–3.0$ L s^{-1}. Igneous rocks can be divided into two large classes: acidic rocks, such as granites, and basic rocks, such as basalts. Metamorphic rocks traditionally are divided into different grades of metamorphism defined by the mineral types that are formed during heating and pressurization. However, there are other characteristics of metamorphic rocks that are important factors in affecting the radon levels in groundwater. The composition of the original material prior to metamorphism affects the amount of uranium in the rock, e.g., metamorphic granites will be very different from metamorphic limestone (marble). Higher grades of metamorphism can be associated with the formation of mineralizing fluids, which can be enriched with uranium. The high radon in the Reading Prong region

TABLE 4.5. Average Radon Concentrations in Groundwater by Aquifer Type

Aquifer type	No. of samples	^{222}Rn (Bq m^{-3})	Reference
Granites			
Maine	136	817,700	42
North Carolina	24	390,800	43
South Carolina	22	298,800	44
Sweden	14	92,000	45
Metamorphic rocks			
Maine			
Sillimanite Zone	35	504,300	42
Chlorite Zone	56	41,000	42
North Carolina			
Gneiss/schist	71	83,000	43
Metavolcanic	21	49,900	43
South Carolina			
High-grade—Monazite Belt	12	53,400	46
Medium-grade	11	118,100	46
Low-grade	7	274,700	46
Sweden			
Gneiss	8	26,000	45
Limestone			
Florida	165	550	39
South Carolina	15	1,300	46
North Carolina	22	3,440	43
Sweden	12	24,000	45
Unconsolidated sand aquifers			
North Carolina Coastal Plain	139	15,760	43
Minnesota (Glacial Drift)	350	11,470	47
South Carolina			
Lower Coastal Plain	15	6,950	46
Middle Coastal Plain	34	9,470	46
Upper Coastal Plain	29	17,340	46

reflects uranium mineralization that occurred during metamorphic events.

Radon in groundwater from the crystalline aquifers of igneous and metamorphic rocks is generally higher than from all other aquifer types. In every study where the average radon content in groundwater is reported by rock type, granites always have the highest levels. Average levels in water from granites, as shown in Table 4.5, are usually 100,000 Bq m^{-3} or greater. Concentrations in individual domestic wells can frequently be as high as 3.7×10^6 Bq m^{-3}.

The radon levels in groundwater from metamorphic rock aquifers range widely (Table 4.5). In a detailed study involving over 2000 wells in Maine, no relationship between radon and metamorphic grade in rock aquifers was found.[42] However, the high-grade metamorphic terrains had the highest individual radon levels, resulting from the development of pegmatites with associated uranium mineralization. In North Carolina, very high-grade metamorphic rock aquifers such as those composed of gneiss and schist had higher radon levels than metavolcanic rocks, which were higher in radon than metasedimentary rocks. In contrast, data from the South Carolina Piedmont showed that water from high-grade metamorphic rocks had lower radon concentrations than water from medium- or low-grade metamorphic rocks.[46] The original composition of the rock is more important than metamorphic grade. In the Piedmont and Appalachian provinces, metamorphic rocks derived from sedimentary rocks will have lower radon concentrations in groundwater than those derived from igneous rocks. However, it is important to realize that radon levels in groundwater from a single rock type or aquifer can vary significantly. For example, radon in eight wells located within 1 km^2 in a granitic aquifer in South Carolina varied from 60,000 to 550,000 Bq m^{-3}.[49]

Radon levels in metamorphic rock aquifers are a function of uranium content and distribution of the aquifer and groundwater flow patterns. The uranium content can vary widely, especially at grades where partial melting occurs. Also, the distribution of uranium within the rock is very important. Groundwater flow in crystalline rocks is through fractures; uranium mineralization frequently occurs along the same fractures.

In metamorphic rock aquifers without uranium mineralization, the radon levels will be a function of the uranium content of the regional rock type, not the metamorphic grade. However, crystalline rock aquifers, in general, tend to have higher radon levels than noncrystalline rock aquifers of similar composition and uranium content because of smaller water-to-rock ratios, slower flow rates, and the tendency for some uranium enrichment along fractures.

4.4.4.2 Sedimentary Aquifers

There are two broad categories of sedimentary aquifers: (1) clastic (sand and sandstone) and (2) carbonate (limestone and dolomite). Clastic aquifers can be composed of loose sand or sand and gravel or sediments that have been consolidated, by cementing agents, into rocks. The mineral composition can vary widely, from pure quartz sand to material that is in essence crushed granite. The aquifers composed of mostly pure quartz are formed from sediments that have been through multiple cycles of weathering and transport, so that only the most resistant minerals remain, e.g., quartz and heavy minerals. Geologists call these types of deposits "clean" sands or sandstones, meaning they include mostly quartz with little clay. These types of sediments have very little original uranium and usually have low radon in groundwater except when they are enriched in uranium by secondary processes. The group of sedimentary rock aquifers of greatest concern for radon are those located in the regions of sandstone-type uranium deposits such as the Colorado Plateau, south Texas, and Wyoming provinces. The uranium in these sandstone aquifers was deposited by precipitation of uranium from groundwater, triggered by reducing conditions. Very little information is available on the radon in groundwater from these aquifers. The nationwide survey results showed no public groundwater supplies in the Colorado Plateau or Texas exceeding 37,000 Bq m^{-1}.[39] Private wells are more likely to have elevated radon in these uranium provinces, but data are not available for analysis.

There are several other types of sedimentary aquifers which have the potential for high radon in groundwater. Arkoses are sediments containing large amounts of a suite of minerals known as feldspars, which are the major component of granitic-type rocks. Arkosic sediments are generally derived from the physical weathering of granites where the sediments have not been transported far away from the granite source rock. Therefore, they usually are not widespread and grade with distance into more mature, quartzose sediments, as sorting and weathering removes all but the resistant quartz grains. These arkosic deposits can subsequently weather in place, with the feldspars turning into clays and the uranium associated with the feldspar adsorbing onto the clay particles. Deposits of this type are most commonly found in the Fall Line aquifers of the east coast of the United States, where a thin layer of the Coastal Plain Province of sedimentary deposits overlies Piedmont rocks. Several studies in North Carolina,[48,43] South Carolina,[49,50] and Georgia[51] have shown that where these Fall Line aquifers occur adjacent to granites in the Piedmont, radon levels in groundwater are elevated, up to 90,000 Bq m^{-3}. In these areas, the sediments in the Coastal Plain were derived from the adjacent granites and originally were arkosic. Deep chemical weathering has altered the feldspars to clays which absorbed some of the original uranium. In similar settings along the Fall Line, where the adjacent Piedmont rocks are composed of metamorphic rocks, the radon content of the groundwater is lower, though still higher than farther downdip in the aquifer. Detailed studies were conducted in South Carolina, where the Coastal Plain aquifers were divided into three groups, upper (Fall Line), middle, and lower

Coastal Plain.[50] The range in radon concentration of the upper and middle Coastal Plain aquifers was similar, but the average concentration of the upper Coastal Plain (17,340 Bq m^{-3}) was twice that of the middle Coastal Plain aquifers (9470 Bq m^{-3}). Lower Coastal Plain aquifers have very low radon, averaging 6950 Bq m^{-3} in South Carolina and 3000 Bq m^{-3} in North Carolina.

The radon concentration in groundwater from carbonate aquifers will be low (Table 4.5). A notable exception is the phosphate mining region in central Florida which has uranium enrichment associated with the phosphate deposits. Concentrations are usually less than 10,000 Bq m^{-3} with average concentrations in North Carolina of 3440 Bq m^{-3} and in South Carolina of 1300 Bq m^{-3}.

4.4.4.3 Radon Distribution in Groundwater

There have been two different efforts to describe the distribution of radon in groundwater. In the nationwide survey by the USEPA, 2457 public drinking water supplies were sampled in a nonrandom manner, representing mostly supplies serving greater than 1000 people and 45% of the water consumed by groundwater users in the United States. These results were used to estimate the mean population-weighted radon levels in both large and small public groundwater systems by state.[52] The results of this study are shown in Table 4.6, as the mean concentration for systems supplying greater than and less than 1000 people. The average concentrations for the United States were estimated to be 8900 Bq m^{-3} for large systems and 28,900 Bq m^{-3} for small systems. The nationwide survey average for the 2457 supplies sampled was 13,000 Bq m^{-3}. Because the radon content of surface water supplies is essentially zero, the average radon concentration in all community water supplies is estimated to be equal to 2000–10,000 Bq m^{-3}.[52]

Although knowledge of the nationwide average radon concentrations is important, it is more important to know where elevated radon may occur. Using data on radon in groundwater from national surveys and detailed studies, a study was conducted to map the distribution of radon in groundwater in the United States. All available radon measurements were plotted on overlays on geologic maps, and the source aquifers were identified. Detailed descriptions of aquifer composition or lithology were obtained from a wide range of literature and supplemented by communication with state water-resource agencies. Radon distribution by aquifer type for available data was used to develop a geologic model to predict the average radon concentrations most likely

TABLE 4.6. Average Concentration (in Bq m^{-3}) of Radon in Public Groundwater Supplies by State and Two Size Categories[a]

State	Population-weighted: <1000 people/system[b]	Population-weighted: >1000 people/system[b]
AL	5,900[c,d]	5,900[f]
AK	3,700[h]	3,700[h]
AZ	4,440[c,d]	11,800[f]
AR	2,780[c,d]	2,780[c,d]
CA	18,500[h]	18,500[h]
CO	14,000[f]	14,100[f]
CT	55,500(MA)	28,500(MA)
DE	3,700[d]	4,660[f]
DC	—[j]	—[j]
FL	37,000[c,d,h]	5,500[f]
GA	40,700(SC)	5,500[f]
HI	1,850[h]	1,850[h]
ID	9,500[f]	9,500[f]
IL	3,700[d]	6,180[f]
IN	3,890[c,f]	3,890[f]
IA	9,250[c,d]	7,400[c,d]
KS	9,250[d]	3,920[f]
KY	9,250[h]	4,070[f]
LA	6,660[h]	6,660[h]
ME	370,000[g]	74,000[g]
MD	25,900(VA)	16,500(VA)
MA	55,500[c,d]	28,500[f]
MI	3,900(IN)	3,900(IN)
MN	7,770[f]	7,770[f]
MS	5,550[h]	3,000[f]
MO	11,100[h]	3,700[c]
MT	18,500[d]	12,100[f]
NB	11,100(SD)	10,700(SD)
NV	20,300[f]	20,300[f]
NH	51,800[c]	43,800[f]
NJ	5,550(DE)	11,100[c]
NM	7,400[c,d,h]	6,660[f]
NY	18,500[h]	4,900[f]
NC	40,700(SC)	10,300[f]
ND	11,100(SD)	5,550[f]
OH	7,400[d,h]	6,250[f]
OK	9,250[d,h]	5,920[f]
OR	11,100[c,d]	9,800[f]
PA	37,000[c,f]	26,600[f]
PR	18,550[h]	7,400[h]
RI	125,800[e]	56,000[f]

TABLE 4.6. (Continued)

State	Population-weighted: <1000 people/system[b]	Population-weighted: >1000 people/system[b]
SC	40,700[e]	10,200[f]
SD	11,100[k]	10,700[f]
TN	3,700[d]	900[f]
TX	5,500[i]	5,550[i]
UT	18,500[h]	13,300[f]
VT	9,250[e]	24,300[f]
Vl	—[j]	—[j]
VA	25,900[c,d]	16,650[c,d]
WA	11,100[c,d]	9,800(OR)
WV	37,000(PA)	26,600(PA)
WI	27,750[c,d]	6,650[f]
WY	32,560[d]	15,400[f]
Guam	1,850[h]	1,850[h]
Am. Samoa	1,850[h]	1,850[h]
US	28,860	8,880

[a] Reference (52).
[b] State abbreviation in parentheses indicates another state that was used as a surrogate.
[c] From Reference (42), Table 9.
[d] From Reference (42), Table 10.
[e] Half the gross alpha-particle activity from Reference (39), Table 3.3.
[f] From Reference (39), Table 3.2.
[g] From Reference (42), Table 7.
[h] Scientific estimate from various sources of data and aquifer type.
[i] From Reference (42), Table 8.
[j] Surface water only.
[k] From Reference (39), Table 3.5.

to occur in aquifers throughout the United States. Five categories of radon distributions were selected and all counties were classified using these categories, as shown in Figure 4.13. Highest concentrations occur in Maine and New Hampshire, where the average levels in groundwater are predicted to exceed 370,000 Bq m^{-3}. These levels are the range of the average levels expected to occur in all groundwater sources, for both public and domestic use. Radon concentrations in the range of 37,000–370,000 Bq m^{-3} are predicted to occur in much of New England, the Piedmont and Appalachian Mountain Provinces, in scattered areas where igneous and metamorphic rocks are utilized as aquifers, and in the phosphate region of central Florida. Very low concentrations, less than 3700 Bq m^{-3}, are expected to occur in the

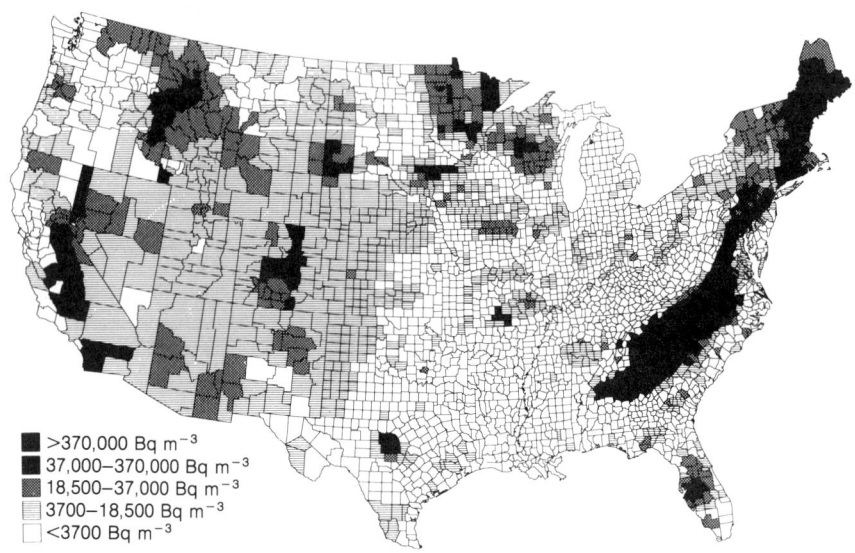

FIG. 4.13. Distribution of radon in groundwater mapped by county, based on available data and aquifer type.

sandstone and sand aquifers which extend from the Appalachian Mountains west to the High Plains, with few exceptions. These ranges of concentrations are the average values expected, and because of the lognormal distribution of radon in groundwater, very few high values are expected for each category. These values are also only qualitative, and the results cannot be rigorously analyzed. However, the map in Figure 4.13 identifies, at a county scale, the likely radon distribution in the majority of wells. Once the results are validated with national survey results which can be statistically analyzed, then the county rankings can be used to estimate the population dose from radon from drinking water.

4.4.5 Radon Contribution from Building Materials

Building materials manufactured from primary raw materials (that is, those products which are not by-products or wastes) have been studied extensively to identify those which have naturally elevated radium concentrations. Of concern are those materials derived from

TABLE 4.7. Summary of Properties Related to Radon in Building Materials[a]

Material	No. samples	^{226}Ra (Bq kg^{-1})	^{222}Rn emanation rate (10^{-6} Bq kg^{-1} s^{-1})	Emanation ratio (%)
Concrete				
USSR	18	66	3.2	3.5
Hungary	~100	13	7.8	28.0
Denmark	4		4.7	
Norway	137	28		1–20
United States	106	9–32	4.3–12.6	13–25
Fly ash (4%)	8	19	10.0	25.0
Alum shale				
Denmark	1		440.0	
Brick (mostly red)				
USSR	12	50	1.6	3.5
Hungary	3	18		2.3
Denmark	2		0.2	
Norway	18	63		1.0
United States	6	45	1.0	1.0
Poland	3	18		5.0
Gypsum				
Denmark	1		0.2	
United States	12	12	6.3	28
Phosphogypsum				
Morocco		925		
United States				
Florida		1222		
Idaho		850		
Poland	3	437		12.0
Fly ash				
Poland	33	96		0.5

[a] Modified from Reference (53).

the earth's crust; wood products are generally not of concern. Each general type of building material is discussed separately.

The components of concrete are generally supplied locally, and there is large variability from region to region. Table 4.7 lists typical values of ^{226}Ra in concrete and its components from around the world. The U.S. data are for samples from ten metropolitan areas. The average ^{226}Ra values are relatively constant, whereas the emanation ratios can vary more widely, from 1 to 30%. Radon flux densities from various building materials are given in Table 4.8. Although concrete is the strongest emanator of radon of all materials, most concretes are still relatively low sources of radon. The exceptions are those which are

TABLE 4.8. Radon Flux Density from Various Building Materials[a]

Material	^{222}Rn flux density (10^{-5} Bq m^{-2} s^{-1})
Ordinary concrete, gravel, and sand from the sea, Danish deposits	35.0
Ordinary concrete, gravel, and sand from pits, Danish deposits	28.0
Ordinary concrete, unknown composition	31.0
Ordinary concrete, unknown composition	38.0
Light-weight concrete, Swedish origin alum shale, old type	858.0
Light-weight concrete, Swedish origin alum shale, new type	363.0
Light-weight concrete, Danish origin, clay based	67.0
Light-weight concrete, Danish origin, clay based	4.0
Expanded clay concrete, LECA	4.0
Bricks, solid type	5.0
Bricks, cavity type	0.2
Chipboard	0.07
Fiberboard	0.1
Gypsum board	0.1

[a] Modified from Reference (54).

composed of materials having high radium content, such as the alum shale used in Sweden from 1930 to 1975.[55]

The practice of disposing of fly ash from coal-generated power plants or phosphate slag by incorporation into concrete is a potential problem. Fly ash can have very high levels of radium, up to several hundred Bq kg^{-1}, but only a few percent can be added to concrete. Also, comparative studies showed that the radon emanation from concrete mixed with fly ash was lower than that for the same concrete without fly ash.[56] However, the use of phosphate slag in concrete in the United States has been studied, and it has been estimated that there are as many as 74,000 homes built with concrete containing up to 740 Bq kg^{-1} of ^{226}Ra.[57] Very few other situations are known where concrete additives have caused potential radon problems.

Varying environmental conditions have been shown to affect radon emanation rates, which could account for the range shown in Table 4.7. Increasing the moisture content of concrete resulted in raising the radon emanation rates by 20–100%.[58] The high porosity of concrete is a likely factor. Just as in porous soils, increases in water content enhance the recoil fraction of radon which is retained in the pore spaces and is able to diffuse into the pore-space air and eventually out of the concrete.

Water-saturated concretes would show decreased radon emanation because of the low diffusion rates in water versus air.

Brick tends to have slightly higher ^{226}Ra concentrations but significantly lower radon emanation rates and ratios than concrete. Red bricks are thought to have higher uranium content, probably because of fixation by iron, than lighter-colored or silica bricks. Because they are more commonly used on the exterior of buildings, their contribution to indoor radon would be even lower. In some places, such as in West Germany, bauxite by-products are used in brick manufacture, resulting in a product with a mean ^{226}Ra content of 280 Bq kg^{-1}.[7]

Gypsum has been evaluated as a radon source, and radon emanation rates have been found to be very low, except where phosphate by-products have been added to make phosphogypsum. In the phosphate milling process, the waste products are enriched in ^{226}Ra. Though no longer used in the United States, phosphogypsum can have over 1000 Bq kg^{-1} of ^{226}Ra and can be a significant source of radon.[53]

Other building materials, such as granite, marble, sandstone, and volcanic rocks, can have a wide range of ^{226}Ra concentrations and radon emanation rates. However, these materials generally are used in limited quantities and would not pose significant population risks except in unusual settings. Exceptions may be in solar houses which use large rock masses for heat storage. Indoor radon levels have been shown to increase by a factor of 2–3 when air was passed through the granitic rock bins in a solar-heated house.[59] Even the use of sand or bricks as the heat-storage material can elevate radon levels in homes, especially in combination with low air-exchange rates.

A comprehensive review and assessment of radon transport through and exhalation from building materials was conducted by the National Bureau of Standards.[53] This work includes a detailed analysis of transport mechanisms and how the microstructural properties and internal characteristics of building materials may affect the transport and exhalation of radon. One of the main conclusions of this review is that only "technologically enhanced" building materials are normally of concern as major contributors to indoor radon concentrations. The radon emanation rates from most building materials would not generate radon levels above 0.0015 WL (4.2×10^{-8} J m^{-3}) for an average home.

Calculation of the contribution of radon from building materials requires knowledge of the radon flux density or exhalation rates from the various components of the structure, which are not quantitatively known. The exhalation process is very complex and dependent on many factors, such as the radium content, macroscopic configuration of the material, moisture content, pressure changes, temperature, and relative humidity.[53] These causal internal and external variables are only qualitatively understood.

The contribution of building materials to indoor radon can be estimated by summing the products of the radon flux density for each type of material, multiplied by its surface area in the structure. See the following equation:

$$Rn_{bm} = \sum_{n=1}^{i} \frac{F_i \cdot A_i \cdot v}{V}$$

where Rn_{bm} = Indoor radon concentration from building materials, in Bq m^{-3}
F = Flux density for material i, in Bq m^{-2} s^{-1}
A = Area of material i, in m^2
v = Ventilation rate, s^{-1}
V = Volume of structure, in m^3

For example, using the range of flux densities typical for most building materials in the United States, the various contributions to a structure with a floor area of 130 m^2, height of 2.5 m, volume of 325 m^3, and ventilation rate of 0.0001 s^{-1} (0.4 h^{-1}) would be as follows:

concrete floor: 130 m^2 × (0.0007–0.0021 Bq m^{-2} s^{-1}) = 2–6 Bq m^{-3}

bricks: 115 m^2 × 0.000048 Bq m^{-2} s^{-1} = 0.1 Bq m^{-3}

gypsum board: 400 m^2 × 0.0000012 Bq m^{-2} s^{-1} = 0.01 Bq m^{-3}

From these sample calculations, it is obvious that normal building materials do not contribute to elevated radon levels. However, problems can arise from certain materials, as shown by the following calculations, using the same conditions as above:

alum shale concrete floor: 130 m^2 × (0.0085–0.07 Bq m^{-2} s^{-1})
= 25–200 Bq m^{-3}
alum shale concrete building: 115 m^2 × (0.0085–0.07 Bq m^{-2} s^{-1})
= 22–180 Bq m^{-3}
phosphogypsum walls: 400 m^2 × 0.012 Bq m^{-2} s^{-1} = 110 Bq m^{-3}
phosphogypsum ceiling: 130 m^2 × 0.0012 Bq m^{-2} s^{-1} = 3.5 Bq m^{-3}

SOURCES

4.4.6 Radon from Fuels

During the 1970s, several studies were conducted to determine the potential of radiological health effects from radon in natural gas, although the presence of radon in natural gas had been known since 1904. The levels of radon in natural gas vary widely, from 150 to 53,700 Bq m^{-3}, but the average values in the United States do not exceed about 3700 Bq m^{-3}.[60] Potential exposures occur when radon is released into homes by the burning of natural gas in unvented appliances and heaters.

The source of such relatively high levels of radon in natural gas is probably the presence of high concentrations of ^{226}Ra in oil-field brines as well as the higher solubility of radon in organic phases. Long-term monitoring of radon in wells has shown that radon in individual wells is relatively constant under steady production conditions.[60] During the processing of natural gas into liquefied petroleum gas (LPG), radon is concentrated in the LPG fraction because of its boiling point. The radon content of fresh LPG ranges from 70 to 48,000 Bq m^{-3}. However, the radon content at the point of use is much less, due primarily to decay during distribution and storage. The average radon content of LPG in distribution lines was estimated to be about 700 Bq m^{-3}.

The contribution of radon from fuels is a function of the amount of fuel burned, the radon content of the fuel, the house volume, and ventilation rate. The average increment in indoor radon due to fuels in the United States is estimated to be 0.6 Bq m^{-3}, or about one-tenth of the natural background level.[60] Higher levels would occur in homes which are close to wellheads or which use fuel derived from wells with high initial radon levels.

4.4.7 Radon Associated with Uranium and Phosphate Mining Activities

Occupational exposure to radon of uranium and other miners has been well documented. The general population also can be exposed to radon from the storage of mining products and disposal of mining wastes; this is primarily the situation in phosphate and uranium mining activities. Exposures can occur via two different pathways: construction of homes on reclaimed land and by radon fluxes from waste piles of phosphogypsum and uranium mill tailings.

Extensive studies have been conducted of the exposure to radon in structures built over formerly mined lands in the phosphate region of Florida.[61-63] Florida generates over 90% of the U.S. phosphate mine rock production. In the mining process, approximately one ton of clay slimes and one ton of sand tailings must be disposed of for each ton of marketable phosphate rock produced. These waste products and the original overburden are returned to the mine in various mixes and combinations of layering which have changed with time. Undisturbed soils in central Florida generally have low ^{226}Ra concentrations (approximately 20 Bq kg^{-1}), which increase gradually with depth through the overburden, reaching a local maximum (approximately 1500 Bq kg^{-1}) in the ore or matrix.[63] Radium concentrations in the soils of reclaimed land are typically an order of magnitude higher than in the original soil and can contain local concentrations similar to the matrix because of poor mixing. Figure 4.14 shows the mean and range of values of ^{226}Ra in soils and the radon fluxes from the various land types.[64] All mining activities tend to increase the radium content and radon flux from the soils. As a result, elevated levels of indoor radon

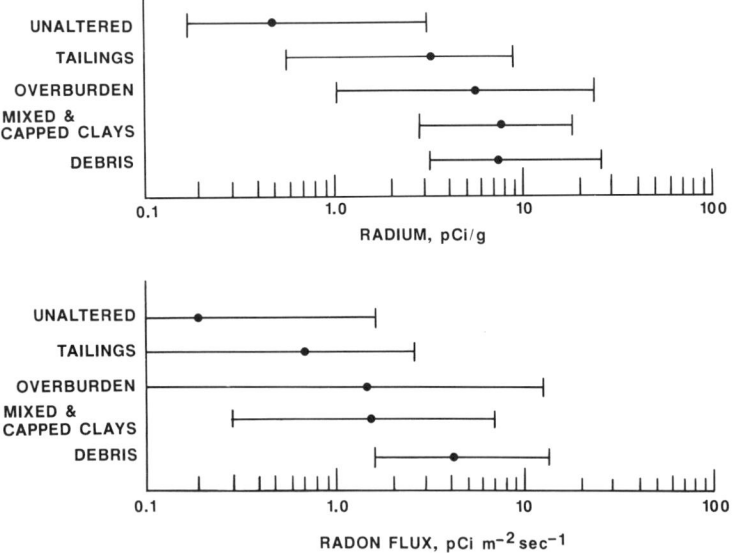

FIG. 4.14. Radium concentration and radon flux for various land types in the phosphate region of Florida.[64]

TABLE 4.9. Soil Radium and Indoor Radon Progeny Summary in the Phosphate Region in Florida[a]

Land type	Surface soil ^{226}Ra (Bq kg^{-1}) $\bar{x}$ (range)[b]	Indoor airborne radon progeny (WL)	
		Slab-on-grade $\bar{x}$ (σ)[b,c]	Crawl space and mobile home $\bar{x}$ (σ)[b,c]
Mined, reclaimed			
Higher-activity overburden and debris	500 (475–540)	0.038 (1.3)	0.005 (1.4)
Tailings	152 (70–230)	0.007 (1.3)	—
Lower-activity overburden	74 (67–89)	0.008 (1.1)	—
Unmined—deposits or fill	70 (52–89)	0.020 (1.4)	0.004 (1.5)
Unmined—nonenhanced	14 (7–37)	0.003 (1.3)	0.003 (1.3)

[a] Modified from Reference (63).
[b] $\bar{x}$, Geometric mean.
[c] σ, Geometric standard deviation.

have been found in structures built on reclaimed land. However, soil radium and surface radon flux were found to be unreliable predictors of the indoor radon levels because of (1) the wide range in distribution (both vertically and horizontally) of radium in the soil beneath the structure, (2) the extent and distribution of cracks in the building slab, and (3) the various ventilation rates.

A land-type classification scheme was developed to characterize the radon potential of the soils for Polk County, Florida, as shown in Table 4.9. The low-activity overburden lands contain mixed overburden from present-day activities. Unmined lands with phosphate deposits occurring relatively near the surface were differentiated from those with thick overburdens which were classified as nonenhanced. The distribution of indoor radon progeny levels for the different land types and structure categories in Polk County, Florida, was estimated as shown in Figure 4.15. Of the structures exceeding 0.02 WL, 75% were located on reclaimed lands. It was estimated that 85% of the structures with slab-on-grade construction built on the high-activity overburden lands would exceed 0.02 WL compared with 4% built on tailings, 0.3% built on low-activity overburden, and 0.1% built on nonenhanced land. Therefore, a significant portion of the problem is associated with lands resulting from practices no longer in use. About 25–30% of slab-on-grade structures built on tailings and low-activity overburden would be in the range of 0.01–0.02 WL, so the need still exists for some restrictions or special construction methods.

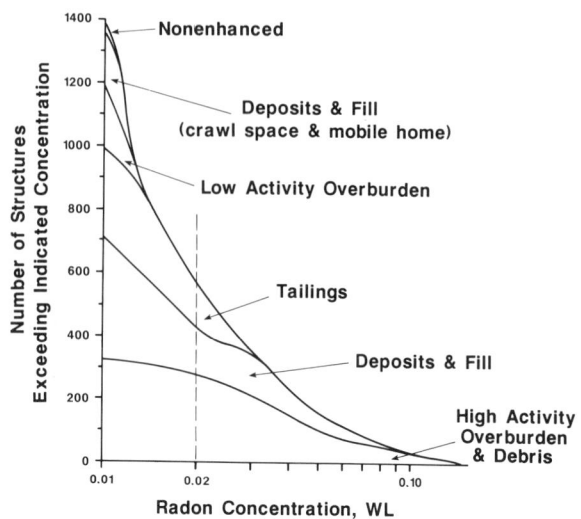

FIG. 4.15. Distribution of indoor radon progeny concentration (in WL) by various land and structure categories in the Florida phosphate mining region.[63]

A second source of radon exposure in the phosphate mining areas of Florida is the disposal of phosphogypsum, a by-product of the chemical processing and phosphoric acid production. The phosphogypsum is stored adjacent to the chemical plants in stacks that are frequently more than 50 m high and cover several hectares. Most of the ^{226}Ra in the feedstock ends up in the phosphogypsum, with concentrations in the range of 900–1200 Bq kg^{-1}. Measurements of the average radon flux from various stacks have ranged between 0.1 and 1.0 Bq m^{-2} s^{-1}, although individual measurements as high as 4 Bq m^{-2} s^{-1} have been reported.[65] Wet areas on the stack have lower average radon fluxes than dry areas, as expected due to slower rates of diffusion of radon through water-saturated pores. Soil cover can reduce the radon flux somewhat, depending on the soil type, thickness, type of vegetation, and integrity of the cover. Actual exposure to radon from phosphogypsum stacks is a function of the source strength (which varies according to operational and closure site conditions), meteorological conditions, and population distribution.

4.4.8 Summary of Contribution of Radon to Indoor Air

Control of indoor radon levels requires knowledge of the relative contributions of radon from various sources. In the 1970s, emphasis

SOURCES

was placed on building materials as sources of indoor radon, especially in areas where technologically enhanced materials, such as alum shale concrete or phosphogypsum, were used in home construction. Monitoring of homes showed, however, that radon emanation from normal building materials could not account for the measured indoor levels, using air-exchange rate estimates. Groundwater was found to be a significant contributor only in selected areas, such as in New England. It was not until the 1980s that soil was identified as an important source of indoor radon, even in areas which were not considered to have elevated uranium concentrations.

The relative contributions of all potential sources of radon are difficult to estimate because of the wide range of distributions possible. Figure 4.16 is a summary of selected data showing the distribution of entry rates in residences from various countries, displayed as cumulative frequency curves.[66] Also shown are the range of potential contributions from various sources. From the curve for the United States, the geometric-mean entry rate for single-family homes is approximately 20

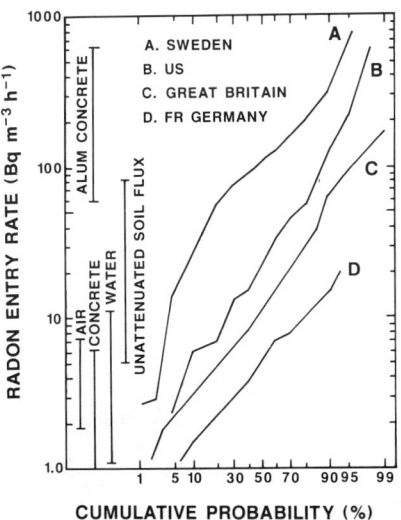

CUMULATIVE PROBABILITY (%)

FIG. 4.16. Cumulative frequency distribution of the radon entry rate determined in dwellings in several countries as the product of simultaneously measured ventilation rate and radon concentration.[66] The bars at the left indicate the range of contributions expected from a variety of sources with the following assumptions: radon concentration in outdoor air, 9 Bq m^{-3}; radon concentration in water, 10^4–10^5 Bq m^{-3}; water usage, 1.2 m^3 d^{-1}; structure, wood frame with floor of 100 m^2 and height of 2.4 m; floor, concrete slab 0.2 m thick; ventilation rate, 0.2–0.8 h^{-1}.

TABLE 4.10. Estimated Contribution of Various Sources to Observed Average Indoor Radon Levels[a]

Sources	Single-family homes (Bq m^{-3})	High-rise apartments (Bq m^{-3})
Soil potential (based on flux measurements)	55	<40
Water		
Private wells	24	
Public groundwater supplies	1.3	1.3
Surface water supplies	0.1	0.1
Building materials	2–4	4
Outdoor air	10	10
Observed average indoor concentrations	55	12

[a] Modified from References (66) and (28).

Bq m^{-3} h^{-1}. The contribution from concrete, the highest common building material source, is estimated to average 2–3 Bq m^{-3} h^{-1}, about one-tenth of the average indoor levels. The potential contribution from soils, with a median of 25 Bq m^{-3} h^{-1}, corresponds closely with the median levels observed in homes.

The relative contributions of various sources to the observed indoor average radon concentrations have been estimated, as shown in Table 4.10. The average U.S. indoor concentration of 55 Bq m^{-3} is based on a systematic analysis of available data, which included 19 data sets totaling 552 houses.[66] Soil is the dominant source in most cases. With average soil-gas radon concentrations of 3700–37,000 Bq m^{-3}, soil gas would only have to contribute a few percent to the indoor air to account for the observed average indoor levels. In areas with high radon in the soil gas, indoor contributions can readily exceed levels at which remedial action may be required.

The contribution of water to indoor radon is a function of the radon content of the water, the volume used, and the efficiency of radon transfer from water to air for different types of water use. A general model for the average incremental indoor radon contribution attributable to water use can be expressed as:

$$Ca = \frac{Cw}{24RV} \sum_{i=1}^{n} e_i w_i$$

where Ca and Cw are the concentrations of radon in Bq m^{-3} in air and water, respectively; R is the air-exchange rate (h^{-1}); V is the dwelling volume (m^3); the set of W_i's represents the average water use for each

TABLE 4.11. Radon Liberated per 37×10^3 Bq m^{-3} in Water[a]

Use	Daily use (m^3)	Transfer efficiency (%)	Radon liberated (Bq)
Shower	0.15	63	3,500
Tub bath	0.15	47	2,600
Toilet	0.36	30	4,050
Laundry	0.13	90	4,330
Dishwater	0.05	90	1,830
Drinking/kitchen	0.03	30	330
Cleaning	0.01	90	330
Total	0.89		16,970

[a] Reference (67).

application (m^3); and the set of e_i's represents the transfer efficiency or fraction of radon released to the air for each application.[67] Table 4.11 shows the transfer efficiencies experimentally determined for the common types of water use in a home, as well as the daily water consumption for each use. Table 4.11 also shows the amount of radon that is expected to be released during the normal use of water which contains 37,000 Bq m^{-3} of radon. Simple multiples can be used to estimate the radon liberated from water of any radon concentration. The actual indoor radon concentrations resulting from water use would be a function of the volume of the structure and the air-exchange rate. Table 4.12 shows the estimated incremental increase in indoor radon per 37,000 Bq m^{-3} in the water for several dwelling types. These results are similar to the earlier estimate that the overall transfer factor for the increase of radon in air resulting from water use is approximately 10^{-4}.[68] Using a long-term average, single-cell model and available data for U.S. housing, the overall transfer factor was calculated to have a geometric mean of 0.65×10^{-4}.[69]

The contribution from water in Table 4.12 ranges widely as a function of water source. Using the estimated radon levels in all public water supplies, from both surface and groundwater sources, of 2000–

TABLE 4.12. Increase of Indoor Radon per 37×10^3 Bq m^{-3} in Water[a]

Dwelling type	Floor area (m^2)	Volume (m^3)	Air exchange (h^{-1})	Bq m^{-3}
Small apartment	63	150	0.25	19
Brick home	140	340	0.5	4
Old frame home	140	340	1.0	2

[a] Reference (67).

10,000 Bq m^{-3},[52] the average contribution from water would be 1.3–6.5 Bq m^{-3}. In domestic wells, with very high radon concentrations, the groundwater can be the dominant source, contributing up to several hundred Bq m^{-3} to indoor radon levels. In most cases, however, soil will be the dominant source of indoor radon in structures.

5

Human Exposure

GEOFFREY G. EICHHOLZ

5.1 INTRODUCTION

Human exposure to radon results from several sources. Those exposures due to technologically enhanced causes such as mining and milling activities will be discussed first and indoor air exposures last.

Since radon is a decay product of uranium and radium, the distribution of these two elements in the lithosphere and in our bodies is important. The kinds of aerosols or particulates to which the radon progeny are attached and the relation to cigarette smoke are also important.

5.1.1 Dietary Uptake of Radium

Radium is a bone seeker, and an appreciable fraction is deposited on bone surfaces and in areas of active tissue turnover. After high intakes, approximately half the initial activity in man is deposited in critical organs and half in the diffuse component. About 70–90% of the radium in the body is contained in bone, the remainder being distributed fairly uniformly in soft tissues.[1] In areas of normal radiation background, ^{226}Ra concentrations in bone range from 2 to 20 pCi kg^{-1} (74–740 mBq kg^{-1}) with an average value of about 8 pCi kg^{-1} (0.3 Bq

---

GEOFFREY G. EICHHOLZ • Nuclear Engineering and Health Physics Program, Georgia Institute of Technology, Atlanta, Georgia.

kg^{-1}). The biological half-life of radium in the body has been estimated to be of the order of ten years.

Dose rates in bone lining cells and bone marrow have been calculated with the aid of the following assumptions: (a) an average retention factor in the skeleton and in soft tissues for ^{222}Rn of 0.33 and 1.0, respectively, and (b) a uniform concentration of radium and its progeny over the total mass of the mineral bone. This leads to total doses of 0.072 mrad y^{-1} (720 nGy y^{-1}) to the red marrow and 0.45 mrad y^{-1} (4500 nGy y^{-1}) to the bone lining cells from radon and its progeny.[1] Organ doses of this order form only a minor contribution to the natural background exposure: the same conclusion can be drawn for most other food pathways involving the uptake of radium.

5.1.2 Uptake of Radon Progeny

Although the typical concern with human exposure to radon arises from its mobility in the environment and the subsequent deposition and decay of its progeny in the lung, there are certain scenarios where radon progeny appears in its own right due to the pervasive character of the atmospheric radon background as discussed in Chapter 4. Lead-210, with its 22.3-y half-life, can accumulate on plants or soils and be ingested or inhaled without the immediate intervention of its radon precursor. If ^{210}Pb is taken up by humans, ingrowth of its progeny can occur. Lead is a bone seeker; it is found in bone mineral with a 70% higher level in cancellous bone than in compact bone. The effective biological half-life of ^{210}Pb, due both to inhalation and ingestion, is about 3300 d. The sum of the contributions from inhalation and ingestion is about 400 pCi (15 Bq).[1]

The main source of ^{210}Pb and ^{210}Po in the atmosphere is ^{222}Rn emanation from the ground. The amount of atmospheric ^{210}Pb produced in this way has been estimated to be 0.6 MCi y^{-1} (22.2 PBq y^{-1}). After precipitation scavenging is allowed for, the average airborne concentrations in the middle latitudes of the Northern hemisphere have been estimated to be 14 fCi m^{-3} (0.5 mBq m^{-3}) for ^{210}Pb and 3.3 fCi m^{-3} (0.12 mBq m^{-3}) for ^{210}Po.[1] Assuming that an adult inhales 20 m^3 of air daily, the intakes of nonsmokers are estimated to be 0.3 pCi d^{-1} (11 mBq d^{-1}) of ^{210}Pb and 0.07 pCi d^{-1} (2.6 mBq d^{-1}) of ^{210}Po.

Tobacco leaves have been known to contain appreciable amounts of these radionuclides, though there seems to be some disagreement whether this arises from surface deposition from atmospheric precipi-

HUMAN EXPOSURE

tation or by way of root absorption. Both nuclides are volatile at the burning temperature of tobacco, so that cigarette smoking will lead to a substantial increase in their intake through inhalation. A cigarette contains about 0.6 pCi (22 mBq) of ^{210}Pb and 0.4 pCi (15 mBq) of ^{210}Po and about 10% of the ^{210}Pb and 20% of the ^{210}Po will enter the lung with the main smoke stream.[2] For a person smoking 20 cigarettes a day, estimated intakes are 1.2 pCi (44 mBq) for ^{210}Pb and 1.6 pCi (59 mBq) for ^{210}Po, subject, of course, to variations depending on the individual's mode of smoke inhalation.

For nonsmokers, consumption of food is usually the most important route for ^{210}Pb and ^{210}Po to enter the body. In cereal products, vegetables, and meat, the typical concentration of ^{210}Pb is 2–5 pCi kg^{-1} (74–185 mBq kg^{-1}), and the ^{210}Po/^{210}Pb ratio is between 0.5 and 1.0. The concentration in milk is lower by an order of magnitude. The ^{210}Po concentrations in the flesh of fish and molluscs are approximately 20 and 500 pCi kg^{-1} (0.74 and 18.5 Bq kg^{-1}), respectively.

The effect of smoking on ^{210}Pb/^{210}Po concentrations in organs and tissues is not as marked as one might expect. The increase is most marked for the lung, exceeding on the average the nonsmokers' levels by factors of 1.5 and 3 for ^{210}Pb and ^{210}Po, respectively. However, despite the fact that the daily intake from smoking is at least 20 times the normal air concentration, the increase in lung concentrations is relatively small, suggesting rapid elimination of ^{210}Po from the lung. There still is considerable uncertainty regarding the most likely location of ^{210}Po deposition in the lung, which does, in fact, depend on its physical form and whether it is carried on an aerosol or smoke particulate. As a result, estimates of the annual absorbed dose to the bronchial epithelium range from 4 to 3000 mrad (0.04–30 mGy).[3] The full impact of ^{210}Pb/^{210}Po intake from smoking in practice depends on smoking habits and the period of duration of smoking.

5.1.3 Atmospheric Radon and Its Progeny

Radon (^{222}Rn) and thoron (^{220}Rn) emanate from soil, water, and uranium-bearing rocks and materials and become dispersed in the air. Being noble gases, they migrate by diffusion and convection without significant interaction with the constituents of air or any airborne particulates. The decay products of airborne radon are produced as free atoms which are temporarily positive ions, since the preceding radioactive decay has carried off electrons, leaving the decay products positively charged. This fact has a strong bearing on their ability to be electrostatically attached to dust particles and other siliceous materials in the air and on the various collection methods used in assaying the atmosphere for radon progeny. The ions tend to form clusters with water droplets or oxygen molecules which form or become attached to aerosols, i.e., suspended submicron-size particulates. Depending on particle size, humidity, barometric pressure and other factors, such aerosols may remain in suspension or settle out gradually on any surface. If inhaled, the clusters and aerosols may deposit in different regions of the respiratory tract, from which they can either be removed by mucociliary action or be transported into the alveolar region. As the radon progeny decay, they emit alpha particles, imparting a radiation dose to the lung region where they are residing. It is the cumulative effect of this alpha dose on lung tissue that is responsible for any health effects observed, primarily lung cancer in the case of severe exposures.

It is evident that the severity of exposure to radon progeny depends on their concentration in the air, their probability of attachment to aerosols, and the particular portion of the respiratory system where they end up. The probability of particulate attachment depends on the nature, size, and concentration of the aerosol as well as on ambient conditions. It is usually expressed in terms of an "attachment coefficient" or attachment fraction, i.e., that fraction of airborne radon progeny that has become attached to respirable aerosol particles. The degree of attachment, then, may vary in different locations. In a mine, for instance, the air may be very dusty and provide plenty of attachment nuclei, while in the open, and after a rain shower, the air may be relatively free of aerosols, resulting in a low attachment rate. Resuspension of soil particles in the air by wind action may be important in that case. Indoors, ventilation and air circulation are most significant in controlling the availability of attachment nuclei or in removing attached radon progeny from areas of potential human exposure as their potential host particulates are swept out.

Two examples may illustrate the effect of aerosol attachment. If a radon progeny concentration, say in a mine, is determined as 1 WL, this represents a radon concentration of 100 pCi L^{-1} only if equilibrium can be assumed. However, any measurement that depends on radon progeny collection typically only collects the attached fraction, and that imperfectly. Corrections must be made for attachment fraction and collection efficiency. If the later radon progeny, e.g., ^{214}Pb, is removed, a value of 1 WL would imply a correspondingly higher initial radon concentration in order to provide the same potential alpha energy release.

A slightly different problem arises if there is a mixture of radon and thoron (^{222}Rn and ^{220}Rn). Generally, the whole thorium series is found in equilibrium with its thorium precursor. However, situations may arise where ^{224}Ra with its 3.6-d half-life may move independently through the environment, and thoron and its progeny will exist in secular equilibrium with it. Under those conditions, 100 pCi (3.7 Bq) thoron per liter of air can result in a progeny concentration equivalent to 13.3 WL (28 × 10^{-5} J m^{-3}). In other words, 1 WL (2.08 × 10^{-5} J m^{-3}) is the concentration of thoron progeny in equilibrium with 7.5 pCi (0.28 Bq) of thoron per liter (280 Bq m^{-3}).

Typical radon concentrations in open air near the ground are about 0.13 fCi mL^{-1} (4.8 Bq m^{-3}), corresponding to a radon progeny concentration of 1.3 mWL (27 nJ m^{-3}), assuming equilibrium. Both lower and substantially higher values have been encountered, the latter particularly near open-pit mines and uranium-bearing rock outcrops. Even plowing a field can increase measured radon levels, as the emanation rate increases through broken ground.

The relative importance of the unattached versus the attached fraction depends on the availability and size distribution of aerosol carriers. The attached radon daughters are mainly deposited in the pulmonary region of the respiratory system, while the unattached progeny are deposited in the upper respiratory tract, though a major proportion of the unattached progeny is removed by nasal deposition.[4] Indoors, the net effect of increased ventilation is to decrease aerosol concentration and residence time of progeny in room air. This leads to a reduction in dose due to a large decrease in the potential alpha-emitter concentration available for deposition in the respiratory system.

The distribution of values of the dose-to-exposure quotient for radon progeny derived from different models is discussed in Chapter 6. It is suggested[1] that for mining conditions, where the unattached fraction is below 10%, a value of 1 rad WLM^{-1} (0.01 Gy WLM^{-1}) can be used in dose assessment. In the case of exposure in homes and outdoor air, a conversion factor of 45 μrad per (pCi − h L^{-1})(1.2 × 10^{-8} Gy m^3 h^{-1} Bq^{-1}) has been proposed, taking into account the lower mean breathing rate.[1] Currently, a commonly used conversion factor for the general population is 1 rem y^{-1} per WLM (2.85 Sv y^{-1} J^{-1} h m^3) of nonoccupational exposure.

5.2 MINING AND MILLING ACTIVITIES

5.2.1 General Characteristics

Not surprisingly, radon levels are highest where the highest uranium concentration occurs in rocks, and historically it was in uranium mines that the potential hazard from radon progeny inhalation was first recognized. However, high radon concentrations are not confined to uranium mines. Any mining activities in uranium-bearing rock, such as in the South African gold mines or the Newfoundland fluorspar mines, may cause the development of significant radon levels in the mine air. Though radon levels tend to be very high in the confined spaces of underground drifts, elevated levels are also found in open-pit mines and around uranium mill tailings.

In the United States there were 7000 underground uranium miners in 1960, about 4000 in 1965, 5000 in 1970, and about 6000 in 1975.[5] Following the adoption of the 4 WLM (1.6×10^{-6} J m^{-3}) per year exposure standard in 1971, a marked reduction in exposure was achieved in the early 1970s. Table 5.1 shows the actual numbers of

TABLE 5.1. Mine Safety and Health Administration Data on Occupational Radiation Exposures for Underground Miners in 1980[5]

Exposure range (WLM)	Number of miners
LLD^a	5,928
LLD–0.1	1,146
0.1–0.2	759
0.2–0.3	550
0.3–0.4	437
0.4–0.5	336
0.5–0.6	339
0.6–0.7	281
0.7–0.8	297
0.8–0.9	299
0.9–1.0	296
1.0–2.0	1,913
2.0–3.0	731
3.0–4.0	159
>4	13
Total	13,484

^{a} LLD, lower limit of detection.

U.S. underground uranium miners involved in the various radon exposure ranges in 1980. The fraction receiving exposures above 4 WLM y^{-1} is a very small percentage of the total.

Nonuranium miners include workers in other metal, nonmetallic mineral, coal, and stone mines. Those potentially exposed to significant (compared to background) levels of radon decay products and gamma radiation are mostly workers in underground metal mines. The number of American nonuranium metal mining workers has generally ranged between 90,000 and 100,000 since 1960. In 1980, about 4% of workers at nonuranium metal mines were assessed for exposure to radon decay products, because these mines contained concentrations of radon decay products in excess of 0.3 WL (6×10^{-6} J m^{-3}). It was assumed that this same number of miners was similarly exposed to elevated levels of gamma radiation. This leads to an estimate of about 4200 U.S. nonuranium miners (compared with 13,500 uranium miners) being exposed to radon decay products and gamma radiation. Based on limited data, it was estimated that the mean annual radon decay-product exposure was 0.2 and 0.3 WLM (0.001 J h m^{-3}) for these potentially and measurably exposed miners, respectively.[5] The mean annual gamma-ray dose was about 150 and 220 mrem (1.5 and 2.2 mSv) for potentially and measurably exposed miners.[5]

For uranium mines the assessment of exposure to radon progeny in air resolves itself into the three distinct problem areas: the air circulating in underground mine workings, occupational exposure in open-pit mines, and elevated radon levels from mining operations that affect surrounding areas. Transfer of radon from rocks adjoining mining activities can be expected to be much greater than from soils or weathered rock because of the extensive fracturing introduced by drilling and blasting and the large amount of freshly broken rock produced in a wide range of sizes.

Uranium mining is generally carried out either by open-pit or underground mining methods. In 1977 there were 251 underground and 36 open-pit uranium mines in operation in the United States. These mines accounted for about 96% of the uranium produced, with each mining method accounting for approximately half of the production. In recent years in-situ solution mining has become widely used and the amount of uranium mined by this method is expected to increase in future years. However, during 1977 this method accounted for only a few percent of the uranium mined in the United States[6], but was employed in two mines in Canada.

In the United States and Canada all of the present uranium mining takes place in western regions. In general, these mines are located in relatively remote areas. Seventy-seven percent of the U.S. uranium production has taken place in the states of New Mexico and Wyoming, but the number of producing mines shrunk drastically in the 1980s. Canadian production, particularly in Saskatchewan and the Northwest Territories, has grown.

5.2.2 Surface Mining

Open-pit mining usually is carried out by excavating a series of terraces in sequence. The mining procedure followed is to remove the topsoil and overburden from above the ore zone and to stockpile these materials in separate piles for use in future reclamation operations. The uranium ore is removed from the exposed ore zone and stockpiled for transport to a uranium mill. Ore stockpiles range in size up to several hundred thousand metric tons of ore. During the mining of the uranium ore, low-grade waste rock is also removed from the pits and stored in a waste stockpile for possible future use.

As the mining progresses, mining and reclamation operations take place simultaneously—pits are mined in sequence and the mined-out pits are reclaimed by backfilling with overburden and topsoil. In some cases the last of the open pits in a mining operation are not backfilled but are allowed to fill with water, forming a lake. Once the pit gets so deep that the cost of removing overburden greatly exceeds the financial return from the ore removal, a mine usually is converted to underground operation. Radioactive emissions from open-pit mining operations are radioactive fugitive dust and ^{222}Rn gas. Because of the diffuse nature of these mining operations, few direct measurements of radon emissions have been made, and emission rates of radon from open-pit mines have usually been estimated using calculational methods.

Model surface mine parameters were obtained from a survey of eight open-pit uranium mines in Wyoming and used to estimate the ^{222}Rn levels.[7] Table 5.2 shows the atmospheric radon emissions derived for a model open-pit uranium mine. The final figure of 1961 Ci y^{-1} (73 TBq y^{-1}) represents the estimated average annual ^{222}Rn emission over the lifetime of the mine. The most extensive set of measurements has been obtained by a team of Argonne scientists at the St. Anthony open-pit mine in New Mexico.[8] Radon fluxes and air concentration near surface mines were monitored over a five-month period. Figure 5.1 shows the correlation of radon flux with barometric pressure. The

TABLE 5.2. Atmospheric Emissions of Radionuclides from the Model Open-Pit Uranium Mine[7]

Source of emission	Radon-222 (Ci y^{-1})
Active open pit	894
New pit being excavated	148
Ore stockpile	103
Subore waste pile	163
Overburden waste pile	148
Refilled pits	391
Increased land area	15
Truck loading and dumping	99
Total	1961

radon release was correlated with surface area for different strata and their ^{226}Ra content. The average ^{226}Ra content of ten samples taken from the ore-bearing region was 102 pCi g^{-1} (3.77 Bq g^{-1}). This average divided by the average flux density for the undisturbed ore-bearing region yielded a specific flux of 0.072 pCi ^{222}Rn m^{-2} s^{-1} for each pCi ^{226}Ra in a gram of ore. Little difference in WL values was found in the mine pit or at some distance from it.

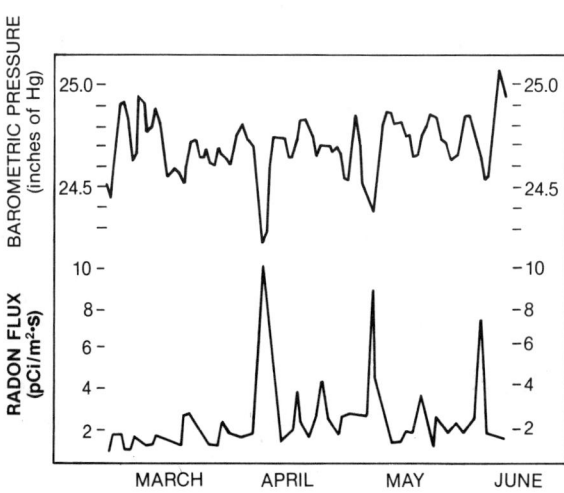

FIG. 5.1. Comparison of radon flux with barometric pressure at an inactive uranium mining pit.[8]

5.2.3 Underground Mines

Because of the irregular nature of ore occurrences, mining activities have to follow the ore body and, consequently, an extensive system of tunnels has to be developed, often at great depth. The mining operation may follow a drift or use a modified room-and-pillar method. The drilling and blasting unavoidably produce large amounts of airborne dust, rock fragments, and fine particulates, and these products liberate a steady flow of radon. Because of the high particulate concentrations in air, a large fraction of radon progeny will be attached to them. This combination poses a distinct health hazard. Reduction in airborne radon-progeny concentrations can be accomplished by water sprays or efficient air circulation; in addition, miners can be protected against dust inhalation by wearing face masks, but that is an unpopular and often counter-effective approach. Water spraying during drilling and mucking operations is of limited utility and merely adds to the general discomfort. This leaves other methods of dust laying, such as aluminum powder and ventilation, as the principal means of exposure reduction. Ventilation in mines is normally required, anyhow, to prevent silicosis, the major occupational disease of hard-rock miners. However, since radon can accumulate in dead-end tunnels and unworked regions of the mine, additional and flexible means of ventilation have to be provided.

It is impractical to sample a worker's breathing zone continuously by conventional methods, and there may be substantial variations in radon levels from place to place. Figure 5.2 is an example of the radiation profile of a mine from the intake air shaft to the exhaust shaft, showing both air flow rate and working-level values. Table 5.3 is an example of a full-shift exposure determination.[9] In this case, the man received 4 WLh of exposure during the shift, at an average level of 0.5 WL (10^{-5} J m^{-3}).

The appreciable reduction in radon exposure due to improved ventilation in underground mines has been documented for New Mexico mines. The median annual exposure was shown to have decreased from 5.40 WLM (0.019 J h m^{-3}) in 1967 to 0.5 WLM (2 mJ h m^{-3}) by 1980. It has remained close to that level since.[10]

5.2.3.1 Surface Effects of Uranium Mines

By ventilating mines as efficiently as possible, increased levels of radon may be conveyed to the surface. The exposure and risk from

HUMAN EXPOSURE

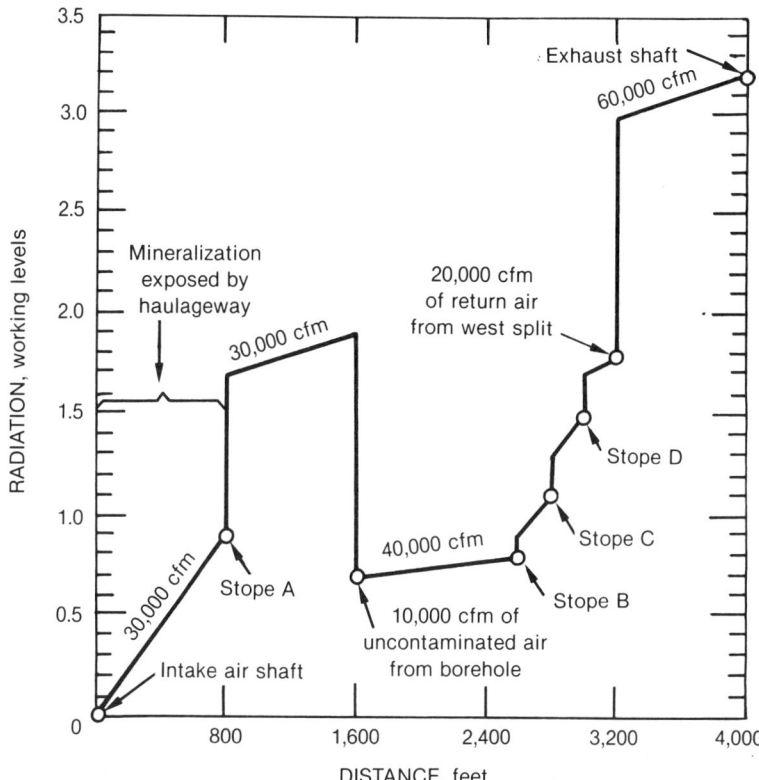

FIG. 5.2. Radiation profile showing working level variations in a mine caused by contamination inflow, radon progeny ingrowth time, and mixing of contaminated and uncontaminated air volumes.[9]

such emissions have been estimated for a model mine operating at 500 tons d^{-1} with an ore grade of 0.2% U$_3$O$_8$ and five vents.[6] Radon emission from mine vents was estimated to be of the order of 6500 Ci y^{-1} and from waste and ore storage another 230 Ci y^{-1}, for a total of 6730 Ci y^{-1} (250 TBq y^{-1}). The risks associated with this are shown in Table 5.4.

A New Mexico study in 1977 found that ^{222}Rn concentrations exceeded the background by 3 pCi L^{-1} (110 Bq m^{-3}) at three of nine locations near uranium mines in the Ambrosia Lake region. The average yearly ^{222}Rn concentration over a two-year period in that area was 4.0 pCi L^{-1} (150 Bq m^{-3}); the highest concentration of 6.4 pCi L^{-1} (240 Bq m^{-3}) was at a trailer court near a mine vent. Measured background

TABLE 5.3. Example of Full-Shift Exposure Calculations for a Mucking Machine Operator[9]

Location	Exposure time (h)	Radon progeny concentration (WL)	Exposure (WLh)
First stope	5.0	0.2	1.0
Second stope	1.5	1.4	2.1
Travelway and haulageway	0.5	1.8	0.9
Surface (lunch)	1.0	0.0	0.0
Totals	8.0		4.0[a]

[a] Average full-shift exposure level = 4.0/8.0 = 0.5 WL; working level months (WLM) of exposure = 4 WL h/170 WL h/WLM^{-1} = 0.023 WLM.

TABLE 5.4. Estimated Risks Due to Radon Venting from a Model Uranium Mine[9]

Source	Individual lifetime risk		Predicted fatal cancers per year in region
	Max. individual	Region average	
Underground mine	9.0×10^{-3}	5.3×10^{-5}	2.7×10^{-2}
Open-pit mine	1.3×10^{-3}	1.6×10^{-5}	8.0×10^{-3}

^{222}Rn concentration averaged 0.5 pCi L^{-1} (19 Bq m^{-3}). A health impact analysis for vented radon from uranium mines has also been performed for individuals in structures near the mine. The results indicate that radon levels will be significantly elevated above background for distances up to 3 km, though the detailed risk levels were made unduly high by assuming horizontal ground level releases and a very high vent release.

5.2.3.2 Solution Mining

Solution mining implies the recovery of uranium from ores through the in situ leaching of the mineral deposit in the existing groundwater by adding chemical reagents. This effectively reverses the natural uranium precipitation process that deposited the uranium mineral in the first place. The resultant uranium-bearing solution is pumped from the mineralized zone to a surface processing plant where the uranium is extracted by conventional recovery techniques. The barren solution from the recovery plant is recycled to the mineralized zone to dissolve additional uranium.[11]

TABLE 5.5. Release Rates of Radionuclides and Annual Population Doses from Airborne Effluents Estimated for Highland Uranium Solution Mining Project after Ref. 12

Radionuclide	Release rate (Ci y^{-1})
^{234}U	4.38×10^{-3}
^{235}U	2.01×10^{-4}
^{238}U	4.38×10^{-3}
^{226}Ra	3.65×10^{-7}
^{230}Th	1.10×10^{-6}
^{222}Rn	1.64×10^{-2}
Solution mining recovery	
^{222}Rn	1.14×10^{2}

	Population dose (man-rem)^a	
Organ	Project effluents	Natural background
Total body	5.35×10^{-3}	1.08×10^{3}
Lung	6.78×10^{-1}	6.74×10^{3}
Bone	2.98×10^{-2}	1.32×10^{3}

^{a} Based on 1970 population of 7490 persons.

In the recovery process, the uranium will stay in solution until extracted as yellowcake and dried. Any dissolved radon, on the other hand, will be driven off and released to the atmosphere. The amount depends, of course, on the ore grade mined and the mill capacity. Table 5.5 shows the predicted release for the Highland mill, which is designed to recover 340 tonnes (750,000 lb) of U_3O_8 per year with a flow capacity of 1200 gal min^{-1} (4543 L min^{-1}). The dose to the lung, the critical organ, was estimated as 0.01% of background dose.[12]

5.2.4 Uranium Milling

After mining the uranium ore, the crushed ore is processed at a mill to extract and concentrate the uranium contained in it. This involves the physical separation of the dispersed uranium-bearing minerals from the "country rock" matrix, mainly silicates, followed by a chemical concentration process. In this process the mineral grains are dissolved and the uranium extracted, usually by ion exchange. It then is recovered, filtered, and dried. The final product is a fairly pure bright yellow uranium compound that is usually referred to as "yellowcake."

The tailings piles form large dikes of finely divided material with a radium content that depends on the original uranium ore grade. As the radium decays, radon can emanate from the pile and the radon progeny may become attached to any fine particulates in the top layer, which could become entrained in air by wind action. As of 1979, it was estimated that there were over 100 million tons of tailings stored at active U.S. uranium mill sites containing over 50,000 Ci (1.85 PBq) each of ^{230}Th and ^{226}Ra and its decay products.[6] Radioactive airborne emissions, mainly from dry pond edges and piles, take place as a result of wind erosion and the diffusion of radon gas. The annual emission rates from tailings disposal areas have been estimated as 0.2–14 mCi y^{-1} (7.4–500 MBq y^{-1}) of ^{238}U, 3–200 mCi y^{-1} (0.1–7.4 GBq y^{-1}) of ^{230}Th, ^{226}Ra, and ^{210}Pb, and 14–8500 Ci y^{-1} (0.5–315 TBq y^{-1}) of ^{222}Rn. They are typically an order of magnitude higher than radon releases from ore crushing and storage areas.[9] The predicted population doses for radon releases from uranium mills are of the order of 600,000 man-rem (6000 man-Sv).

Tailings dike failures have occurred on several occasions.[13] Their main effect is the release of radium-containing water into surface watersheds. Radon release is usually inconsequential. On the other hand, airborne releases may result in ambient radon levels well above normal background; some examples are shown in Table 5.6, where values as high as 282 pCi L^{-1} (10 kBq m^{-3}) on a pile are shown and up to 42 pCi L^{-1} (1.6 Bq L^{-1}) at some distance.[14] For the large, inactive tailings pile at the Vitro mill near Salt Lake City, it has been estimated that a 3×10^9-kg mass of contaminated material emanated radon at an average rate of 550 pCi m^{-2} s^{-1} (20.4 Bq m^{-2} s^{-1}).[15] Surface air concentrations of ^{222}Rn on the pile varied from 1.2 to 4.5 pCi L^{-1}, falling off to background-equivalent level at a distance of about 1 km. Though these levels are small in this particular location, the tailings pile may increase the radon exposure, in person-WLM, by about 10%, as shown in Table 5.7, and contribute a small, finite increase in the probability of lung cancer.

Radon release from tailings piles can be reduced or controlled by various means, such as earth layers, plastic sheeting, incorporation of tailings in asphalt or concrete, or chemical removal or recovery of radium. Only earth covers are seriously considered for the purpose, as the other methods either degrade when exposed to sunshine or involve prohibitive costs.[16] Figure 5.3 shows the reduction in radon flux obtained for different types of soil. Most clayey soils will retain some

TABLE 5.6. Ambient Radon Levels near Uranium Tailings Piles[14]

Type of study	Range of radon concentrations (pCi L^{-1})^a	
	On pile	Within 1 km
Several measurements at four piles	1–34	0.5–3.4
Intensive study at one pile for a year	8–66	0.8–7.2
Several samples at two piles in one month	3.2–8.4	0.4–14
Few measurements at numerous inactive piles over a few days	12–282	0.2–42
Several samples at one pile over a 1-month period	4.5–7.7	1.7–7.8

^{a} The range of results presented for multiple-pile studies are from measurements made at different piles as opposed to a range for any particular pile.

residual moisture, of the order of 9–12%; complete drying out would destroy much of the radon attenuation and erosion control potential. Three meters of soil are the prescribed cover layer and efforts are being made to provide such cover for all active and inactive U.S. tailings piles.

A rather special situation has existed in some towns in mining areas, where mill tailings were used for landfills and subsequently built on. This situation was documented extensively for Grand Junction, Colorado,[17] but similar conditions have also been observed at Elliott

TABLE 5.7. Extrapolated ^{222}Rn Exposures Attributable to Vitro Tailings Pile for Medium- and High-Density Population Projections[15]

Exposure source	Population	Integrated population exposure (person-WLM y^{-1})
Tailings	1,850,000	15,500
Background^a	1,850,000	116,000 ± 65,000
Tailings	3,000,000	24,400
Background^a	3,000,000	188,000 ± 105,000

^{a} 0.25 ± 0.14 pCi L^{-1}.

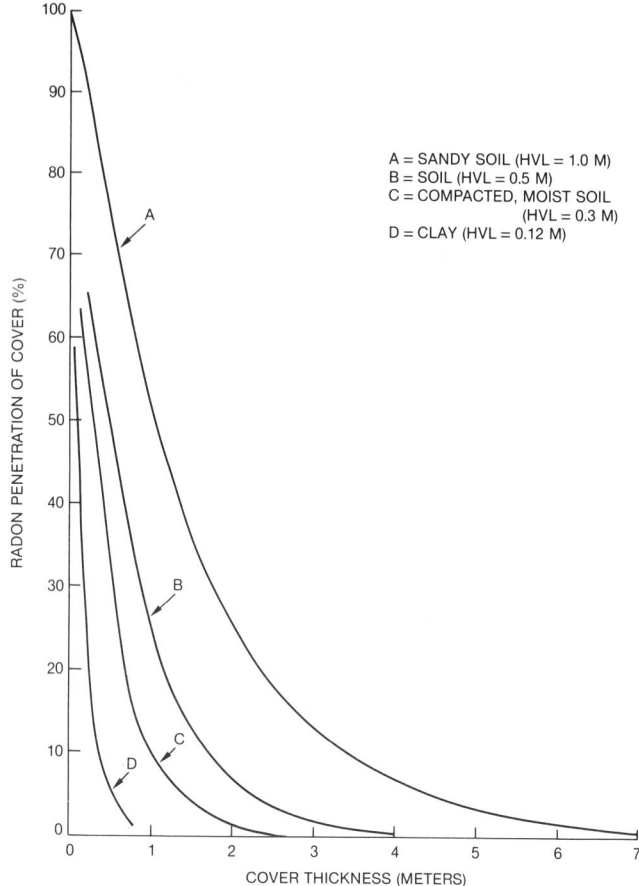

FIG. 5.3. Radon penetration of cover as a function of cover thickness.[16]

Lake, Ontario. At Grand Junction, gamma exposure rates up to 104 rad h^{-1} (1 Gy h^{-1}) were found inside the high school. It was shown that the proportion of buildings built on tailings that exceeded the 0.01 WL (2 × 10^{-7} J m^{-3}) level was much greater than the fraction that exceeded the gamma exposure limit.[18] The general indication was that diffusion of radon through the basement structure was the predominant cause of radon exposure, rather than any diffusion from tailings materials used in structural components.

5.2.5 Phosphate Mining and Milling

Uranium ores are not the only minerals with high radium–radon levels. As was shown in Chapter 4, many other minerals have substantial uranium contents. In some cases, such as with the South African gold ores, recovery of the uranium as a by-product is economically feasible. Phosphate shale, mainly mined for fertilizer production, is also such a mineral. In the United States alone, 130×10^6 tons (120 Tg) of phosphate rock were mined in 1973, yielding 38×10^6 tons of marketable rock. About half of the marketable rock is converted to fertilizer, the other half being used to produce other commodities, such as phosphoric acid. Mining and processing phosphate ores redistributes ^{238}U and its decay products among the various products, by-products, and wastes of the phosphate industry. The dispersal of wastes in the environment as well as the use of phosphate fertilizers in agriculture and of chemical gypsum as a building material are possible sources of exposure to the public.

The average activity concentrations range up to 130 pCi g^{-1} (4.8 Bq g^{-1}) for ^{238}U and ^{226}Ra, which are generally found to be in radioactive equilibrium, and up to 4.4 pCi g^{-1} (0.16 Bq g^{-1}) for ^{232}Th. The activity concentrations of ^{238}U vary widely from area to area, but they are usually much higher than those of ^{232}Th, which are comparable to those observed normally in soil.

The products and by-products of the phosphate industry contain various fractions of the uranium, radium, and thorium originally present in the rock, and these fractions in turn depend on the manufacturing process. In the wet process the marketable ore is combined with sulfuric acid, and phosphoric acid and gypsum result. This mixture is called normal superphosphate. By separation, phosphoric acid is obtained, and gypsum is sent to a waste pile. Most of the uranium and much of the thorium is transferred to the phosphoric acid during the separation process, whereas most of the radium remains with the gypsum by-product.[1] Consequently, most of the radon evolution is associated with utilization of the gypsum by-product.

Phosphate mining is a major U.S. industry. Central Florida mining activities account for about 91% of total U.S. production, and considerable attention has focused on the environmental impact on that area.[19–22] Phosphate shale also occurs in Georgia.[23]

The radon impact from phosphate mining is of three kinds: increases in the airborne radon concentrations at or near active and inactive mine sites, increases in indoor radon levels in houses on mine sites or tailings, and increases in gamma-ray exposures and radon concentrations in houses built with phosphate-slag-bearing materials.

5.2.5.1 Phosphate Mine Sites

The extraction of phosphates from mined ores results in tailings that consist of sand and slimes, which contain approximately 10% and 45% of the initial radium content, respectively. Most of the remainder goes with the marketable rock, is separately processed in fertilizer production, usually elsewhere, and ends up in the process slag. The disturbed land underlying the mine site will still contain uranium-bearing rock. Overburden, sand tailings, dewatered clays, and often part of the leached zone are mixed during the reclamation of phosphate-mined lands. These materials, with their different ^{226}Ra concentrations, combine to become a source of ^{222}Rn emanation. Measurements of ^{222}Rn emanation rates at various reclaimed land sites in Florida obtained values ranging from less than 0.1 to 25.0 pCi m^{-2} s^{-1} (4–1000 mBq m^{-2} s^{-1}). A mean emanation rate of 8 pCi m^{-2} s^{-1} (0.3 Bq m^{-2} s^{-1}) is believed to be a reasonable estimate.[24]

The estimated ^{222}Rn release from phosphate production is shown in Table 5.8.[24] It is evident that radon release from reclaimed land accounts for about two-thirds of the total amount released. It has been estimated that 33% of the reclaimed land exceed the 0.05 WL level and 41.7% the 0.01 WL level.[22]

For a "model" phosphate mine and mill, handling 10^6 tons of ore

TABLE 5.8. Estimated Release of ^{222}Rn from Phosphate Production[24]

Source	Estimated annual release (kCi y^{-1})	(TBq y^{-1})
Mining	0.1	3.7
Beneficiation		
Slime	—	—
Sand tailings	9.2	340
Rock	2.4	100
Processing		
Wet	3.4	126
Thermal	0.2	70
Reclaimed land	36	1330
Total	51.3	1970

per year, mining a 4-m seam below 3-m overburden, radon fluxes of 0.3, 1.6, and 4 pCi m^{-2} s^{-1} (11, 60, and 148 mBq m^{-2} s^{-1}) have been estimated for the undisturbed area, the tailings, and the removed overburden, respectively. The maximum radiation exposure levels are given as 0.22 mWL, for a population dose of 4.9 person-WL,[25] not really significant compared to normal background levels. Correspondingly low risk levels have been estimated for other phosphate mining areas.[23,26]

5.2.5.2 Indoor Radon Levels near Phosphate Mines

Radon diffusing through the floor, or more likely through cracks and openings in the floor, will cause the buildup of radon progeny inside a building. Direct gamma-ray exposure from radon progeny in the soil seldom contributes significantly to total exposure.

Radon progeny levels were evaluated by the U.S. Environmental Protection Agency (USEPA) and the Florida Department of Health and Rehabilitative Services (DHRS) in Polk County, Florida, and the results are shown in Table 5.9.[32] It was found that substantially higher WL levels were observed in homes on reclaimed or mineralized land. It was also found that higher levels occur in houses with slab floors than those with a crawl space underneath. Although these levels are somewhat lower than those observed in houses built on uranium mill tailings, it is clear that a number of the houses surveyed require remedial action according to the Surgeon-General's guidelines (10CFR712). For exam-

TABLE 5.9. Summary of the Florida Survey Results[(27)]: The Indoor Radon Decay Product Level Distribution by Number of Structures as Monitored by the USEPA and the Florida Department of Health and Rehabilitative Services (DHRS)

	Number of structures in the range (% of total)		
Level (WL gross)	EPA[a] ($N = 22$)	DHRS[b] ($N = 111$)	Composite ($N = 133$)
Greater than 0.05	5 (23)	3 (2)	8 (6)
0.03–0.05	3 (14)	9 (9)	12 (9)
0.01–0.03	4 (18)	22 (20)	26 (20)
Less than 0.01	10 (45)	77 (69)	87 (65)

[a] U.S. Environmental Protection Agency.
[b] Florida Department of Health and Rehabilitative Services.

ple, in Florida an estimated 640 structures in Polk County had radon progeny levels above 0.02 WL (4×10^{-7} J m^{-3}), with several hundred similar situations occurring in adjacent counties.[28]

5.2.5.3 Use of Phosphate Slag in Building Materials

One of the waste products of phosphate production by the thermal process is a slag in which most of the residual radium is concentrated. One of the major industrial uses of this slag has been as a lightweight aggregate in the production of concrete building blocks. Consequently, in regions adjoining areas of phosphate production one can find many buildings with elevated levels of uranium, radium, and radon progeny in their walls. The consequences of using phosphogypsum in houses for walls and ceilings has been estimated to be a gamma dose rate of about 7 rad h^{-1} in air (60 krad y^{-1} or 0.6 kGy y^{-1}) and radon levels of about 0.15 pCi L^{-1} (5.5 Bq m^{-3}).[19,29,30]

5.2.6 Other Mineral Extraction Facilities

Since uranium minerals occur widely dispersed in the earth's crust, they are found to accompany many other minerals that are being mined commercially. Consequently, elevated radon levels have been found in various mines, such as Newfoundland fluorspar mines, South African gold mines, Swedish lead–zinc mines, and Norwegian magnetite mines.[25,31] Generally speaking, such mines show the same relationship between lung cancer incidence and working level values, for nonsmokers, and the same age dependence, as are found in uranium mines.[32] The discovery of significant radon concentrations in many of these mines has brought a realization of the need for better ventilation and for monitoring of radon levels at many mine sites that considered themselves outside any radiation concerns.

The occurrence of uranium and radium in many metallic minerals has also led to the appearance of detectable radon levels at mills and smelters where radon is released by grinding, dissolution, or melting operations. For instance, zirconium extraction from zircon sands results in 150–1300 pCi g^{-1} (5.6–48 Bq g^{-1}) of soluble ^{226}Ra and progeny in the chlorinator tailings and 87–154 pCi per dry gram (3.2–5.7 Bq g^{-1}) in the clarifier sludges; no radon concentrations were reported above the tailings.[33] Other industries with significant radionuclide emission are the zinc, aluminum, and copper industries.[18]

5.2.7 Thoron Production in Mines and Mills

Because of the short half-life of thoron, ^{220}Rn, its mobility is limited and the impact of its progeny is of minor consequence, apart from the rest of the thorium chain. However, during mining and milling of thorite, monazite, and other major thorium ores, thoron levels could reach working level concentrations comparable to those involving action levels with ^{222}Rn. Thoron emanation from rocks and tailings can be inhibited by a relatively thin (approximately 10 cm) layer of sand or soil.[34]

In the Elliott Lake, Ontario, uranium district, radon and thoron progeny were encountered at comparable concentrations. To assess the potential impact of thorium mining and milling, the impact of a model open-pit thorium mine capable of producing 1600 tons of 0.5% ThO$_2$ ore per day was investigated for some actual production sites.[35] It was deduced that the lung dose due to radon inhalation completely dominates the dose commitment and is not greatly different for two selected sites.

5.3 RADON IMPACT FROM FOSSIL FUEL COMBUSTION

The fossil fuels, oil, natural gas, and coal are considered among the most serious sources of air pollution. This is primarily because of their release of sulfur dioxide, nitrous oxides, fly ash, and particulates, all major ingredients of smog. However, as with all other resources that are recovered from the earth's crust, one may find radon dissolved in gas and oil, and substantial concentrations of uranium-bearing minerals are slag- and dust-forming impurities in coal.

5.3.1 Natural Gas

Natural gas fields contain substantial concentrations of radon. Table 5.10 lists some of the wellhead concentrations reported. Some radon is lost in subsequent transmission and by decay in transport. Pipeline samples in the New York area averaged 20 pCi L^{-1} (740 Bq m^{-3}), but have varied considerably with the supply source.[36] The concentration

TABLE 5.10 Radon Concentrations in Natural Gas at Various Wells[1]

Location of well	Radon concentration (pCi L^{-1})	
	Average	Range
Borneo		
Ampa field	—	1.5–3.2
Canada		
Alberta	62	10–205
British Columbia	473	390–540
Ontario	169	4–800
Federal Republic of Germany	—	1.0–9.6
Netherlands		
Slochteren	—	1.1–2.8
Other fields	—	3.7–44.7
Nigeria		
Niger delta	—	0.9–2.9
North Sea		
Leman field	—	2.0–3.8
Indefatigable field	1.8	—
United States		
Colorado, New Mexico	25	0.2–160
Texas, Kansas, Oklahoma	<100	5–1450
Texas Panhandle	—	10–520
Colorado	25.4	11–45
Project Gasbuggy	15.8	—
California	—	1–100
Kansas	100	—
Wyoming	10	—
Gulf Coast (Louisiana, Texas)	5	—
California, Louisiana, Oklahoma, Texas	—	1–120

of radon remaining in the gas when it reaches the consumer will depend on many factors, such as gas processing, storage time, and separation of various fractions; the heavier fraction is bottled as liquefied gas. Much of the radon will appear in the liquefied gas sold by retailers and constitutes a potential source of exposure in homes. Seasonal variations result from variations in consumption which give rise to variations in storage time. Minimum concentrations occur during winter when previously stored gas is consumed. The large variations between retailers presumably reflect different storage times and variations in processing. Large tanks had five times the concentration that small tanks had. This result is consistent with the operational practices of the retailers.[37]

Natural gas is used widely for space heating and cooking, resulting in the release of radon and radon progeny to the indoor air. Several models have been developed to estimate human exposure from this source.[29,30,38] Most assume that no radon progeny is transported in the natural gas, as it would plate out on pipeline surfaces. This means that the natural gas will release only radon, and the equilibrium among its progeny dispersed in the home will depend on the release rate and the ventilation rate. Assuming a radon concentration of 20 pCi L^{-1} (740 Bq m^{-3}) and that 0.8 m^3 of gas are consumed in a 230-m^3 house with an air change of one space volume per hour, an average in-house radon concentration of 0.0028 pCi L^{-1} (100 mBq m^{-3}) occurs from use of unvented kitchen ranges.[38] The contribution from use of space heaters is more difficult to estimate, because their use is both more seasonal and also more prolonged at any one time. In the United States many of the cooler regions are at some distance from gas fields, and this results in somewhat lower radon concentrations at the heater. It should also be remembered that ventilation rates are likely to be low under cold winter conditions in well-insulated homes. Taking a maximum instantaneous radon level of 4.5 pCi L^{-1} (170 Bq m^{-3}) for a 45°C (80°F) indoor/outdoor temperature differential and the most severe heating requirement of 10,000 degree days (one degree day = one day at one degree temperature differential), a maximum yearly average radon concentration of 0.28 pCi L^{-1} (10 Bq m^{-3}) has been estimated.[38] This is well within the range of normal indoor radon levels.

5.3.2 Oil

Petroleum contains trace quantities of uranium, thorium, and their progeny. No environmental distribution of radioactive material has been observed around oil-fired power plants. Presumably, any trace radon would be driven off in the refining process and any radiological impact would be insignificant.

5.3.3 Coal-Fired Power Plants

It is well known that coals contain varying concentrations of uranium, thorium, and their decay products. Large coal-fired power stations burn pulverized coal, and the combustion products in both gaseous and particulate form are discharged into the atmosphere. Because of its high ash content, typically 5–10%, there is a large particulate emission from combustion of coal, even though larger fly ash particles are removed by precipitators or scrubbers. Typical ^{226}Ra concentrations in coal and coal residues range from 0.2 to 14 pCi g^{-1} (7.4–520 mBq g^{-1}).[1] A substantial fraction of the ^{226}Ra will appear in the fly ash. Estimates of the flow of fly ash through the stack per megawatt-year of electricity produced a range of from 0.7 to 30 tons. For a representative value of 10 tons per megawatt-year, the activity of ^{226}Ra, ^{228}Ra, and ^{232}Th would amount to 10^{-5} Ci (370 kBq), while for ^{210}Po it would be 10^{-4} Ci (3.7 MBq). The activity of ^{222}Rn released per megawatt-year has been estimated at 10^{-3} Ci (37 MBq) on the assumptions that ^{222}Rn is in radioactive equilibrium with its ^{226}Ra precursor in coal, that the average concentration of ^{226}Ra in coal is 0.3 pCi g^{-1} (11 mBq g^{-1}), and that all the radon is discharged during combustion of the coal.[1]

Though predicted, no significant radon levels have been found near coal-fired power plants. Model calculations for current coal-fired plants and "new" plants with improved off-gas treatment predict emissions of 0.67 Ci y^{-1} (25 GBq y^{-1}) ^{222}Rn, and 0.51 Ci y^{-1} (19 GBq y^{-1}) ^{220}Rn, and 1.9 Ci y^{-1} (70 GBq y^{-1}) ^{222}Rn and 1.6 Ci y^{-1} (59 GBq g^{-1}) ^{220}Rn, respectively.[25] This implies a tripling in radon emission for "new" plants, while all other radioactive emissions are significantly reduced by the improved treatment. In any case, the risks from radon are not considered significant in comparison with those from much lower concentrations of radionuclides of the uranium and thorium series in particulate materials.[25]

The emanation rates of radon from coal ash from pulverized coal furnaces have been investigated by Barton and Ziemer.[39] The emanation rates increased, as expected, with decreasing particle size, i.e., larger specific surface area; but they also increased with moisture content up to a level of 20% moisture by weight; beyond this value they fell off. Although the contribution of ash piles and slag piles to the total environmental radon inventory may be minor, there may be some benefit in keeping their surface damp.

5.4 AIRBORNE RADON

5.4.1 The Ambient Background

The radon exposures discussed in the preceding sections are often described as being "technologically enhanced" and, therefore, to some

HUMAN EXPOSURE

extent controllable. There is, however, a significant radon contribution to the general "natural" radiation background whose controllability is much more problematic and which, in fact, has certain aspects of inevitability. Most of this radon exposure results from inhalation of the radon progeny attached to aerosols that are found in the ambient air around us. As we have seen, the radon concentrations vary geographically, depending primarily on the uranium content of soil and near-surface rocks.

Exposure to radon progeny in normal environmental situations differs from that in mining atmospheres in three respects: 1) the median particle size of the carrier aerosol is smaller (0.1 μm versus about 0.2 to 0.4 μm for mines), 2) the fraction of ^{218}Po that exists as free ions is larger (0.07 versus 0.04 in mines), and 3) the exposure is continuous rather than part-time. All three of these factors increase the bronchial dose per unit exposure. Compensating for these effects, the environmental concentration of radon progeny is usually lower than in an underground situation.

Some measured and inferred outdoor radon concentrations in the United States are presented in Table 5.11.[40] Comparable results have been found for continental areas of other countries. The values shown are only approximate, as it must be remembered that the rate of radon emanation from the soil varies diurnally and seasonally, and radon progeny measurements will depend on aerosol concentrations and meterological dispersion conditions. The emanation rate is lower when the ground is cold or frozen or when it is wet following rain. Variations with altitude and soil conditions have been discussed in Chapter 4.

A reasonable average level for radon concentrations near ground level seems to be of the order of 150 pCi m^{-3} (5.55 Bq m^{-3}). In Chapter 4 the range was suggested to be 4–15 Bq m^{-3}. This is a useful concentration to bear in mind for outdoor conditions when evaluating indoor concentrations. Much higher and somewhat lower levels can obviously be encountered at various locations and at different times of year, even in the absence of specific mineral sources or mining activities.[41]

The doses and associated risks resulting from these radon releases will be discussed in subsequent chapters. However, it is important to view the impact of any enhancement of radon progeny concentrations against this background. The principal setting where such enhancement is being observed is inside buildings, mainly in private homes.

A special situation is found in underground caves, where high radon levels may be encountered from radium-bearing rocks or mineral waters. Some European spas still pride themselves on their radium-

TABLE 5.11. Some Measured Outdoor ^{222}Rn Concentrations in the United States and Antarctica[40]

Location	Radon concentration (pCi m^{-3})	
	Range	Mean
Illinois	50–1000	—
New York	20–500	130
New Mexico	—	240
New York	40–230	120
New York	100–200	170
Ohio	170–1040	480
Illinois	70–300	—
Florida	20–300	—
Washington, D.C.		122
Massachusetts	10–43	20
California		90
New York	15–200	100
Ohio	70–850	270
Ohio		260
Washington, D.C.		47
Illinois		25
California	2.5–10	6
Tennessee		17
Alaska		3
Puerto Rico		0.1
Washington		2
Little America		2.5
South Pole		0.5

bearing mineral waters and their supposed curative properties. Others provide rock galleries, where visitors can breathe radon emanating from the rock walls. Radon concentrations in Slovenian Karst caves are in the kBq m^{-3} range, with some ranging as high as 6500 Bq m^{-3}.[42]

Related problems may arise in many caves that are open to the public, such as the Carlsbad Caverns in New Mexico, where concentrations as high as 36–50 pCi m^{-3} (1.3–1.9 Bq m^{-3}) have been measured on some days.[43] Approximately a million persons visit the caves each year and, though the individual dose for a 3–4-h underground visit would not be high, a significant population dose may result. Also, the occupational exposure to park personnel should be limited below 4 WLM y^{-1}.[26]

In the open, radon and its progeny will be subject to meteorological

HUMAN EXPOSURE

transport processes that determine their dispersion and precipitation (see Chapter 4). Depending on the presence and concentration of airborne aerosols the attached fraction will vary. Following a heavy rain, which precipitates aerosols, a larger fraction will remain unattached. Thus the aerosols control the behavior of the radon progeny in the air and in the lung and the unattached fraction may vary between 3 and 25%.[40] Many of the aerosols arise from suspended soil particles, which typically are silicates and may be electrically charged due to atmospheric processes. The charged particles readily attract the charged radon and thoron progeny.

Aerosol concentration and size distribution vary widely, but 0.1 mg m^{-3} with a median diameter of 0.5 μm may be considered a fair description of outdoor urban particulates. In rural regions aerosol concentration levels might drop to about a quarter of that value. Number concentrations of the order of 10^4 to 10^6 particles cm^{-3} are common, so that less than one airborne particulate in a million carries natural radioactivity.[40] Indoor concentrations are somewhat lower and the particles tend to be smaller.

Atmospheric aerosols below 0.5 μm in radius are composed largely of ammonium sulfates and other ammonium compounds, which rapidly grow by coagulation to about 0.1 μm in size, and have a mean radius of about 0.15 μm in the troposphere.[44] Additionally, long-lived radon progeny suspensions arise from soil dust, forest fires, and other pollutant sources. Coagulation of particles larger than 0.3 μm in radius is slower and more heavily controlled by precipitation scavenging. Most of the radioactivity is attached to smaller particles, which have a mean tropospheric residence time of 4–6 days.[44] This difference in particle size and residence time must be taken into account when one extrapolates data from the exposure conditions prevailing in underground mines to those prevailing out-of-doors or inside houses.

5.4.2 Indoor Radon

5.4.2.1 Introduction

The accumulation of radon and its progeny inside houses was first observed at Grand Junction, Colorado, in 1971, where it was ascribed to the use of uranium mill tailings as land fill.[17] Subsequent work on the development of methods of controlling the inflow of radon into houses from the underlying ground drew attention to the potential for radon progeny buildup in poorly ventilated spaces, and it became gradually more obvious that similar conditions could develop in any building that was exposed to significant radon inflow.[18,45,46] Since then, it has been overwhelmingly evident that radon progeny exposure inside houses, particularly well-insulated ones, may constitute a major portion of the natural-background exposures to the general population and may involve frequently much higher population exposures than those so carefully monitored and controlled due to effluents from nuclear power plants.[48]

Four sources of interior radon concentrations can be identified:

(a) ambient radon levels in the outside atmosphere, primarily from soil emanation;
(b) seepage of radon from underlying soil or rock through the building foundations;
(c) radon release from water use, primarily for cooking, heating, and washing; and
(d) radon emanation from uranium- and radium-containing structural materials such as floors and walls.

Entrance of outdoor air through open windows or air-conditioner intakes is difficult to control, but the levels are rarely high enough to cause concern; through the use of open windows, adequate ventilation ensures that interior radon levels are not greatly above the exterior ones. A calculational model has been developed to permit estimates of the internal dose from inhaled radionuclides and the external photon dose from airborne and surface-deposited radionuclides. The model considers air ventilation and the deposition of radionuclides on inside surfaces, as well as the shielding effect of the building against exterior radiation sources.[47] While this model does not specifically address the radon problem, the approach remains generically valid. Assuming a steady-state condition in which the radionuclide concentration outside a building, C_v, is constant with time and that the radionuclide concentration inside the building, C'_v, is increased by air ventilation into the building but is decreased by air ventilation out of the building, deposition of radionuclides on inside surfaces, and radioactive decay, then the radionuclide concentration inside the building is described by the first-order differential equation:

$$\frac{dC'_v}{dt} = \lambda_v C_v - \lambda_v C'_v - \left[\frac{V_{df}S_f + V_{dw}S_w + V_{de}S_c}{\Omega}\right]C'_v - \lambda C'_v \quad (5.1)$$

where t = time, λ_v = air ventilation rate (in units of time^{-1}), V_{df}, V_{dw},

HUMAN EXPOSURE

V_{dc} = deposition velocities (in units of length and time) on floor, walls, and ceiling, S_f, S_w, S_c = surface areas of floor, walls, and ceiling, Ω = volume of building, and λ = radioactive decay constant.

Assuming zero radionuclide concentration inside the building at time zero, by integrating Equation (5.1) the reduction factor is given by:

$$\text{RF}_{in}(t) = \frac{C'_v(t)}{C_v} = \frac{\lambda_v}{\lambda_a}(1 - e^{-\lambda_a t}) \qquad (5.2)$$

where

$$\lambda_a = \lambda_v + \frac{(V_{df}S_f + V_{dw}S_w + V_{dc}S_c)}{\Omega} + \lambda$$

By setting $(1 - e^{-\lambda_a t}) = 1$, which is accurate for times greater than a few hours in most assessments of chronic releases to the atmosphere, the time-independent reduction factor becomes:

$$\text{RF}_{in} = \frac{\lambda_v}{\lambda_a} \qquad (5.3)$$

Note that the steady-state internal dose reduction factor is just given by the ratio of two time constants—λ_v describing the increase in indoor air concentration and λ_a describing the decrease in concentration.

Recommended air ventilation rates for single-family housing units are λ_v = 0.5–1.5 h^{-1}. Therefore, for radionuclides with half-lives greater than about one day, the reduction factor is independent of the specific radionuclide and depends only upon the air ventilation rate, deposition velocities on inside building surfaces, and building geometry.

Calculations were performed for the following values of input parameters[47]:

$$a = 2, 5, 10 \text{ m}$$

$$\lambda_v = 0.2, 1.0, 5.0 \text{ h}^{-1}$$

$$V_d = 0.0001, 0.01, 1.0 \text{ cm s}^{-1}$$

The range of values for V_d was based on theoretical and experimental results for dry deposition processes and experiments on iodine vapors; and they are less well known for radon progeny.

The internal dose reduction factor decreases (i.e., the protection provided by residence inside the building against increases in radionuclide inhalation) with decreasing building radius, air ventilation rate, and increasing deposition velocity. The reduction factor varies by about three orders of magnitude for the range of parameter values given above.

The model predicts significant reductions in inhalation dose for both short-lived radionuclides (half-lives less than $1/\lambda_v$) and radionuclides occurring in particulate form that have deposition velocities greater than about 0.01 cm s^{-1} generated outside the building.[47]

5.4.2.2 Soil Radon Seepage to Indoor Air

In the majority of houses where elevated indoor radon levels have been found, radon has percolated from the underlying soil through the foundations of the buildings. In most cases this is a diffusion process that is driven by a concentration gradient in the soil and aided by a pumping action due to fluctuations in the barometric pressure. The pathways may be varied, as shown in Figure 5.4, but usually involve cracks or penetrations through the basement slab or the existence of a crawl space under a frame structure. The radon concentrations inside the house consequently are a complex result of soil emanation rates, entrance diffusion rates, and prevailing ventilation rates inside an otherwise closed structure. Since the natural ventilation rate of houses is a linear function of both wind speed and indoor/outdoor temperature difference, internal radon concentrations correlate with emanation rates from concrete foundations and soil if proper allowance is made for variations in the ventilation rate.[47]

Ventilation rates in larger structures can be determined by the use of sulfur hexafluoride (SF$_6$) gas, which can also be used to identify radon entry paths and seepage flow rates. After injection into the air, SF$_6$ concentrations can be monitored by collecting air samples with 10-mL syringes followed by analysis of the samples in a gas chromatograph with an electron-capture detector. A more difficult determination is that for the attachment coefficient, in view of the smaller-size aerosols prevailing in indoor air. This has been discussed in Chapter 3. For reasons of simplicity, the USEPA has routinely assumed 50% equilibrium for the inhaled fraction for assessment purposes.

Radon entry into basements, especially in residences, has been the

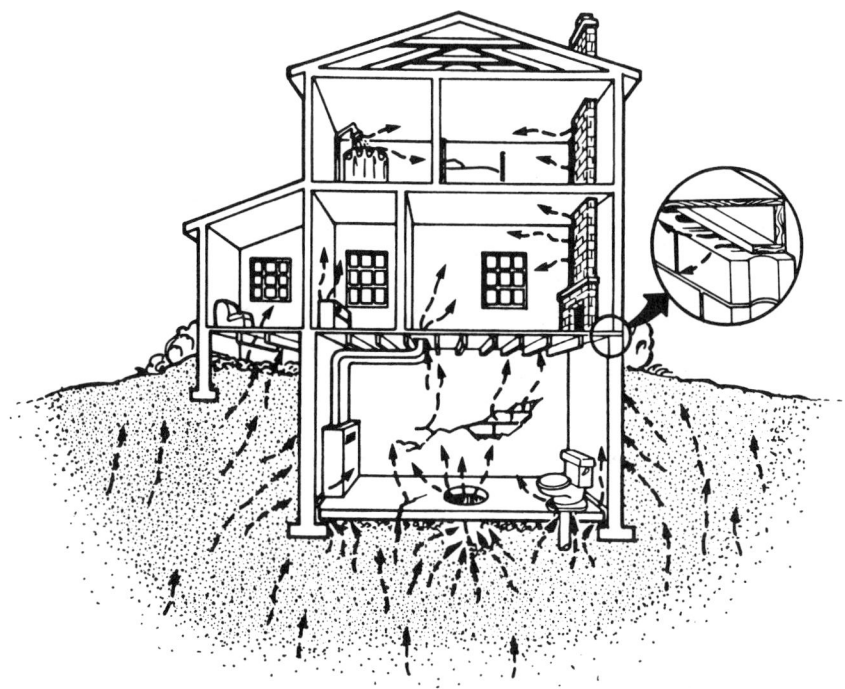

FIG. 5.4. Sources and entry routes of indoor radon.

subject of extensive investigations over the past several years. It has been shown that this is by no means an uncommon occurrence or confined to limited geographical areas. Radon concentrations in 87 dwellings in Great Britain showed mean emanation rates of 0.54 pCi $L^{-1} h^{-1}$ (0.02 Bq $L^{-1} h^{-1}$) and a high value of 5.5 pCi $L^{-1} h^{-1}$ (0.2 Bq $L^{-1} h^{-1}$) where there were exposed stone walls.[49] Table 5.12 shows the results of measurements of the effect of different ventilation rates on radon progeny concentrations and disequilibrium. Assuming a mean ventilation rate of one room-volume change per hour and an occupancy factor of 0.8, a mean cumulative population-exposure rate was projected of 0.144 WLM y^{-1} (5.8 × 10^{-8} J m^{-3}) from exposure inside buildings and roughly 0.15 WLM y^{-1} if open-air exposure is included.[49]

For seven typical houses in the Oak Ridge, TN, area the radon progeny levels in basements ranged from 0.006 to 0.09 WL with a mean of 0.016 WL (3.3 × 10^{-7} J m^{-3}).[50] Some houses exceeded the "remedial guideline" of 0.015 WL by factors of 2 to 6. The measurements also

TABLE 5.12. Radon Progeny Concentrations within a Room Emanating 0.54 pCi $L^{-1} h^{-1}$ for Various Ventilation Rates[47]

Ventilation rate (h^{-1})	Concentration (pCi L^{-1})			Working level (WL)
	^{218}Po	^{214}Pb	^{214}Bi, ^{214}Po	
0.1	5.07	4.76	4.55	0.0471
0.2	2.64	2.35	2.15	0.0230
0.5	1.10	0.85	0.69	0.0081
1.0	0.57	0.37	0.26	0.0035
1.5	0.39	0.23	0.15	0.0022
2.0	0.31	0.17	0.11	0.0016

clearly indicated an increase in radon and progeny as a result of simple energy conservation measures.[50,51] Ventilation rates are significantly lower in newer houses than in pre-1950 ones, with a corresponding increase in radon concentrations from an average of 23 Bq m^{-3} in older houses to 58 Bq m^{-3} in newer ones.

An extensive survey of 21 residences in the Northern New Jersey–Long Island, New York, region has shown diurnal variations, particle-size distribution of attached radon progeny, the uncombined fraction, and radon exhalation from cellar floors.[52] The particle size of attached aerosols peaked fairly sharply around 0.1 μm. Figure 5.5 shows the distribution of radon concentrations for cellar and first-floor levels versus ambient concentrations. They defined a quantity called "working level ratio," WLR, as

$$\text{WLR} = \frac{100 \times \text{working level}}{\text{radon concentration (pCi } L^{-1})}$$

The working level ratio effectively measures the radon-progeny equilibrium. For the houses measured, average WLR values were 0.50 ± 0.10 for cellars, 0.61 ± 0.12 for first floors, 0.66 ± 0.18 for second floors, and 0.78 ± 0.12 outside, indicating a substantial inflow by ventilation rather than by basement seepage. Annual doses were calculated using the formula

$$\text{Annual dose (mrad)} = 42 C_{Rn} + 25{,}000 \text{WL}$$

where C_{Rn} = mean annual radon concentration, in pCi L^{-1}, and WL = annual mean working level value. The first term in this equation gives the dose due to unattached ^{218}Po, assumed to be 0.07 of the radon

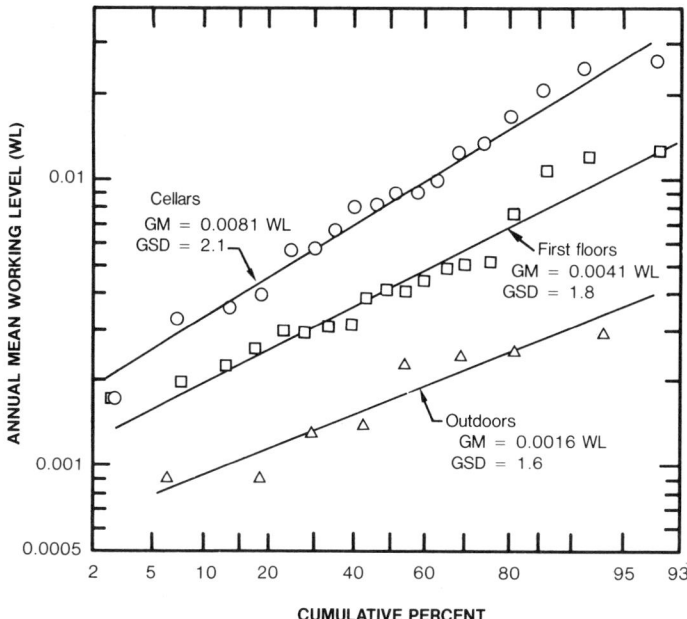

FIG. 5.5. Distributions of annual mean working level measurements for homes.[52]

concentration. On this basis, annual mean lung doses ranging from 42 to 420 mrad (0.42 to 4.2 mGy) were calculated.

Some of the houses in that study were located in the region often referred to as the "Reading Prong" and tended to show dose values in the upper portion of the range. The Reading Prong is an area on the eastern foothills of the Appalachians that stretches northeasterly from near Reading, Pennsylvania, through Northern New Jersey towards the lower Hudson Valley. This area has become the center of intensive studies ever since a number of unusually high radon concentrations were found, typically in new houses with an "energy-efficient" design. Table 5.13 presents the results of a large screening survey, using damage-track detectors, of 14,230 homes in the Reading Prong. It is seen that over 60% had radon-progeny concentrations above a tentative guidance value of 0.02 WL (4×10^{-7} J m^{-3}).

The general evaluation approach on houses with high radon levels has been to correlate these levels with high soil concentrations, penetrations through the cellar floor, and cracks in the foundations. To validate tests of this type, measurements must be conducted over long

TABLE 5.13. Radon Progeny Screening Results from the Alpha Track Mail Program Conducted by the Pennsylvania Department of Environmental Resources[a]

Exposure level (WL)	Number in the range	Percent of total[b]
Above 1.0	81	0.6
0.500–0.999	134	0.9
0.100–0.499	1585	11.1
0.050–0.099	2276	16.0
0.021–0.049	4651	32.1
0.010–0.020	3448	24.2
0.001–0.009	2145	15.1

[a] Bureau of Radiation Protection.
[b] 14,230 homes were included in the survey.

periods to allow for diurnal and seasonal variations.[53–55] Crawl spaces and seasonal ventilation rates have to be allowed for, but even so a survey of 453 homes of physics professors in 42 states found low correlations with such features and concluded that geographical factors predominate.[56] The results showed a median concentration of 39 Bq m^{-3} (1.05 pCi L^{-1}), with the highest at 559 Bq m^{-3} (15.1 pCi L^{-1}). An earlier compilation of radon levels in the United States, shown in Table 5.14, confirms the widespread occurrence of high radon levels in homes.[57] A very high proportion exceed the 0.01 WL (2×10^{-7} J m^{-3}) level, and even a guidance value of 0.02 WL (4×10^{-7} J m^{-3}) is exceeded by a large number.

It may easily be concluded that high radon levels in houses are fairly common and certainly not confined to the Reading Prong or to areas built on mill tailings. If one adopts an average ^{222}Rn concentration indoors of 1.1 pCi L^{-1} (40 Bq m^{-3}) and a working level ratio of 0.5, this corresponds to an average concentration of 5 mWL. However, significant fractions of houses, such as those measured in Chicago, are much higher, ranging up to 130 mWL.[55] It has been shown that a mean value of 5 mWL and an 80% occupancy factor translate into an indoor exposure contribution of 0.205 WLM, with 20% spent outdoors in a 1-mWL ambient environment; this adds up to a total exposure of 0.22 WLM (8×10^{-4} J h m^{-3}).[58] Depending on the risk model adopted, the estimates for lung cancers in the U.S. population due only to radon exposure are in the range of 11,000–22,000 "excess" cases per year. Of this, about 1000 cases per year may be ascribed to energy-efficient insulation of homes.[58] (See Chapter 9 for a more complete risk analysis.)

With some indoor radon levels in the Reading Prong area recorded as high as 1000 pCi L^{-1} (37 kBq m^{-3}), 250 times the recommended

HUMAN EXPOSURE

TABLE 5.14. Indoor Radon and Radon Progeny Decay Product Levels[57]

House location	Approximate number of homes	Average radon concentration (pCi L^{-1})	Average WL	Measurement condition	Percent above 0.01 WL	Percent above 0.02 WL
NY/NJ						
Basements	18	2.0	0.01	YR[a]	39	17
First floors	18	1.0	0.007	YR	17	0
Grand Junction, CO	29	(1.1)[c]	0.007[c]	YR	25	0
Florida						
Phosphate	100	(2.5)	0.015	YR	25	25
Background	29	(0.5)	0.0033	YR	3	0
Background	28	(0.065)	0.004	YR	4	0
Background	13	0.8[b]	0.004[c]	YR	14	14
Butte, MT	56	(3.3)	0.02	YR	75	38
Anaconda, MT	16	(2.6)	0.013	YR	56	25
Alabama and neighbor states						
P slag, 1st fl.	5	(2.8)	0.017	Jan–May	40	40
P slag, basement	17	(3.6)	0.018	Jan–May	76	35
Control, basement	5	(2.8)	0.014	Jan–May	60	40
San Francisco region	25	0.3	(0.002)	Grab	0	0
Energy-efficient homes	17	4	(0.027)	Grab	76	35
Soda Springs, ID	100	1.4[b]	0.006[c]	Grab	25	2
Illinois	22		6 houses > 10.0 pCi L^{-1}			
			9 houses > 5.0 pCi L^{-1}	(41% > 0.03 WL)		

[a] YR, Year-round average under occupied conditions (air pump integrated measurements).
[b] Geometric mean.
[c] Values in parentheses are not direct measurements but are calculated using a characteristic radon decay products/radon equilibrium ratio of 0.5 for basements and 0.61 elsewhere.

action level of 4 pCi L^{-1} (150 Bq m^{-3}), and a large proportion in the 100–250 pCi L^{-1} (3700–9250 Bq m^{-3}) range, the scientific and regulatory communities are faced with a technical and ethical dilemma. There is no epidemiological evidence that natural radon levels of the order of 4 pCi L^{-1} (150 Bq m^{-3}) result in any discernible health effects. Even at substantially higher levels, no such evidence has been found, nor probably is there any statistical feasibility for such a determination. Yet the radiation protection community is naturally reluctant to "accept" radiation exposures of that magnitude. They represent substantial lung dose commitments, that are an inevitable part of the natural radiation background, whereas strenuous and extravagant efforts are being made to control and reduce population-dose levels, orders of magnitude smaller, that may arise from routine effluents from nuclear power plants. Because of the severe economic and political implications of translating hypothetical risks into remedial action, the exposure to radon and its progeny will test the commonsense, sense of proportion, and the willingness to accept certain risks of both the regulators and the affected population.

5.4.2.3 Water Sources

Radon and its progeny can reach man either through their presence in water or through the decay of dissolved radium in water. Radon can dissolve in water when the water contacts uranium- or radium-bearing minerals. The subsequent pathways of radon to man can vary: radon may be ingested from drinking water and decay in the stomach; it may boil off on heating or cooking and be inhaled in the vicinity; or it may enter a house via humidifiers or showers, where it is driven off in water spray.

The average concentration of radon in U.S. public drinking water supplies is in the range of 200 to 600 pCi L^{-1} (7.4–22 kBq m^{-3}).[58] There are many private and municipal wells with radon concentrations exceeding 20 pCi L^{-1} (0.74 kBq m^{-3}) and concentrations in water of the order of 1000 pCi L^{-1} (37 kBq m^{-3}) are not unusual. Some Austrian spas reported having radon concentrations between 0.5 and 120 nCi L^{-1} (18.5–4400 kBq m^{-3}), leading to airborne concentrations of the order of 1 pCi L^{-1} (37 Bq m^{-3}) in the open air and 10 pCi L^{-1} (370 Bq m^{-3}) in room air.[5]

While the effect of ingestion of ^{226}Ra from drinking water has received a great deal of attention, radon ingestion has been of less concern. A series of measurements on subjects with both full and empty stomachs showed that radon diffuses from the stomach in a short time, with a biological half-life of the order of 30–40 min.[48] Because of this short residence time, only radon and its short-lived progeny ^{218}Po cause a significant dose to the stomach. Residual radon remains in the stomach for about an hour before being discharged to the small intestine, though some discharge of water from the stomach begins within a few minutes of ingestion.[59] Hursh's calculations showed that the stomach must be considered the critical organ for ingestion and that the total decay of the radon progeny, ingested as such, contributes a negligible dose to the stomach walls compared with that from ingested radon.

A series of radon dose measurements have been conducted for users of radon-rich drinking water.[59] The stomach doses derived were 240 and 380 rem mCi^{-1} ^{222}Rn (65 and 103 mSv MBq^{-1}) for subjects with full and empty stomachs, respectively. For waters with a radon concentration of the order of 100 pCi L^{-1} (4 kBq m^{-3}), this translates into annual doses of about 2.6–4.2 mrem y^{-1} (26–42 μSv y^{-1}) to the stomach, assuming a daily raw water consumption of 0.3 L. This would normally be a minor addition to the natural background exposure. Using the currently recommended quality factor of 20 for alpha particles, the dose equivalents convert to stomach doses of 480 and 760 rem mCi^{-1} ^{222}Rn (130 and 205 mSv MBq^{-1}) ^{222}Rn. Ingestion doses of 400–800 rem mCi^{-1} (108–216 mSv MBq^{-1}) ^{222}Rn to the stomach and 4–14 rem mCi^{-1} (1–4 mSv MBq^{-1}) ^{222}Rn to the whole body have been recommended as reasonable.[60] For a 1000 pCi L^{-1} (37 kBq m^{-3}) concentration, the annual dose equivalents were estimated as 100 mrem (1 mSv) to the stomach and 2 mrem (20 mSv) to the whole body. Even under conservative assumptions this ingestion dose is small compared with the inhalation dose to the bronchial epithelium from the radon transferred to respirable air indoors from use of the same water, though far from trivial.[60]

The appearance of radon in air from water supplies, particularly indoors, will depend on radon concentrations, water temperature, and hydraulic conditions, such as pressure drop, turbulence, and air bubble formation. For the state of Maine alone, radon concentrations in well water were found to vary by as much as a factor of 20 between various rock types, with an arithmetic mean for the state of 10,000 pCi L^{-1} (370 kBq m^{-3}) for 2000 samples; public utilities averaged about 2000 pCi L^{-1} (75 kBq m^{-3}), though ranging as high as 11,000 pCi L^{-1} (400 kBq m^{-3}). Table 5.15 shows radon analyses for water supplies averaged for each state.[61] It is evident that a mean concentration of 1000 pCi L^{-1} (37 kBq m^{-3}), assumed above, is not unreasonable for private wells, while public groundwater supplies are much lower at 130 pCi L^{-1} (4.8 kBq m^{-3}).

TABLE 5.15. Radon in Water: Results by State and Source[61]

	Geometric mean (pCi L^{-1}) of radon concentration (number of samples)	
State	Private well	Public ground water supply
AL	120 (22)	70 (182)
AR	230 (2)	12 (22)
AZ	—	250 (124)
CA	43 (6)	470 (15)
CO	—	230 (76)
DE	—	30 (72)
FL	6000 (34)	30 (327)
GA	2100 (2)	67 (225)
IA	—	220 (85)
ID	—	99 (155)
IL	—	95 (314)
IN	—	35 (185)
KS	—	120 (47)
KY	1500 (10)	32 (104)
MA	1000 (8)	500 (212)
ME	7000 (24)	990 (71)
MN	1400 (1)	130 (233)
MO	ND^a (2)	24 (138)
MS	—	23 (104)
MT	4300 (8)	230 (71)
NC	15 (29)	79 (404)
ND	—	35 (133)
NH	1400 (18)	940 (52)
NJ	—	300 (38)
NM	59 (14)	55 (171)
NV	—	190 (57)
NY	1500 (4)	52 (292)
OH	—	79 (165)
OK	—	93 (83)
OR	450 (18)	120 (69)
PA	910 (16)	380 (105)
RI	6500 (69)	2400 (575)
SC	1100 (28)	130 (384)
SD	4200 (2)	210 (155)
TN	ND (2)	12 (98)
UT	—	150 (195)
VA	560 (42)	350 (284)
VT	210 (23)	660 (71)
WI	730 (40)	150 (278)
WY	—	330 (32)
US	920 (434)	130 (6298)

^{a} ND, not detected above background levels.

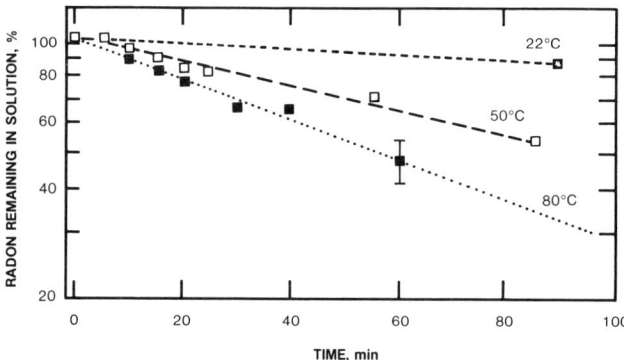

FIG. 5.6. Loss of radon from a 750-mL water sample at three constant temperatures.[53]

The basic question that arises relates to the impact of airborne radon from water sources in relation to other environmental sources of radon. The suggested environmental conversion factor is 0.7 rad WLM^{-1} (2 × 10^4 Gy J^{-1} h^{-1} m^3) for the bronchial epithelium.[60] However, actual air concentrations will depend on the humidity of the surrounding air, room temperature, and air circulation. These factors have given rise to a wide range in conversion factors for radon concentrations in air resulting from a given radon concentration in free-flowing tap water. Most concentration ratios are of the order of 10^{-4} pCi L^{-1} in air per pCi L^{-1} in the water.[61]

A number of experiments have been conducted to determine the radon transfer from water to air. Figure 5.6 shows the results from radon loss on heating. At room temperature, relatively little radon is lost, so that radon will be largely retained in the water in distribution systems or water tanks. On the other hand, vigorous boiling will remove radon rapidly, as seen in Figure 5.7. On that basis, radon transfer to the atmosphere was investigated for various domestic water uses. Table 5.16 shows the estimated radon releases per day for a water supply containing 1000 pCi L^{-1} (37 kBq m^{-3}) of ^{222}Rn for an American family of four. The figures seem to be on the high side in assuming use of both showers and tub baths and a rather liberal use of toilet flushes. Figure 5.8 shows two examples of observations on airborne radon concentrations in homes following the use of water in showers and other domestic activities. In both cases the water was tap water from a public utility supply, containing still 1500–2000 pCi L^{-1} (56–74 kBq m^{-3}) of radon.

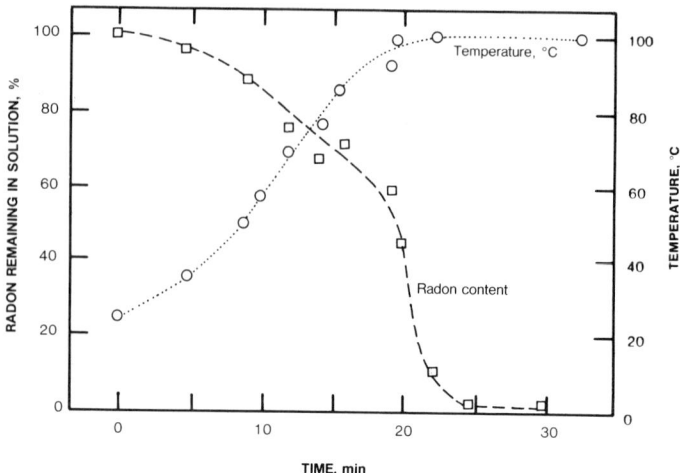

FIG. 5.7. Loss of radon as water is heated and boiled.[53]

It is obviously difficult to convert such measurements into an estimate of lung cancer risk from this source, since the inhaled radon progeny concentration will vary strongly with ventilation conditions, aerosol concentrations, and the effect of water droplets near the point of release. In most cases this source represents an added increment to radon-progeny concentrations arising from other causes, which are similarly affected by air circulation conditions. It has been suggested that 10,000 pCi of radon per liter of water contribute approximately 1 pCi to a liter of indoor air (the range is 0.17 to 3.5).[60] Assuming radon

TABLE 5.16. Radon Liberated in the Use of Domestic Water Containing 1000 pCi L^{-1} of Radon

Use	Daily consumption (L)	Transfer efficiency (%)	Radon liberated (pCi)	(kBq)
Showers	150	63	94,500	3.5
Tub baths	150	47	70,500	2.6
Toilet	365	30	109,500	4.1
Laundry	130	90	117,000	4.3
Dishwasher	55	90	49,500	1.8
Drinking and kitchen	30	30	9,000	0.33
Cleaning	10	90	9,000	0.33
Total	890		459,000	17

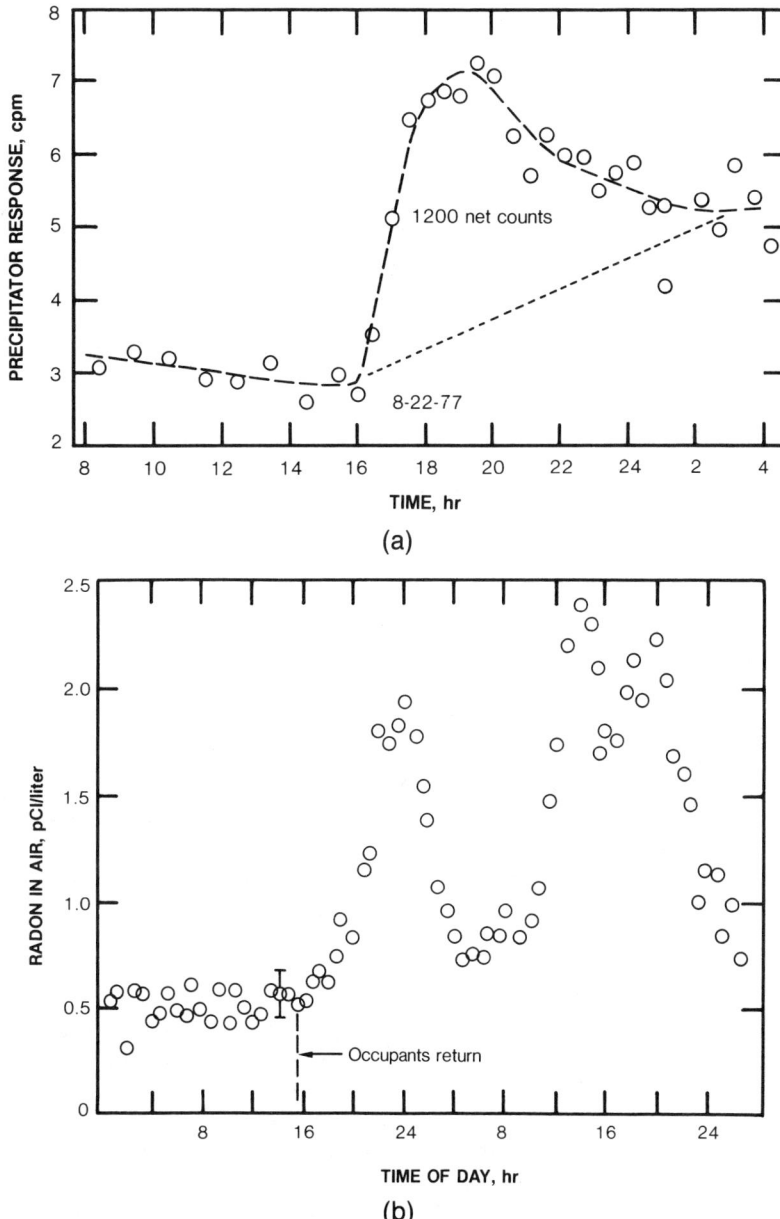

FIG. 5.8. (a) Determination of integrated radon exposure in an apartment following the use of 250 L of water in a shower[53]; (b) atmospheric radon levels in an apartment using water containing 1500 to 2000 pCi L^{-1} of radon.[53]

release rates comparable to those shown in Table 5.16, it has been estimated that radon in drinking water may cause some 30–600 lung cancer cases a year in the United States. This can be compared with 5000–20,000 cases of lung cancer estimated to be caused by radon from the soil seeping into homes.

5.4.2.4 Building Materials

Building materials cannot always be clearly isolated as a source of radon, because often local building materials are derived from similar rock to that underlying the houses. However, particularly in high-rise buildings and those with granite facings, such as many government buildings in Washington D.C., radium-bearing structural materials can be identified as a source of human exposure. Initially this was seen primarily as a gamma-ray exposure problem, because the gamma-emitting radon progeny are trapped in the solid; concern has been expressed that tight sealing of wall surfaces would result in a significant, and possibly hazardous, increase in external gamma exposure from the trapped radionuclides.[62] Curves have been developed comparing the gamma fluxes from sealed and permeable concrete walls, showing substantial differences in exposure with wall permeability.[63]

More important, probably, is the radon diffusion out of radium-bearing building materials. This diffusion is a function of pressure, temperature, porosity, and radon concentration.[64] For various building materials, porosities range from less than 0.1 to slightly above 0.3 and radon diffusion lengths range from 0.1 to 0.3 m, corresponding to interstitial diffusion coefficients of 2×10^{-8}–20×10^{-8} $m^2\ s^{-1}$.[54] To some extent the radon emanation from walls and floors can be controlled by surface coating, painting, or wallpapering.[62,65]

6

Dosimetry

DOUGLAS J. CRAWFORD-BROWN

6.1 INTRODUCTION

A man walks into the doctor's office, raises his arm and says, "Doc, it hurts when I do this." The doctor says, "Don't do that." An old vaudeville joke may seem an odd start to a chapter on dosimetry, but hidden within this joke is a kernel of truth concerning the nature of dosimetry and why we bother with it. The field of radiation protection is faced with past experience concerning the uranium miners and other groups exposed to radon and its progeny. If we look carefully, there are lessons to be learned from these experiences, lessons which can guide us in determining how to act in future situations involving radiation which might differ in some way from the past experiences. For example, new experiences may differ from the past in regard to the amount of radioactivity inhaled, the kind of radioactivity inhaled, the size of the particles inhaled, or even the ages and health of the people involved.

The past experience with the uranium miners suggests that they were at increased risk of developing lung cancer. What lessons can we learn from this experience? On the surface, all we can state is that working in uranium mines seems to produce lung cancer ("Doctor, it hurts when I go into uranium mines"), and therefore we conclude that we shouldn't go into uranium mines (Doctor: "So don't go into uranium mines"). This conclusion, however, is drastic and gives no guidance as

---

DOUGLAS J. CRAWFORD-BROWN • Department of Environmental Science and Engineering, School of Public Health, University of North Carolina, Chapel Hill, North Carolina.

to how we should act in other cases where people are exposed to the hundreds of other radionuclides present in the environment. This cause of cancer, therefore, must be attributed to some property of the experience in mines which is shared by other experiences with radiation. An ideal doctor would not tell the patient just to stop raising his arm, but would look for the cause of the pain within the muscles, joints, or nerves inside the arm or in the belief of the patient.

We are seeking, therefore, some aspect or property of the uranium miners' experience which can be labeled as the cause of their lung cancer and which might be shared by others exposed to radiation of any kind. The relationship between this property and the probability of developing lung cancer then should be the same regardless of which radionuclides or radiations produced the property.

This property in traditional dosimetry is the dose equivalent[1] delivered to the cells responsible for the cancer of interest (here, lung cancer). Determining the dose equivalent requires that two quantities be specified: the absorbed dose (D) and the qualify factor (Q). The absorbed dose is the average density of energy (ergs per gram) absorbed by the cells of interest and specifies the macroscopic distribution of energy in an organ or tissue. The quality factor (Q) is related to the linear energy transfer (LET) of the radiation and specifies the microscopic distribution of the energy. The product of D and Q is the dose equivalent (H), and in traditional dosimetry the dose equivalent is perceived as a unique indicator of risk.

If the dose equivalent delivered to the radiobiologically important cells is computed for the uranium miners and divided into the probability of developing lung cancer in the mining group, the risk per unit dose equivalent can be extrapolated to any other experience. The only requirement is that the dose equivalent delivered under the new experience be computed for the same cells. In the case of gamma-ray, x-ray, and many beta-particle radiations, it is sufficient to determine the average dose equivalent delivered throughout an organ, since the radiation energy is deposited fairly uniformly for these radiations. For alpha-particle-emitting radionuclides, however, care must be taken to ensure that the dose equivalent is computed for precisely the cellular subpopulation believed to produce the biological effect (here, lung cancer), since the dose delivered by alpha particles is highly nonuniform in an organ. When dealing with alpha emissions, it is quite common to find that one cellular subpopulation in an organ receives a dose of zero, while another subpopulation in the same organ receives hundreds of rads.

This highly nonuniform nature of the dose equivalent delivered

DOSIMETRY

to different cells in an organ such as the lung has in recent years led to suggestions that dose equivalent is not a sufficiently detailed quantity to use in extrapolating risks.[2] A new field, microdosimetry,[3] is rising to replace the old system of dosimetry. Microdosimetry examines the microscopic structure of the energy deposited by radiations in a cell. This structure is summarized by quantities such as the specific energy, Z, and its distribution in an organ. The specific energy distribution, $f(Z,D)$ is the probability that a small volume with a diameter on the order of 1 μm or less will achieve an energy density of Z, given that the organ as a whole has received a dose of D. The mean of $f(Z,D)$ simply is D, a fact which will be employed later in this chapter.

Essentially all of the past studies concerning exposure to radon have utilized the concept of dose or dose equivalent. Because of this history, the present chapter will focus on dose and dose equivalent as the quantities of most concern in risk assessments for inhaled radon and progeny. It will be assumed that the probability of developing lung cancer following exposure to airborne radon and progeny is related to the dose equivalent delivered to a distinct subpopulation of cells in the lung, called the critical cells. For a look at how microdosimetric quantities such as hit probabilities[4] could be used in performing the same task, the reader might examine any of a number of reports.[5-8]

Computing the dose equivalent to a cellular subpopulation in the lung following inhalation proceeds in several distinct stages:

1. It first is necessary to determine where the inhaled radon or progeny will deposit in the lung. This calculation is performed through the use of lung deposition models which mimic the movement of radionuclides while they are carried by the air during breathing. The deposition will be affected by the size of any aerosol particles to which the radionuclides (here, the progeny) are attached, the degree to which the radionuclides are free or unattached, the volume of air breathed, and the age of the individual. The deposition model must be capable of predicting both the amount of the radionuclide deposited and the distribution of deposition throughout the various subregions of the lung.

2. Once it has been estimated where the radionuclide deposits and to what degree, it is necessary to predict how the radionuclide will move within the lung. Because of transport processes in the lung, a radioactive particle deposited at one location may actually decay elsewhere. In the case of radon progeny, we will be focusing on the upper passageways of the lung, where the movement of particles is dominated by movement on the mucociliary blanket. The problem at hand, then, is to take the pattern of deposition in the lung and transform it over to the pattern

describing where the radionuclides are carried. This requires a mathematical model describing how the mucociliary blanket moves within the lung.

3. By knowing where the radionuclides are deposited and then carried by the mucociliary blanket, as well as the radiological half-life of the radionuclide, it is possible to determine where the radionuclides actually decay within the lung. For the radon progeny, the decays of interest are the alpha-particle decays, since they contribute most of the dose.

4. The last step involves using the information on where each alpha particle is emitted and calculating the resulting dose equivalent to the critical cells in the lung. This step clearly requires that the location of the critical cells be determined. The calculation of dose equivalent requires that depth–dose curves be generated. These curves describe the dose, and dose equivalent, delivered to cells located at various depths away from the walls of the lung passageways. The depth–dose curves themselves are produced by using stopping power equations, which estimate the stopping power of the alpha particles as a function of distance from the point of emission. The stopping power also is used to specify the quality factor to be assigned at each depth.

The result of these four steps for the problem of radon is to take an estimate of the concentration of radon and progeny in air and the degree of attachment to aerosol particles and convert this to an estimate of the dose equivalent delivered to a critical subpopulation of cells in the lung. If both this dose equivalent and the probability of lung cancer have been computed for a group of uranium miners, a risk factor (risk per unit dose equivalent) can be calculated. This risk factor then is assumed to apply to any new experience involving any mode of irradiation to the same subpopulation of cells. The role of dose equivalent will have been to act as a conceptual, and computational, framework uniting a whole range of experiences, both past and future. To paraphrase our original vaudeville joke, then, consider the following:

A man walks into the doctor's office and says, "Doc, it hurts when I go into the uranium mine." The doctor says, "So don't let your dose equivalent exceed X." OK, the joke's not so funny anymore, but it is more scientific.

6.2 ANATOMY AND MORPHOLOGY OF THE LUNG

The first step in this discussion is to delineate the structure of the lung. A visual tour of a system to be modeled is necessary to ensure

DOSIMETRY   177

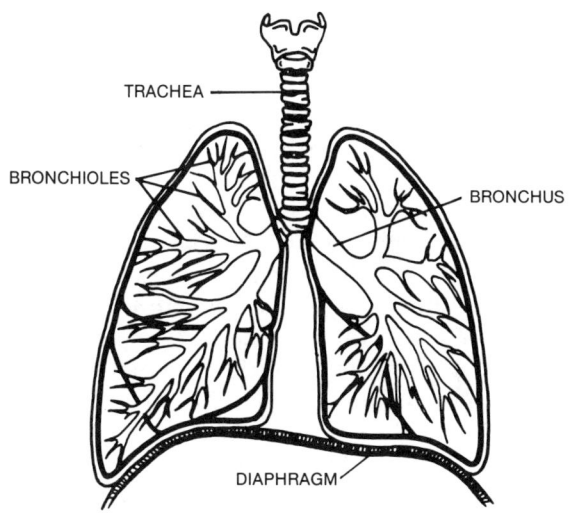

FIG. 6.1. A general depiction of the lung from the trachea down to the level of the distal bronchioles. (Reprinted from *Biology*, by C. F. Herreid, © 1978, with permission from Macmillan Publishing Company, Inc., New York.)

that the model contains all aspects of the system likely to affect the model predictions. We'll start at the top of the lung near the nose or mouth (called the proximal region of the lung) and proceed on down through the lung passageways to the alveolar sacs (called the distal region of the lung). A general picture of the lung can be found in Figure 6.1.

There are several different levels of detail available in examining the lung. At the lowest level of resolution lies the Task Group Lung Model (TGLM) published by the International Commission on Radiological Protection.[9] This model breaks the lung into three regions, which are labeled the nasopharyngeal (NP), the tracheobronchial (TB), and the pulmonary (P). The level of detail provided by this simple approach is insufficient for performing deposition calculations, so the TGLM model will not be discussed further. The lung can, however, be pictured generally as consisting of three compartments: a conducting zone, in which gas is transported but not exchanged; a transitional zone, in which there is both transport and exchange; and a respiratory zone, in which only gas exchange with the blood occurs.

More detailed models must account for the actual branching structure found in the lung. Air enters into the lung either through the nasal passage or the mouth. These two passages join directly beneath the tonsils in a zone termed the oropharynx. From this point, the air passes into the trachea and then through a series of bifurcating tubes or passageways. Each passageway splits to produce two or more passageways, with each split resulting in new passages whose diameters are smaller than the original. The passageways themselved are grouped into loose categories called generations, with each generation characterized by a certain range of values for the tube diameters and lengths.

The shapes of the internal chambers of the nose and mouth are complex and not very amenable to mathematical modeling. Deposition in these parts of the lung, in conjunction with the pharynx, usually is determined through direct empirical measurements. Models, therefore, usually begin with the trachea, which in an adult is a thin-walled, flexible cylinder approximately 12 cm in length and with a diameter of 2 to 2.5 cm. Its cylindrical shape is maintained by cartilage rings which encircle it. The flexibility is due to the presence of fibroelastic tissue which spans the space between the incomplete cartilaginous rings. These rings, in turn, are surrounded by a dense matrix of connective tissue.

The inner surface of the trachea, as in most of the bronchial passages, consists of pseudostratified columnar epithelial cells coated by cilia (see Figure 6.2). This layer lies above a layer of basal cells which, in turn, rest on a thick basement lamina. The basal cells are the source of new epithelial cells and divide when a depletion of the columnar cells occurs. The differentiated basal cells then migrate into the columnar layer and mature. Throughout the layer of columnar cells there may also be found goblet or mucus-secreting cells, whose role is the

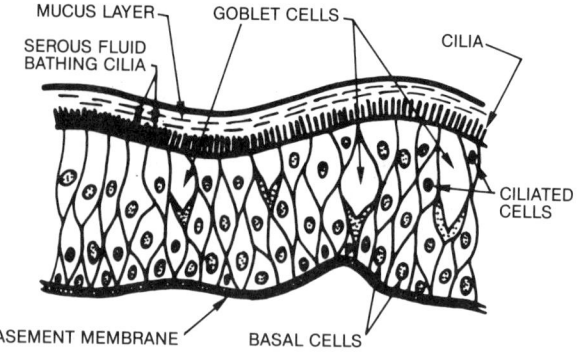

FIG. 6.2. A depiction of the epithelial layer in the tracheobronchial generations of the lung (shown in cross section).

secretion of mucin. The mucin is a polysaccharide which combines with water to form a lubricating solution called mucus. This mucus is moved along the passageways towards the esophagus, propelled by the beating of the cilia on the surface of the passages.

The trachea divides into two main bronchi, each of which enters either the right or left side of the lung at the hilar region. These main bronchi divide into two smaller bronchi on the left side and into three on the right. This branching produces five distinct lobes in the lung, two on the left side of the lung and three on the right. These divisions then give rise to further divisions and smaller bronchi, which continue to divide into several orders of bronchioles. The bronchiole subdivisions continue until the terminal bronchioles are reached. At this most distal level, the TB or conducting zone ends and the terminal bronchioles branch to form the respiratory bronchioles, alveolar ducts, and alveolar sacs, in which carbon dioxide and oxygen are exchanged with the bloodstream through a thin layer of membranes.

The cartilaginous rings associated with the trachea disappear at the level of the bronchi and are replaced by a series of cartilage plates which encircle the bronchi completely. The plates retain the cylindrical shape of the tubes during respiration. They become irregularly distributed as one moves toward the more distal parts of the lung and are replaced by smooth muscle. When the adult bronchiole falls below 7 mm, the cartilage disappears.

The cellular structure of the lung (pseudostratified columnar cells followed by basal cells and a basement membrane) remains fairly constant until reaching the terminal bronchioles. Thus, all of the passageways above the terminal bronchioles are coated by a layer of mucus responsible both for humidifying the inhaled air and transporting deposited material out of the lung. Mucus flow, which has been reviewed by several authors,[10,11] is propelled by the beating of the cilia, which consist of an arrangement of fibrils that causes bending of the cilia by interaction between adjacent fibrils.[12] This bending presumably is activated by the breaking of ATP near the basal bodies and results in the tips of the cilia exerting a force in the direction of the esophagus. The tips stick into the bottom of the mucus layer, so the mucus layer moves with each beat from the cilia. There is some coupling between the goblet and ciliated cells, since a minimum amount of mucus is needed for ciliary action.[13]

The upper passageways down to the terminal bronchioles serve simply to humidify and filter the air and to raise the air temperature to that of the body. The functional unit of the lung begins with the respiratory bronchiole and continues to the alveoli. By the time the respiratory bronchioles have been reached, the lung passages will have divided through many generations (up to 20 depending upon the particular anatomical model). As a result, the adult lung will possess as many as 300×10^6 alveoli,[14] yielding roughly 40 to 100 m^2 of surface through which gases may be exchanged with the pulmonary circulatory system.[15] The alveolar surface is covered by two primary types of cells joined by tight junctions which prevent passage of material between cell walls. The first is a highly specialized cell known as the squamous or Type I alveolar epithelial cell. The Type II alveolar cell, or granular pneumocyte, acts as the source of the surfactant layer which aids in maintaining the flexibility and spherical shape of the alveolar sacs.[16] These cells also divide to form Type I cells when damage occurs to the lung and the Type I population declines.

6.3 REDUCING THE ANATOMY TO A MATHEMATICAL FORM

For purposes of modeling deposition and movement of radionuclides, the descriptive features must be formalized into distinct quantities which can appear in mathematical functions. These functions require that the modeler specify the number of generations to be assumed, the radii and lengths of tubes in each generation, the branching angles and the angles of the tubes with respect to the vertical, and the number of tubes in each generation. Certain simplifying assumptions typically are employed to make the mathematical modeling easier.

It must first be decided how to simplify the branching scheme. Two primary methods have been used in the past to represent the branching. The first is to develop a dichotomous branching scheme in a symmetric tree. This approach assumes that each generation of the lung divides its passageways into two identical subpassages, which in turn constitute the next generation. The second approach utilizes nonsymmetric branching, in which each generation divides into subpassages which differ in dimensions. While anatomical measurements suggest that the latter approach is more realistic, the symmetric schemes are easier to deal with computationally. Regardless of the general scheme adopted, there are available several different sets of data detailing the physical characteristics of the generations.

The most familiar and frequently used anatomical models were

TABLE 6.1. Anatomical Dimensions of the Adult Human Lung

Region of lung	Volume (cm³)	Number per generation
Findeisen–Landahl model		
Mouth	20	1
Pharynx	20	1
Trachea	25	1
Main bronchi	10	2
Lobar	4	12
Segmental	5	100
Subsegmental	7	770
Terminal bronchioles	50	60,000
Respiratory bronchioles	30	150,000
Alveolar 1	100	3,000,000
Alveolar 2	600	40,000,000
Alveolar 3	2000	100,000,000
Weibel model		
Mouth	20	1
Pharynx	20	1
0	30.52	1
1	11.12	2
2	4.11	4
3	1.50	8
4	3.23	16
5	3.29	32
6	3.54	64
7	4.04	128
8	4.45	256
9	5.15	512
10	6.25	1,024
11	7.45	2,048
12	9.58	4,096
13	11.67	8,192
14	16.20	16,384
15	22.41	32,768
16	30.56	65,536
17	42.00	131,072
18	61.00	262,144
19	93.00	542,288
20	140.00	1,048,576
21	224.00	2,097,152
22	350.00	4,194,304
23	591.00	8,388,608
24	3150.00	300,000,000

developed by Weibel,[17] Davies,[18] Landhahl,[19] Findeisen,[20] Horsfield,[21] and Yeh and Schum.[22] The Weibel, Findeisen, and Landahl models have received the widest attention in the dosimetry of radon progeny due to their simplicity. The Landahl model actually is a modified version of the Findeisen model and at times is referred to as the Findeisen–Landahl model. Values for the various parameters typical for adults are displayed in Table 6.1 for the Weibel and Findeisen–Landahl models. The Weibel model has proven to be the more anatomically correct of the two and currently is in wide use.

Commonly employed asymmetric models are those developed by Yeh and Schum[22] and by Raabe et al.[23] These models differ from the others in that nonsymmetric branching is assumed and the data can yield probability distributions for the branching angles, tube diameters, etc. While this nonsymmetric approach increases the complexity of calculation, these models are the most anatomically correct of those available and are ideal for calculations designed to assess variability of deposition. As a result, these models are coming into wider use with the advent of easier access to computers capable of running Monte Carlo codes.

All of these models regard the tubes of the lung as rigid cylinders with fixed dimensions, at least in generations above the terminal bronchioles. The alveolar sacs typically are modeled as spheres whose radii vary throughout the breathing cycle. Reported diameters range from 150 to 300 µm in adult humans.[17] The trachea usually is designated as generation 0, with the generation numbers increasing as one moves to the more distal regions of the lung. The mucal layer (see Figure 6.2) is assigned a thickness of 7 µm for the mucus itself and an additional 7 µm for the watery layer separating the mucus and the upper layer of epithelial cells in the tracheobronchial region. Considerations of the surfactant layer in the pulmonary, or alveolar, region are not significant for purposes of dosimetry.

6.4 CHANGES IN ANATOMY WITH AGE

Unfortunately, there is little information available for developing mathematical models of the lung at very young ages. There are, however, some data which can be used in conjunction with what appear to be reasonable assumptions to yield estimates of anatomical parameters at young ages. A primary assumption employed in developing age-dependent models of the lung is that relative generational volumes in the tracheobronchial region should remain invariant during growth.[24] As a result, any generational volume (i.e., the total volume of air in any specific generation) at a given age may be defined as a fixed fraction of the total tracheobronchial volume. This latter volume, in turn, often is taken to be given by the anatomical deadspace for the lung.

Available data indicate that the number of airways in each tracheobronchial generation may be assumed to be constant with age for generations located above the respiratory bronchioles. Below this level, the number increases with age up until age 8, after which it remains constant. (Horsfield, however, states that this number of channels may not have stabilized by age 8.[15]) The age-dependent changes in the number of airways in each generation below the terminal bronchioles is displayed in (Figure 6.3).

The methodology for estimating airway diameter, length, and number for each generation has been reviewed by Hofmann et al.[25] Age-dependent generational volumes, $V_g(A)$, at age A for the space above the respiratory bronchioles are calculated from the relation

$$V_g(A) = V_g V_{DS}(A)/V_{DS,\ adult} \tag{6.1}$$

where $V_g/V_{DS,\ adult}$ is the fraction of the total anatomical dead space

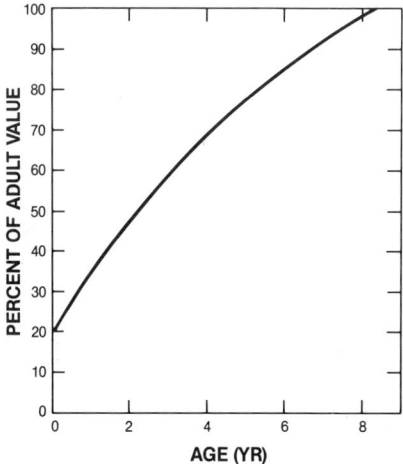

FIG. 6.3. The number of airways in each generation of the pulmonary region of the lung as a function of age. The results are given as the fraction of the adult value.

(DS) represented by generation g and $V_{DS}(A)$ is the age-dependent volume of this dead space. Values for $V_{DS}(A)$ may be found in Table 119 of ICRP Publication 23.[26]

Since the generational volume is dependent on both the diameter and length of the airways in a region, and since few data are available on either of these parameters as a function of age, a simplifying assumption usually is introduced.[25,27] This approximation asserts that the ratio of the airway radius to length remains constant with age, an assumption which derives from the belief that growth occurs due to the random division of cells in a cylindrical wall. If $\alpha_g(AD)$ is the value of this ratio for the adult, $N_g(A)$ is the number of passages in generation g at age A, and $V_g(A)$ is the volume of generation g at age A, then the radius of tubes in generation g at age A may be given as

$$R_g(A) = [\alpha_g(AD)V_g(A)/\pi N_g(A)]^{1/3} \tag{6.2}$$

The length then is given as

$$L_g(A) = R_g(A)/\alpha_g(AD) \tag{6.3}$$

TABLE 6.2. Age-Dependent Values for Total Surface Area (S) and Radii (R) for the Generations of the Weibel Model Tracheobronchial Region[b]

Generation	α^a	S (cm^2)/R (cm) at age (years)				
		0	2	8	16	Adult
Mouth	0.143	4.2/0.31	7.2/0.41	18.5/0.65	32.4/0.86	41.3/0.97
Pharynx	0.500	2.8/0.47	4.8/0.61	12.2/0.98	21.4/1.31	27.2/1.47
0	0.075	6.3/0.27	7.6/0.30	27.4/0.57	48.2/0.76	67.8/0.90
1	0.128	3.4/0.19	5.8/0.24	14.8/0.39	26.1/0.52	36.5/0.61
2	0.218	1.9/0.13	3.1/0.16	7.9/0.26	13.9/0.35	19.8/0.41
3	0.368	1.1/0.09	1.8/0.11	4.5/0.18	7.9/0.24	10.7/0.28
4	0.177	2.6/0.068	4.5/0.089	11.4/0.14	20.1/0.19	28.7/0.22
5	0.164	3.7/0.055	6.0/0.070	15.4/0.11	27.0/0.15	37.6/0.18
6	0.156	4.7/0.043	7.9/0.055	20.3/0.089	35.7/0.12	50.6/0.14
7	0.151	6.7/0.036	11.0/0.045	28.5/0.073	50.0/0.097	70.4/0.11
8	0.145	9.1/0.029	15.2/0.037	38.4/0.059	67.5/0.078	95.8/0.093
9	0.143	12.7/0.024	21.0/0.031	53.7/0.049	94.3/0.065	133.7/0.077
10	0.141	18.1/0.020	30.8/0.026	78.3/0.041	138/0.055	193/0.065
11	0.140	26.1/0.017	43.3/0.022	112/0.035	196/0.046	273/0.055
12	0.144	38.3/0.015	64.4/0.019	164/0.030	288/0.040	404/0.048
13	0.152	54.0/0.013	90.6/0.016	231/0.026	407/0.035	597/0.041
14	0.161	82.7/0.011	139/0.015	356/0.024	625/0.031	876/0.037
15	0.165	127/0.010	215/0.013	550/0.021	967/0.028	1359/0.033
16	0.182	192/0.009	323/0.012	826/0.019	1452/0.025	2038/0.030

[a] Independent of age, equal to the ratio of radius to length.
[b] S (cm^2)/R (cm).

The age-dependent surface area calculated for generation g at age A, $SA_g(A)$, may be found in Table 6.2. Also included in this table are the calculated values for the channel radii and lengths as a function of age. All values are specific to the Weibel anatomical model A.

Available data indicate that the alveolar volume remains directly proportional to the number of airways in the pulmonary region during the first eight years of life. This suggests that volumetric changes with age in this region may be due simply to the effect of increasing number of pulmonary airways. Therefore, it usually is assumed that the linear dimensions of airways or alveolar sacs in generations below the terminal bronchioles are constant up to age 8. After this age, the dimensions are calculated according to the methodology detailed above for the TB region, with the exception that the values for anatomical deadspace are replaced by the total alveolar volume. Age-dependent values for the latter quantity may also be found in ICRP Publication 23.[26]

6.5 MODELING DEPOSITION

Having specified the physical characteristics of the lung, the next step in lung dosimetry is to estimate the number of particles deposited in each generation of the lung. Several processes contribute to the deposition of atoms, molecules, and aerosol particles in a cylindrical tube or a sphere. The most important mechanisms are impaction, sedimentation, and diffusion (or Brownian motion), with some minor contribution from deposition of charged particles by the image force resulting from the rearrangement of charge on the surface of lung passages. The relative contributions from the three main processes of deposition depend upon the size distribution of any inhaled aerosols and on the unattached fraction, factors which must be considered in specifying the atmosphere to which people inhaling radon and progeny are exposed. There are many different approaches to modeling deposition by these three processes, with the following discussion detailing those equations utilized by the present author in past studies of lung deposition.[27-29]

Consider first the probability that an aerosol particle of diameter d (cm) and particle density ρ (g cm^{-3}) will be deposited by impaction. In general, impaction can be viewed as a case in which the momentum of a particle is too large to permit it to navigate a turn and, hence, it crashes into the wall of a generation near the bifurcation. Let $P_I(n)$ be the fraction of such particles which impact in the nth generation, given that they have made it to this generation. In that case, the following empirical equation holds:

$$P_I(n) = 150\rho d^2 V_{n-1}/(R_n + 150\rho d^3 V_{n-1}) \tag{6.4}$$

where V_{n-1} is the linear velocity of the air in generation $n-1$ and R_n is the radius of the nth generation. Values for V may be found by dividing the volumetric flow rate in the lung by the total cross-sectional area of the airways in a generation.

Sedimentation occurs due to the gravitational force which pulls a particle downwards while it moves through the lung. Following the discussion on impaction, let $P_s(n)$ be the probability that a particle deposits in the nth generation by sedimentation, given that the particle has arrived at this generation. The following relation then holds:

$$P_s(n) = 1 - \exp[-0.8\ U_t t_n \cos(\theta_n/R_n)] \tag{6.5}$$

where t is the time during which the air is present within the nth generation, θ_n is the average angle of inclination of the airway with respect to the horizontal, and R_n is the channel radius. Values for t_n may be computed by dividing the generational volume by the volumetric flow rate for the lung associated with the breathing pattern. The terminal settling velocity, U_t, is given by Stokes' equation:

$$U_t = Cg\rho d^2/18n \tag{6.6}$$

where n is the viscosity of air, g is the gravitational acceleration, and C is the Cunningham correction factor which corrects Stokes' law for particles whose diameters are small compared to average molecular distances. Alternative formulations have been given by Millikan[30] and Findeisen.[20]

Several formulations are available for calculating the probability of deposition by diffusion. The general approach was laid out by Gormley and Kennedy in 1949,[31] who studied the deposition of particles by diffusion while moving through cylinders. Their theory states that the deposition fraction in a generational tube of length $L(n)$ under a volumetric flow rate $V(n)$ through the tube and for a particle diffusion coefficient D_Δ is

$$P_D(n) = 4.07h^{2/3} - 2.4h - 0.446h^{4/3} \ldots \tag{6.7}$$

for $h < 0.0156$ and

$$P_D(n) = 1 - 0.819e^{-7.314h} - 0.0975e^{-44.6h} - 0.0325e^{-114h} \ldots \quad (6.8)$$

for $h > 0.0156$. In these expressions, $h = (\pi/2)L(n)D_\delta/V(n)$. Davies,[32] Ingham,[33] and Thomas[34] have each recalculated the coefficients in the Gormley–Kennedy equations. The diffusion coefficient usually is taken to be in the area of 1.3×10^{-5} cm^2 s^{-1} for nuclei and 5.4×10^{-2} cm^2 s^{-1} for free ions.

The discussion above focused on an assessment of the fraction of particles depositing in a generation, given that the particles make it to the beginning of that generation. The total probability, $P(n)$, of a particle or atom depositing in the nth generation then is given by

$$P(n) = 1 - [1-P_I(n)][1-P_s(n)][1-P_D(n)] \quad (6.9)$$

Not all of the particles or free atoms inhaled, however, find their way to the nth generation. Some of them will have deposited in more proximal airways. The fraction, F_n, of inhaled particles surviving to reach the nth generation is equal to

$$F_n = 1 - P_{\text{NP}} - \sum_{n=0}^{n-1} P(n) \quad (6.10)$$

where P_{NP} is the probability of the particles being removed by the nose and pharynx (if nose breathing is assumed) or by the mouth and pharynx (if mouth breathing is assumed).

Deposition in the NP region is less amenable to mathematical modeling due to the geometrically complex shapes encountered in this region. As a result, empirical determinations of P_{NP} must be used and extrapolated to nonadult ages. Previous studies have indicated that deposition in the NP region depends upon whether nose or mouth breathing is used, with breathing through the nose resulting in larger values of P_{NP}. This deposition is approximately the same during both inhalation and exhalation.

Pattle[35] studied the deposition of monodisperse methylene blue particles and found the empirical relation

$$P_{\text{NP}} = -0.62 + 0.475 \log(d^2F) \quad (6.11)$$

where d is the effective aerodynamic diameter of the particle and F is the inhalation rate (liters per minute). Experimental results confirm the validity of the relation.[9,36–42] Of particular importance is the study of George and Breslin,[42] which looked at radon progeny partially attached to an aerosol with an activity median aerodynamic diameter (AMAD) of approximately 1 μm. They found P_{NP} to be approximately 2% for the attached atoms and 62% for the unattached.

Modeling deposition in the mouth also is complicated by the underlying geometry. As in the case of nose breathing, P_{NP} for mouth breathing is obtained from direct measurement. The ICRP[9] used the data of Dennis[43] for deposition in the mouth, but the more recent data by Stahlhofen et al.[44] are more complete. Figure 6.4 shows P_{NP} for mouth breathing plotted as a function of log d^2F, where d and F are defined as in Equation (6.11).

Determining $P_{NP}(A)$ at other than adult ages is complicated by the purely empirical nature of P_{NP} and the lack of empirical information on deposition in children. Because of these features, simplifying assumptions must be introduced in order to determine P_{NP} at young ages for both nose and mouth breathing. The use of one set of assumptions in generating age-dependent estimates of P_{NP} has been reported,[29] in which attention is restricted to particles yielding an adult value for P_{NP}

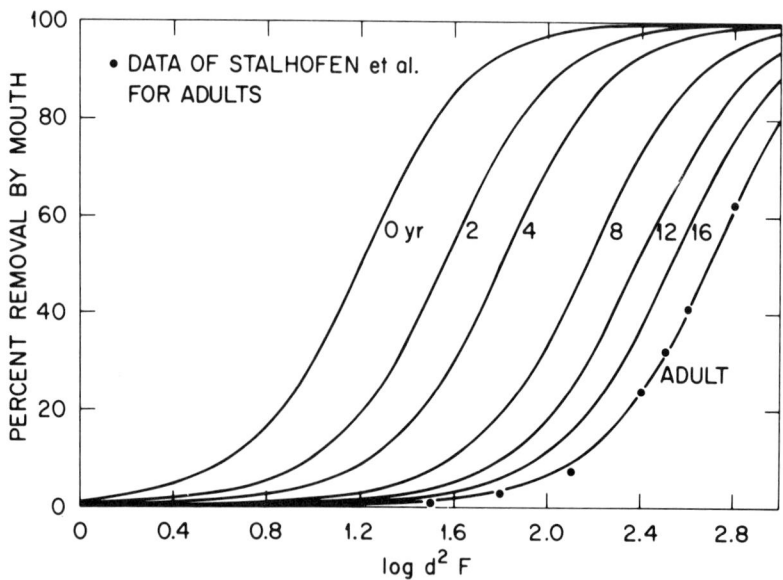

FIG. 6.4. Deposition fraction in the mouth for inhaled particles as a function of age (yr), particle diameter (d), and volumetric flow rate (F). Particle diameters are in μm and F is in units of L min^{-1}.

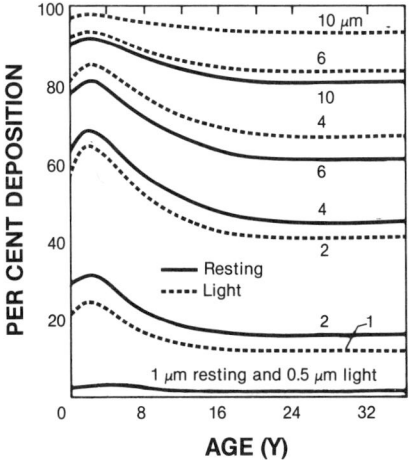

FIG. 6.5. Deposition fraction for inhaled monodisperse particles in the NP region of the lung. Results are presented for various particle diameters and for resting and light activity. Only particle sizes yielding NP deposition fractions greater than 1% are shown.

larger than 1% and due primarily to impaction. The method uses semiempirical relations[45–48] giving the probability of impaction as a function of tube diameter, bend angle, particle size, particle density, and flow rate. For a detailed discussion of the model, the reader should consult the original article.[29] Representative results are given here in Figures 6.4 and 6.5 and can be used for estimating $P_{NP}(A)$.

6.6 MOVEMENT ON THE MUCOCILIARY BLANKET

The result of deposition calculations is a physical description of where inhaled particles or free atoms deposit in the various generations of the lung. This deposited material usually is assumed to be distributed uniformly on the surfaces throughout each generation. Once deposited, the material in the TB region will move on the mucociliary blanket, and the material in the P region transfers either to the bloodstream, lymph nodes, or the TB region. Since essentially all of the dose delivered to cells in the TB region occurs from material initially deposited in that region, movement of material from the P region will not be explored here. The kinetics of movement of radonuclides from the P region has been reviewed by the ICRP.[9,49]

The rate of movement of materials on the mucociliary blanket has not been measured for most generations of the lung. As a result, it is not possible at present to state accurately the values which should be assigned to rates in most generations. The existing data concerning transport within the TB region are confined to the trachea and main bronchi[50-53] and indicate that a typical packet of mucus requires between 10 and 30 minutes to traverse the trachea in adults. Values in other generations can be determined by assuming a constant mucal blanket thickness throughout the lung and employing a mass balance equation.[54] The calculation of transport rates typically is presented in the form of transit times for each generation, given as the average length of time required for a differential element of mucus to pass through a generation. Selected values from references[27-29,54,55] are presented in Table 6.3.

In general, let $T(n)$ be the transit time for mucus through the nth generation of the lung, let λ be the radiological decay constant for the radionuclide in question, and let $R(n)$ be the number of atoms of the radionuclide deposited initially in generation n. The number of atoms disintegrating in the nth generation, $N(n)$, then is given by the relation

$$N(n) = R(n)\{1 - [1 - e^{-\lambda T(n)}]/\lambda T(n)\} + R(n+1)[1 - e^{-\lambda T(n)}]$$

$$\times \text{ts}\,[1 - e^{-\lambda T(n+1)}]/\lambda T(n) + \sum_{i=n+2}^{m} \Big\{ R(i)[1 - e^{-\lambda T(n)}][1 - e^{-\lambda T(i)}]$$

$$\times [1 - e^{-\lambda T(i)}] \exp\left[\sum_{j=n+1}^{i-1} \lambda T(j)\right] \Big/ \lambda T(n) \Big\} \qquad (6.12)$$

For the case of the radon decay chain, the initially deposited atoms may consist of any and all of the progeny of radon. Each atom is treated separately in using Equation (6.12), and care must be taken to account for serial decay through the entirety of the decay chain supplying alpha-particle emissions. Clearly, Equation (6.12) can be used directly to yield the number of ^{218}Po decays in each generation from the pattern of deposition of ^{218}Po in the generations. This requires only that λ be set equal to that of ^{218}Po. The same is true for the number of ^{214}Bi decays from deposited ^{214}Bi, with λ being specific to ^{214}Bi. For atoms of ^{214}Pb deposited in the generations, the number of atoms of ^{214}Pb decaying in each generation can be computed. This number in each

DOSIMETRY

TABLE 6.3. Adult Regional Transit Times for Mucus in the Lung (Minutes)

Region	Hague and Collinson	Altshuler et al.	Crawford-Brown
0, trachea	9.6	8	24.0[a]
1, main bronchi	3.8	6	13.0
2, lobar	7.6	11	7.0
3	3.04		3.8
4, segmental	25.4	37	10.2
5	21.4		13.3
6	18.0		17.9
7, subsegmental	Long	82	24.9
8	Long		33.9
9	Long		47.3
10, terminal		1980	68.1
11			96.1
12			142.8
13			201.5
14			309.9
15			480.8
16			721.1

[a] Utilizes a tracheal velocity of 5 mm min^{-1}.

generation then can be set equal to $R(n)$ for ^{214}Bi and the number of ^{214}Bi decays in each generation (but resulting from initial deposits of ^{214}Pb) can be computed from Equation (6.12). The same process, but with one additional iteration, can be used to compute the number of ^{214}Bi decays occurring in each generation as a result of deposition of ^{218}Po atoms. A standard assumption then is used which distributes the alpha decays in any generation uniformly over the surface area of that generation. Generational surface areas already have been provided in Table 6.2.

Age-dependent values for the various transit times needed for Equation (6.12) are not available at present. Since mucal velocities depend on the density of cilia, frequency of cilia strokes, density of goblet and serous cells, and blood supply,[56] for which data are not available, extrapolation of transit times from adult data to children cannot be placed on any formalized, theoretical foundation. Instead, certain rather arbitrary assumptions must be utilized. For example, one can assume that clearance rates from the three regions of the ICRP model remain invariant with age, implying that the transit times through each generation do not change with age.[25] By contrast, one also could assume that tracheal velocities remain constant with age, as suggested by the similarity of velocities between species.[27–29] The methodology

TABLE 6.4. Age-Dependent Transit Times for the Weibel Model

Generation	Transit time (min) at age (years):				
	0	2	8	16	Adult
0	7.3	8.0	15.2	20.6	24.0
1	4.0	6.1	8.3	10.9	13.0
2	2.2	3.2	4.4	5.8	7.0
3	1.3	1.9	2.5	3.3	3.8
4	3.1	4.7	6.4	8.4	10.2
5	4.2	6.3	8.6	11.3	13.3
6	5.4	8.4	11.3	15.0	17.9
7	7.8	11.6	15.8	21.0	24.9
8	10.5	16.0	21.4	28.3	33.9
9	14.7	22.1	30.0	39.6	47.3
10	21.0	32.5	43.6	57.8	68.1
11	30.3	45.7	62.1	82.2	96.7
12	44.3	67.9	91.3	120.9	142.8
13	62.5	95.6	128.8	170.7	201.5
14	95.8	146.3	198.0	262.5	309.9
15	147.1	226.9	305.9	405.8	480.8
16	221.8	341.1	459.7	609.6	721.1

for computing transit times for each generation at a given age then proceeds as described earlier, assuming an age-invariant mucus layer thickness and no changes in thickness as the generation number changes. These times are shown in Table 6.4.

6.7 DOSE CALCULATIONS

By far the simplest approach to estimating the lung dose is to average the absorbed energy over the entire mass of the lung. For cases involving gamma- or X-ray- or beta-emitting radionuclides, the dose to any particular subpopulation of lung cells is approximately the same as the average lung dose. This feature arises because of the long range of these radiations in lung tissue. For alpha-emitting radionuclides decaying in the alveolar region, this same approach is valid since the Type I and Type II cells are uniformly distributed in space and the alpha particles typically cross through several alveolar sacs in losing their energy. For cases in which only the average lung dose is needed, the dose, $D(A)$, in rads at age A is given by

$$D(A) = \sum_n \frac{N_n(A)\mathrm{EF}(A) \times 1.6 \times 10^{-8}}{M(A)} \quad (6.13)$$

where $N_n(A)$ is the number of decays in the nth generation at age A, $F(A)$ is the age-dependent absorbed fraction for the lung (defined as the fraction of the radiation energy absorbed within the lung), E is the emission energy for the radiation (in MeV), and $M(A)$ is the age-dependent lung mass as given, for example, by ICRP Publication 23[26] in units of grams. When computing doses specific to the alveolar region, the mass should include the mass of the blood supply in this region, since this mass will absorb part of the emission energy and contribute to $F(A)$.

Computing doses to specific cell subpopulations in the TB and NP regions is more complicated due to the potential for very large inhomogeneities in dose. For example, when dealing with alpha particles, the epithelial cell doses may be quite a bit higher than the doses to basal cells. It becomes necessary to calculate depth–dose curves describing the dose to cells located at the different depths within the walls of the various generations (see Figure 6.2). This approach is necessary for the dosimetry of radon progeny since most of the significant dose comes from alpha emissions.

Fortunately, alpha particles travel in a fairly straight line due to their large mass. Alpha particles, as in the case of all charged particles, lose their energy through several kinds of collisions. These interactions normally are separated into three broad groups: interactions with individual electrons (most common), interactions with nuclei, and interactions with the entire atom (coupled system). The latter two modes of dissipation are negligible for alpha particle energies of concern in the dosimetry of radon progeny.

The total stopping power (dE/dx) can be given by the relation

$$\frac{dE}{dx} = \frac{4\pi Z^2 e^4 NZ}{mv^2} \left\{ \ln\left[\frac{2mv^2}{I(1-\beta^2)}\right] - \beta^2 - \gamma - \sum_i \frac{C_i}{Z} \right\} \quad (6.14)$$

In this relation, Z is the atomic number of the element (tissue composition), v is the alpha velocity, N is the density of atoms in the absorbing medium (tissue), I is the mean excitation and ionization potential of the absorbing atoms, m is the rest mass of the alpha, e is the electronic charge, β is the fraction of the velocity of light represented by the velocity of the charged particle, γ accounts for the polarization effect (small for alphas of the energy important here), and Ci/Z represents shell corrections for a material of atomic number Z. At low velocities encountered towards the end of an alpha's path, the charge will drop below +2 due to charge fluctuations as the alpha attaches electrons from the surrounding medium.[57,58]

Equation (6.14) applies to the contribution to the total stopping power from each separate element in tissue. In radon dosimetry, it usually is assumed that the Bragg additivity rule[58] holds true, so that the contribution for each element is calculated and then the contributions are summed after weighting by the relative atomic densities (N). Various authors[59-62] have used this approach to develop stopping powers in lung tissue for a wide range of alpha particle energies. The table published by Walsh[60] is provided in Table 6.5.

There are a number of ways in which the depth–dose curves may be generated for alpha particles in the TB and NP regions. In the following discussion, one approach[27-29] will be described. In the approach used here, doses are computed to hypothetical spherical volumes located at any given depth within the cylinder wall. The volumes themselves are considered to be 1 μm in diameter and composed of tissues with a density of 1 g cm^{-3}. For a given generation of the lung, the dose of a 1-μm diameter sphere located a distance X into the wall also is equal to the mean of the specific energy distribution for all spheres located at that distance.

In general, let S_n be the areal density of alpha particle emissions for the radionuclide of interest on the walls of the generation. Figure 6.6 gives a depiction of the geometry assumed in all of the following depth–dose calculations. If N_n is, as defined earlier, the number of decays in generation n, then S_n is simply N_n divided by the total surface area of that generation (see Table 6.2). The decays are assumed to be distributed uniformly over the surface of that generation. It then is necessary to compute the flux, ϕ, of alpha particles reaching the target volume. The units for this flux are alphas per unit area. Consider the flux contribution from a differential source element, dA, located a distance T from the target volume center. The flux at this point from dA is simply

$$\phi_n = S_n dA/4\pi T^2 \quad (6.15)$$

where the index, n, refers to the generation number.

Now let $S(T)$ be the stopping power of an alpha which has gone a

TABLE 6.5. Stopping Power versus Depth of Penetration for Alpha Particles in Tissue

Depth (μm)	Exit energy (MeV)	Stopping power (MeV cm^{-1})
0.0	8.00	
2.5	7.84	670
5.0	7.68	675
7.5	7.51	690
10.0	7.34	700
12.5	7.16	710
15.0	6.94	725
17.5	6.81	735
20.0	6.63	750
22.5	6.44	765
25.0	6.75	780
27.5	6.06	800
30.0	5.86	820
32.5	5.65	840
35.0	5.44	860
37.5	5.23	885
40.0	5.01	915
42.5	4.78	945
45.0	4.53	980
47.5	4.30	1010
50.0	4.05	1080
52.5	3.78	1120
55.0	3.50	1180
57.5	3.20	1250
60.0	2.89	1320
62.5	2.56	1450
65.0	2.20	1590
65.5	2.12	1620
67.0	1.87	1770
67.5	1.78	1810
68.0	1.69	1890
68.5	1.60	1910
69.0	1.50	1995
69.5	1.40	2050
70.0	1.30	2160
71.0	1.08	2400
71.5	0.96	2500
72.0	0.83	2620
72.5	0.69	2790
73.0	0.55	2890
73.5	0.41	2790
74.0	0.28	2480
74.5	0.18	1920
75.0	0.08	1200
75.5	End of range	

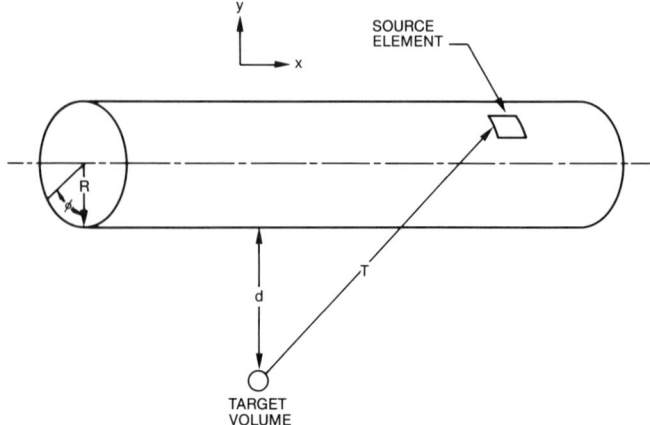

FIG. 6.6 Irradiation geometry for decays occurring on the surface of the tubes in the generations of the TB region. The parameters d, R, and T refer to the distance of the targets in the wall, the radius of the tube, and the distance from source element to target volume, respectively.

distance T through the intervening tissue (see Table 6.5). The total energy delivered to a sphere of diameter $2r$ located at the point of interest is

$$E = S(T)(2/3)(2r) \tag{6.16}$$

The factor of 2/3 arises from the fact that the average chord length for an alpha through a sphere is 2/3 times the diameter. Equation (6.16) assumes that the entire energy transferred from the alpha to the medium in passing through the sphere is absorbed in the sphere. This is a reasonable approximation since the diameter of the sphere is much larger than the range of delta rays produced by the alpha particles in the radon decay chain, which travel a distance of less than 0.1 μm in tissue.[3]

Since the target volume has a cross-sectional area equal to πr^2 and a mass of $(4/3)\pi r^3 \rho$, where ρ is the density of tissue, the dose, D_n, is equal to

$$D_n = \frac{E}{M}(1.6 \times 10^{-8}) = (1.6 \times 10^{-8})\phi S(T)/\rho \tag{6.17}$$

DOSIMETRY

The total dose (in rads) to the target volume then is obtained by the integration of Equation (6.17) over all surface source elements, i.e.,

$$D = \int_{S_A} \frac{S_A S(T)(1.6 \times 10^{-8}) \, dA}{4\pi T^2} \tag{6.18}$$

For a cylindrical tube, it may be shown that[63]

$$T = [X^2 + (d + R - R \cos \theta)^2 + R^2 \sin^2 \theta]^{1/2} \tag{6.19}$$

where X is the distance along the x-axis measured from the normal line connecting the cylinder and target volume (see Figure 6.6), R is the radius of the tube in a generation, d is the distance from the cylinder surface to the target volume measured along the normal, and θ is the angular coordinate for the surface source element. In this case, Equation (6.18) becomes

$$D = \int_x \int_\theta \frac{S_A S(T)(1.6 \times 10^{-8}) \, dx \, R \, d\theta}{4\pi T^2} \tag{6.20}$$

In radon dosimetry the contribution to the dose at a depth d typically is broken into two components. The first is called the "near wall" contribution and accounts only for the energy deposited in the target by alphas which travel through tissue alone. The second, or "far wall," contribution accounts for alphas which must first pass through the air space within the tube before reaching the "near wall" and, hence, entering the tissue. The effect can be accounted for by separating the value of T in Equation (6.20) into two components, one for the distance in air and the other for the distance in tissue (which will be specific to a given source element). The distance in air then is converted to an equivalent distance in tissue which would produce the same energy loss. This conversion may be accomplished by noting that the tissue equivalent pathlength in 1 mm of air is roughly 1.1 μm[64] when the alpha energy varies from 1 to 10 MeV.

It often is necessary to also compute the dose-averaged stopping power at each depth in tissue, since this quantity will be useful in specifying the quality factor at this depth. To compute this quantity, let S_i be the stopping power of the alpha emitted in source element i when it has reached the target volume and let D_i be the dose contributed to this target volume by the ith element. The dose-averaged stopping power then is

$$S = \frac{\sum_i S_i D_i}{\sum_i D_i} \quad (6.21)$$

For alpha particles, the dose-averaged stopping power and dose-averaged LET are approximately the same.

Some assumption also must be made as to how the alpha emissions are distributed within the mucus layer. This distribution depends upon the solubility of the progeny in mucus and whether the progeny remains attached to aerosols. Various authors have used different approaches to this problem, although the approaches can be divided into two broad categories. The first[54] assumes that the decays occur only on the surface of the mucus. The second[65] assumes that the decays are distributed in a linear fashion within the mucus, with the density going to zero at the interface between the mucal layer and the epithelial cells. Kirkichenko[66]

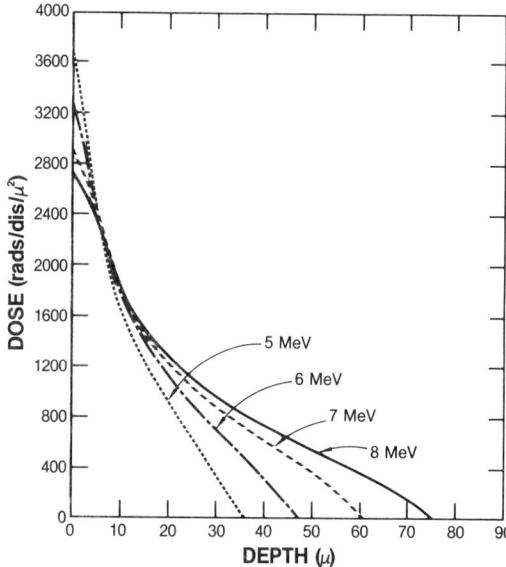

FIG. 6.7. Depth–dose curves for alpha particles emitted on the surface of cylinders with walls composed of tissue. Separate curves are shown for alpha energies of 5, 6, 7, and 8 MeV, with a tube radius of 0.05 cm.

DOSIMETRY

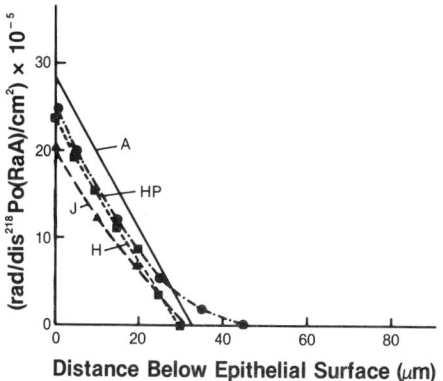

FIG. 6.8. A comparison of dose factors for the alphas from RaA as a function of the depth into the various generations near the proximal end of the TB region. References are given in the text of this chapter. (Reprinted from NCRP Report 78 with permission from the NCRP.)

measured the distribution in rabbits and dogs and found that the radon progeny are partially mixed in the mucal layer. Fortunately, the difference in depth–dose curves generated under the two approaches is small.

Several papers have published depth–dose curves for alpha emissions in the lung passages.[28,55,62,67] Representative results are given in Figure 6.7. A summary of results published by the NCRP[67] is provided in Figure 6.8. The curves are not a function of age.

6.8 CRITICAL CELLS

From the methods outlined in the previous sections, it is possible to compute the deposition in each generation for a given aerosol size or free atom, the movement on the mucus blanket, the density of decays in each generation, and the depth–dose curves for these generations. The question then arises as to the dose which should be assigned to the cells which produce bronchial cancers following inhalation of the radon progeny. An answer to this question requires that the location of these cells within the cylinder walls be specified.

One of the general principles of radiobiology is the assumption that radiosensitive cells primarily are those which have not differentiated and are still capable of differentiation and division. The cells of the bronchial epithelium are differentiated and lost by desquamation into the mucus layer. They then are replaced by division of the basal cells, suggesting that the basal cells should be regarded as the most radiosensitive when arguments are developed from first principles.[54,65,69] As early as 1918, it had been suggested that the basal cells were the source of small-cell carcinomas, which arise in uranium miners.[68] Many of the past studies on the dosimetry of radon progeny have focused on doses to the basal cell layer in each generation.

It is not clear, however, that the basal cells are, in fact, the progenitors of interest in radon dosimetry. In recent years, both K cells[67,70] and cells distributed within the epithelium[71] have been implicated as possible sources of bronchial cancer. These conclusions have been supported by recent epidemiological studies on the uranium miners which show that approximately 35% of the bronchial tumors in the Czechoslovakian miners were epidermoid tumors[72] and that this fraction depends upon the dose, age at exposure, and assumed latency period.[67] Clearly, there is a need for more detailed information on the cells to be considered critical in the induction of cancer.

Until more information is available, however, a common practice is to compute doses to basal cells. Unfortunately, there is little information available on the location of these cells in adults,[54,73] and the data are nonexistent for children. Some of the available data are

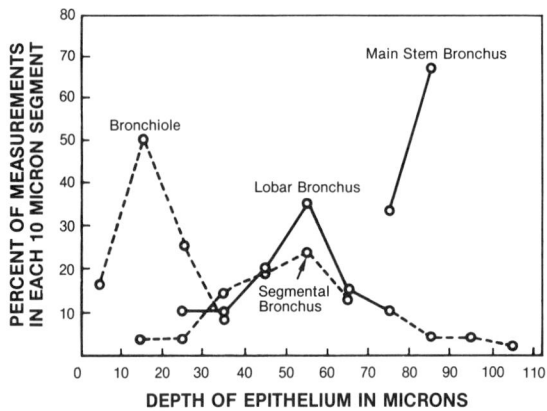

FIG. 6.9. Mean basal cell depths within the various generations of the TB region of the lung. [Data reprinted from Reference (73) with permission from *Health Physics*.]

DOSIMETRY

reproduced in Figure 6.9. From these data, it appears that the mean basal cell depth in a generation is proportional to the average channel radius in that generation. This proportionality also was noted by the ICRP.[26] Using this relation, Crawford-Brown[28] employed a mean basal cell depth of 0.01 times the channel diameter and a triangular distribution with a minimum and maximum basal cell depth varying by 50% around the mean value. An advantage of this approach is that it provides basal cell distributions in all generations and at all ages, since the distribution is an explicit function of channel radius, $R(n)$. Care must be taken to account for the mucal thickness and the height of the basal cell nuclei above the basement membrane (usually 7 μm).

6.9 PHYSIOLOGICAL PARAMETERS

The deposition equations for particles and free ions in the lung are functions of the rate at which air flows through the lung passages. Air flow can be summarized by three main factors: the total volume of air taken in during each breath (called the tidal volume), the breathing rate (breaths per minute), and the amount of time assumed for breath-holding between inhalation and exhalation.

There is a large amount of variability in these factors, especially when age is considered. Deposition calculations, therefore, usually use

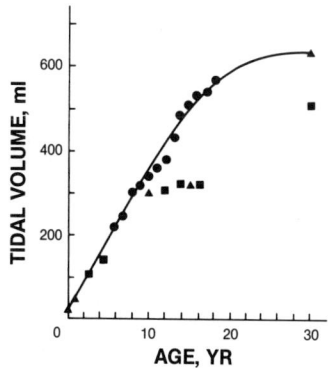

FIG. 6.10. Changes with age in the tidal volume for breathing. The data are from Hofmann et al.[25] (Reprinted with permission from the authors.)

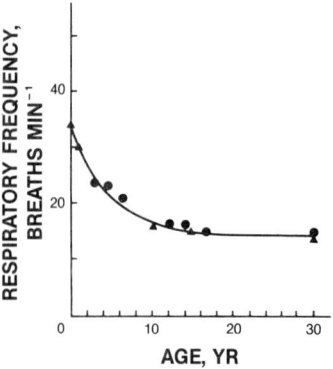

FIG. 6.11. Changes with age in the breathing frequency. The data are from Hofmann et al.[25] (Reprinted with permission from the authors.)

average values for normal, healthy people. In addition, since increased activity usually requires more oxygen, separate calculations often are performed to simulate conditions of rest and light activity. A review of these factors can be found in the report by Raabe,[74] and suggested values have been provided by the ICRP.[9,26] No attempt is made here to review possible values to be used for the tidal volume and breathing rate. Instead, typical age-dependent values are presented in Figures 6.10 and 6.11.

There also are a number of possible approaches to specifying the amount of time to be assigned to breath-holding. During this period of time, the air flow ceases, and only diffusion and sedimentation occur. A fairly standard assumption is that both the inspiration and expiration periods occupy $\frac{3}{8}$ of the total breathing cycle, with $\frac{1}{8}$ of the cycle being assigned to each of the two periods of breath-holding before and after inspiration.

6.10 PREVIOUS CALCULATIONS OF DOSE FROM RADON AND PROGENY

Given the number of models and parameters which must be chosen in computing cellular doses following inhalation of radon and progeny, it should come as no surprise to find that a wide range of approaches have been adopted by various authors to develop a conversion factor between integral exposure (in units of WLM) and the dose to radiosensitive cells. This section describes some of the most pertinent past studies. The studies can differ in their choice of: (1) relative concentrations of the radon progeny, (2) aerosol size distribution, (3) fraction of the progeny unattached, (4) deposition models, (5) mucociliary transport rates, (6) anatomical models, (7) depth–dose curves, or (8) location of critical cells. With such a large number of choices, it is a little surprising that the results are fairly close together.

By far the most widely cited papers in this area are those of Altshuler et al.,[54] Jacobi,[65] Haque and Collinson,[55] Harley and Pasternack,[75] Chamberlain and Dyson,[76] and Hofmann et al.[25] The papers by Harley and Pasternack[75] and Hofmann et al.[25] also present results as a function of age at the time of inhalation. In addition, the National Council of Radiation Protection and Measurements (NCRP)[67] and Walsh[77] have published excellent reviews of these various approaches and detailed discussions of methodologies specific to radon and progeny. All of the studies assume that the dose from the radon itself is negligible, since radon is an inert gas which will not attach to the walls of the generations.

It is not possible in this chapter to give a detailed description of the assumptions employed in each of the past studies. For this information, the reader should consult the original papers or the reviews by the NCRP and Walsh cited above. Instead, only a few summary results will be presented here, given in the order in which the studies appear in Table 4.1 of the NCRP report.[67]

Chamberlain and Dyson[76] report a conversion factor of 0.0038 Gy WLM^{-1}. The study by Altshuler et al.[54] finds a conversion factor of 0.035 Gy WLM^{-1} for mouth breathing, which drops to 0.019 Gy WLM^{-1} for cases of nose breathing. A similar model was used by Jacobi,[65] who found the conversion factor for mouth breathing to be 0.031 Gy WLM^{-1}.

Haque and Collinson[55] used the Weibel model for calculating deposition and they found a conversion factor of 0.11 Gy WLM^{-1} for mouth breathing and 0.08 Gy WLM^{-1} for nose breathing. A later study by Harley and Pasternack[62] used the same anatomic model but the mucus clearance times of Altshuler et al.[54] They found conversion factors ranging from 0.0023 to 0.011 Gy WLM^{-1} under conditions of nose breathing and a range of atmospheric conditions. The reader should note that the above studies computed doses to the third to fifth generations of the Weibel model, as this is believed to be the critical

TABLE 6.6. Relative Age-Dependent Dose Rates for Exposure to Airborne Radon Progeny[a]

Age (years)	Mean basal cell dose rate[b]	Mean dose rate[b] to TB region
0	2.60	2.00
2	2.40	1.75
5	3.10	2.50
10	2.40	2.25
15	1.65	1.75
Adult	1.00	1.00

[a] Assumes all ages exposed to the same atmosphere.
[b] Ratio of value at the indicated age to the value for the adult.

section of the lung for induction of bronchial cancers following inhalation of radon progeny.

There have been two principal studies concerning conversion factors at other than adult ages. The most complete and detailed is that by Hofmann and Steinhausler[78] and Hofmann et al.[25] They also studied the effect of age on the dose to the pulmonary region and found that this dose remains a factor of at least three to five below that to the TB region at all ages. A summary of their results may be found in Table 6.6. It will be noted that the dose rate delivered either to the epithelial cells or to the basal cells rises dramatically at very young ages.

Harley and Pasternack[75] also studied the effects of age on lung doses from radon progeny at several ages using the Yeh–Schum morphometry.[22] Their results suggest a slightly smaller increase of the conversion factor at young ages. Their conversion factors averaged over the second generation range from 0.007 Gy WLM^{-1} for adults to 0.017 Gy WLM^{-1} for ten-year-olds and 0.013 Gy WLM^{-1} for one-year-olds.

The NCRP report[67] also gives a good description of the extent to which conversion factors are affected by changes in the various assumptions which go into the calculation of adult doses. They consider ten variables, which they divide into two broad groups:

A. Physical characteristics
  1. Fraction of unattached RaA
  2. Daughter product equilibrium
  3. Particle deposition models
  4. Particle size distribution
  5. Method of computing dose

B. Biological characteristics
   6. Breathing pattern
   7. Bronchial morphometry
   8. Mucociliary clearance rates
   9. Location of target cells
   10. Mucus thickness

The effect of each of these factors, given by the extent to which the conversion factor is changed from a mean value when the factors are varied over their range, is displayed in Table 6.7.

TABLE 6.7. Variation in Dose Conversion Factor (Gy WLM^{-1}) Due to Various Factors

Factor	Variation[a]
1. Unattached RaA	
4%	−10%
20%	+30%
2. Daughter equilibrium ratios	20%
3. Deposition models	A few percent[b]
4. Particle size	
AMAD = 0.125 μm	0
AMAD = 0.05 μm	+100%
AMAD = 0.17 μm	−20%
5. Calculation of dose to a particular cell	A few percent
6. Breathing pattern	
Normal	0
Total active pattern	+20%
Total resting pattern	−25%
Mouth versus nose	+35%
7. Child versus adult morphology	+60%[b]
8. Mucociliary clearance	
Normal	0
Complete stasis	+10%
9. Mucus thickness	
Normal (14 μm)	0
No mucus	+20%
Twice normal	−30%
10. Location of critical cells	
Average (45 μm)	0
Shallow (22 μm)	−80%

[a] Variation about the conversion factor when the mean value is employed.
[b] Appears to be an underestimate according to discussion in this chapter.

6.11 MISCELLANEOUS FACTORS AFFECTING CALCULATIONS OF LUNG DOSES

There are a number of additional considerations which are not considered in existing modeling efforts for radon exposures but which may affect these calculations if included. The first involves the effect of turbulent flow on deposition in the lung passages.[74,79-81] The impaction, sedimentation, and diffusion equations given earlier assumed that laminar flow held throughout each generation. Turbulent flow appears to occur only in the generations including the trachea and main bronchi and affects deposition by transporting particles to the tube walls in these generations. The impact on lung deposition calculations for inhalation of radon progeny by the general public should be minimal.

Another unresolved problem revolves about the possibility that particles may grow while passing through the lung. The temperature and humidity of air entering the lung is quickly brought up to normal body temperature and very high (99%) relative humidity, with the humidity stabilizing most rapidly.[74] Particles which are inhaled may, therefore, absorb water and change their effective aerodynamic diameters as they move into the lung. This feature is not, at present, accounted for in existing models of lung deposition and the impact on calculations is not known.

Recent work[82] indicates that for large-diameter (>1 μm) particles, deposition may occur much more frequently at the bifurcations, or divisions, separating generations. Once again, the effect of this feature on calculations of dose from radon progeny is not known at present. It is interesting to note, however, that the K cells implicated earlier as possible critical cells for producing lung cancer tend to be found at bifurcations rather than distributed throughout the generations.[67] Since the mucus flow may also be slowed at these bifurcations, it seems clear that more detailed studies of deposition at bifurcations and calculations of doses at these locations may be necessary.

A final additional consideration arises from the possibility that there may be a causal connection between doses to epithelial cells and doses to basal cells. Recent radiobiological evidence[83] suggests that cells transformed into a precancerous state require one or more cell divisions before they manifest their potential to produce cancer. The resulting cell death in the epithelial layer then would stimulate cell division in the basal cells. The dose to the basal cells then would be related to the force of induction in the lung, and the dose to epithelial cells would be related to the force of promotion.

6.12 A COMPARISON OF MODEL PREDICTIONS AND EXPERIMENTAL DATA

During the past decade, several attempts have been made to check deposition and mucociliary clearance calculations through a comparison with experimental measurements. The experimental deposition data in humans have been restricted to regional (TB or P region) deposition, rather than deposition within the various generations. In recent years, however, there have also been studies of deposition in hollow lung casts at various ages, allowing a comparison of calculations of deposition in the generations of the lung. In addition, limited data allow an assessment of the extent to which model predictions of mucociliary clearance from the entire TB region are correct.

Model calculations[29,74] for the generations of the TB region have been summed to yield total TB deposition and P deposition for different aerosol AMAD and geometric standard deviations. A comparison with the averages of experimental data[44,84–89] then was performed. Some results for the TB deposition, P deposition, and total lung deposition are displayed in Figures 6.12, 6.13, and 6.14.

Models appear to have problems at particle diameters of 0.1 μm and below. In this region, the model calculations systematically yield estimates below the measured values. The underestimates arise primarily in the pulmonary deposition, so this should have little impact on radon progeny calculations. These results do indicate, however, that there are unresolved problems in modeling pulmonary deposition, perhaps due to the fact that existing models do not account for turbulent deposition due to the mixing of inhaled air with the normal alveolar volume. Essentially no data are available at very young ages, although recent studies[82] using lung casts from 5- to 12-year-old children indicate that the deposition fractions at least increase as age decreases, which is a prediction from the theoretical studies.

Finally, Crawford-Brown and Eckerman[29] compared the mucociliary calculations with the data on TB clearance.[51] The data consisted

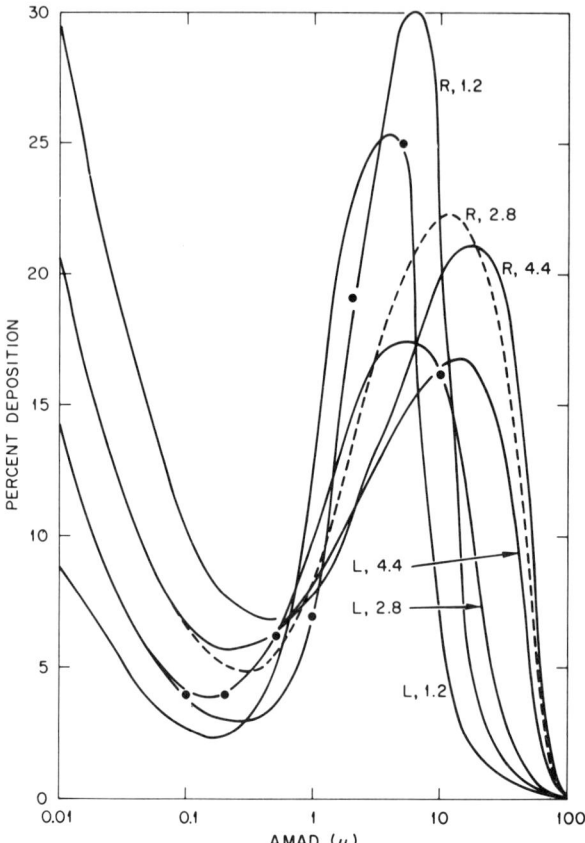

FIG. 6.12. A comparison of model predictions and experimental results for the deposition fraction in the total TB region. Deposition fractions for both resting (R) and light activity (L), and for three different values of the geometric standard deviation characterizing the distribution of particle sizes, are displayed.

of measurements of retention of radioactive aerosols in the TB region following inhalation. A model[27] was used to assign particles to the various generations of the lung and then calculate the activity remaining in the entire TB region as a function of time after inhalation. The comparison is presented in Figure 6.15 and indicates that the model represents clearance well, at least during the first five hours of clearance.

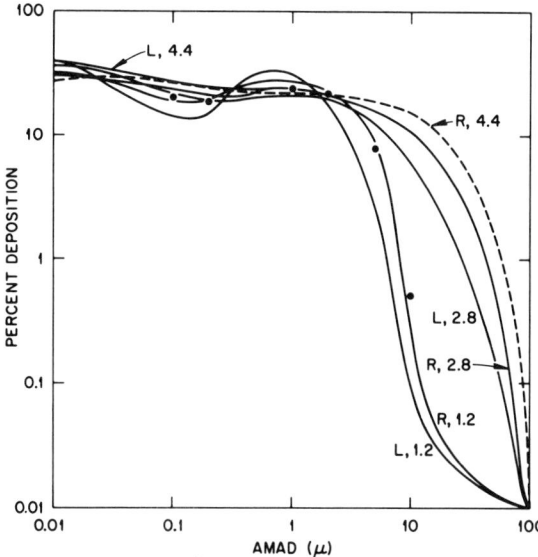

FIG. 6.13. A comparison of model predictions and experimental data for the deposition fraction in the pulmonary region of the lung. Deposition fractions for both resting (R) and light activity (L), and for three different values of the geometric standard deviation characterizing the distribution of particle sizes, are displayed.

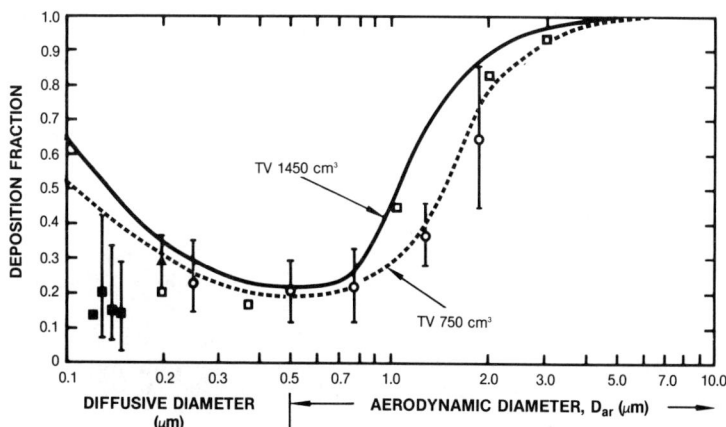

FIG. 6.14. A comparison between calculated and measured deposition fractions for the entire lung under conditions of nasal breathing. (Reproduced from an unpublished report by O. Raabe, with permission from the author.)

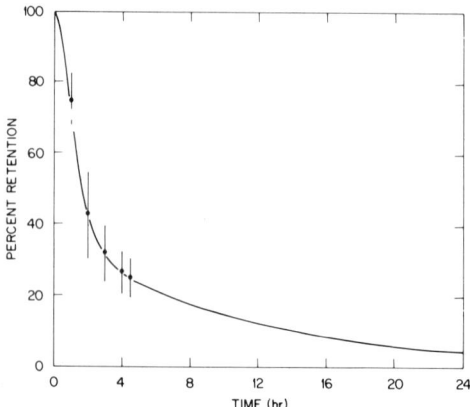

FIG. 6.15. A comparison between predictions of the fraction of deposited material remaining in the TB region of the lung following inhalation versus the experimental data of Yeates et al.[51]

6.13 DOSIMETRY OF INGESTED RADON

Aside from exposures to radon progeny in air, the public also is exposed to radon in water supplies. It has been suggested, however, that risks from ingestion are usually small compared to those arising from inhalation under most conditions.[90]

The dosimetry of radon for ingestion is much simpler than that for inhalation since the radon distributes fairly uniformly throughout the organ. The dose rate from the alpha-particle emissions, which produce most of the dose, in the radon decay chain in an organ then is:

$$D = CE(1.6 \times 10^{-8})/\rho \qquad (6.22)$$

In this equation, D is the dose rate in rads per unit time, C is the density of alpha emissions (alphas per unit volume per unit time), E is the energy (MeV) per alpha decay from Rn and includes the energy of all alphas in the decay chain (19.2 MeV), and ρ is the density of tissue (mass per unit volume). This equation assumes that the radon progeny decay at the site of their production. The total dose then is equal to the integral of the dose rate over some time interval.

Measurements and calculations of tissue doses from ingested radon

DOSIMETRY

have been confined to cases of adult exposures. These reports agree in the finding that stomach doses are much higher than doses to other organs, due primarily to the fact that the stomach is the point at which the radon enters the body. The radon then moves out of the gastrointestinal (GI) tract and into the bloodstream, from which it diffuses into the various tissues and organs. Most of the radon (more than 95%) then passes out of the body through exhalation since the bloodstream will carry the radon to the lung, where it diffuses into the alveolar spaces. A recent study[5,91] has examined the dosimetry of the radon after it enters the alveolar spaces and is exhaled. Several groups[92-94] have studied the retention and concentration of radon in the adult body following ingestion by using direct measurements. Their results indicate that an intake of 1.0 Bq of radon would produce an average dose equivalent to the entire body of between 2.7×10^{-10} and 3.8×10^{-9} Sv if a Q of 20 is assumed. These probably are overestimates of the dose equivalent delivered to tissues other than the GI tract, since most of the decays measured in the studies probably occurred in the GI tract. Doses to the stomach are higher, probably averaging approximately 1×10^{-8} Sv Bq^{-1}. In addition, these past studies indicate that

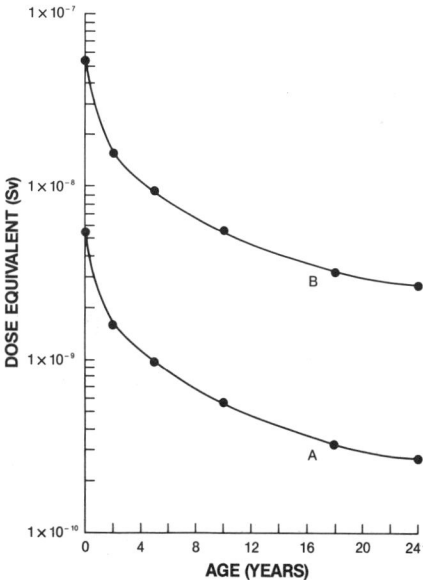

FIG. 6.16. The dose equivalent per Bq of ingested ^{222}Rn as a function of age. The adult value assumed is 2.7×10^{-10} Sv Bq^{-1} for curve A and 2.7×10^{-9} Sv Bq^{-1} for curve B.

the biological removal half-time for radon in the body is on the order of 30 to 70 minutes,[92,94,95] depending upon whether the stomach is empty or full. The longer half-time is associated with a full stomach.

There have been no direct measurements of these factors at other than adult ages. In a recent paper, however, a theoretical model for estimating the effect of age has been developed.[91] The theory was founded partially on the finding[95] that the half-time for removal of radon from the body was about the same for humans and rats. It then was hypothesized that the removal half-time was proportional to the ratio of lung surface area to the mass of soft tissue, which remains fairly constant with age. The dose equivalent delivered to body tissues other than the stomach is displayed in Figure 6.16 as a function of age. The dose equivalent to the stomach should be higher at all ages by a factor of 20 above curve A in this figure.

6.14 DISCUSSION AND CONCLUSIONS

While dosimetry of the lung is a complex field, the base of information is growing more complete on a daily basis. A great deal of data is being collected on lung deposition in hollow casts of the human lung, with some of the most recent efforts being directed towards developing a lung model which is physiologically realistic. These experimental models would, ideally, simulate the temperature and humidity found in lung passages as well as the morphometric properties. It has only been in the past few years that lung casts for young children have been obtained. These casts should go a long way towards determining whether the theoretical calculations of deposition in children are correct in their predictions. This is an important issue, since existing theoretical studies suggest that young children may represent a very sensitive population with respect to inhalation of radon progeny. As a result, regulatory standards designed to control the risk of airborne radon and progeny may be dominated by considerations of the lung doses to young children.

Perhaps we should try to peer a bit into the future and guess where future research efforts might be most fruitfully located. Clearly there is a need for much more information on the role of age in deposition and mucociliary clearance. Since there are distinct ethical problems associated with making direct measurements in young children, it seems likely that modeling efforts will continue to rely on theoretical deposition models normalized to experimental data using lung casts. Regardless of the age of the individual, however, there is a paucity of detailed information on transit times through generations other than the trachea and main bronchi.

Another large area of uncertainty at present concerns the location of the critical cells, both within the various generations and within the walls of each generation. The prediction of dose depends upon which generation is chosen and which depth assigned to the critical cells. It also seems plausible that the actual quantity of interest for predicting the probability of lung cancer could involve the combination of doses to several cell subpopulations in the lung. Fortunately, it is not necessary to have identified the "ultimate" quantity of interest if the only requirement is to extrapolate risk factors from one level of exposure to radon progeny down to another level, everything else being constant. All that would be required is that the calculated dose be directly proportional to this "ultimate" quantity.

Going through a detailed calculation of doses to basal (or other) cells may appear to be a complicated process just to get a risk factor. It is, however, a necessary process if the experience of the uranium miners and other groups exposed to radon progeny is to prove useful in telling society how to act with respect to the many other radionuclides for which we have no epidemiological experience. Without a reduction to estimates of dose to critical cells, each radionuclide would need to be studied in isolation, an impossible task given the large number of radionuclides in the environment. With the use of dosimetry, we can predict that avoiding the lung cancer incidence seen in the uranium miners can be accomplished not just by restricting access to the mines, but by regulating any activities likely to produce the same dose equivalent to critical cells. As is true with all theories, the theory of dosimetry allows an immense range of past and potential experiences to be explored under a common conceptual framework.

7

Health Effects

FRED T. CROSS

7.1 INTRODUCTION

The radon-related radiation dose to respiratory tissue and any resulting biological effects come from the inhalation of existing radon progeny in the air rather than from inhaled radon and its subsequent decay products. Once inhaled, radon gas quickly finds its way to the bloodstream. It is a chemically inert gas, and only a small fraction of that inhaled will be absorbed by the blood and not exhaled. Further, because the half-life of radon is relatively long compared to breathing time, only a small amount of it will decay while in the lung. The relatively short half-lives of the immediate radon progeny (excluding ^{210}Pb and its subsequent decay products) preclude any appreciable radiation dose to nonrespiratory tissue or even to pulmonary lymph nodes. In addition, the longer-lived progeny will not grow in fast enough, or they will be substantially cleared before an appreciable radiation dose accrues to any tissue of the body. This is particularly true at environmental radon-exposure levels and is supported in both human and animal studies.

Acute and subacute early effects, as well as late effects, can be expected following exposure of the respiratory tract to radon progeny. High concentrations of radon decay products in the lungs of animals can result in profound structural and functional changes that may produce lifespan-shortening, pulmonary emphysema, pulmonary fibro-

---

FRED T. CROSS • Biology and Chemistry Department, Battelle Pacific Northwest Laboratory, Richland, Washington. Work supported by the U.S. Department of Energy under Contract DE-AC06-76RLO 1830.

sis, and lung cancer. Many of the more than 40 distinctive cell types of the respiratory tract could be affected.[1] Low concentrations of radon decay products primarily increase the risk of lung cancer. The nature and magnitude of biological effects that may occur following inhalation of radon decay products will depend on many factors, such as fractions deposited in the respiratory tract and their retention times, translocation to other tissues, and rate of excretion by the body.

Inhaled short-lived radon decay products will, to a large extent, decay at their deposition site. Consequently, the tissues in the nasopharynx, the tracheobronchial tree, and the pulmonary region receive the majority of the radiation dose. The dose to the bronchi generally predominates in humans. These sites contain precursor or stem cells that are particularly sensitive to the cytotoxic and carcinogenic properties of alpha-emitting radon progeny. They may be more sensitive to carcinogenesis because of exposure to other environmental agents (such as cigarette smoke) that may increase cell division.

The mechanisms of early damage to respiratory tissues from inhaled radionuclides have been addressed in more detail by the International Commission on Radiological Protection (ICRP).[1] With regard to environmental radon, we will be concerned more with the carcinogenic process than with those effects more prevalent at high radiation exposures, such as early death, pulmonary emphysema, and pulmonary fibrosis.

7.2 MULTISTAGE THEORY OF CARCINOGENESIS

The multistage model[2-5] is one of the most popular theories for explaining temporal patterns in carcinogenesis. This model predicts an increase in cancer incidence as a function of time since exposure to a carcinogen. In general, the model proposes that a malignant tumor arises from a single cell which has undergone a series of heritable changes. A carcinogen may act on any or all of the stages in the cell division process leading to cancer. Those affecting the first stage are commonly referred to as initiators, while those affecting later stages are called promoters or progressors. Initiators are characterized by long latency periods between initial exposure and death, which often exceed 20 years in humans. Promoters, on the other hand, usually have shorter latency periods since fewer stages must occur before a malignant cell is produced.

Initiation involves alterations in cellular deoxyribonucleic acid (DNA), which contains the genetic code and transmits the hereditary pattern. Both chemicals and radiation may affect DNA, although their mechanisms may differ. Much of the DNA damage is repaired within the cell. The damage that is not repaired before cell division takes place usually produces mutations. The precise molecular mechanisms that transform cells with altered genetic information into cancerous cells are only poorly understood. However, the variation in the activity of DNA repair enzymes is thought to partly account for differences in cancer susceptibility among individuals.

The two-stage model of carcinogenesis postulates transitions from a normal cell (NC) to an intermediate cell (IC) to a malignant cell (MC). The model has quantifiable transition rates to describe the growth characteristics of the NC and IC populations.[6] The two stages can be interpreted as reflecting mutations, at the same gene locus, on homologous chromosomes; possibly an "anti-oncogene," a particular type of "cancer gene," is the gene involved. Oncogenes are thought to play a major role in cell proliferation, while anti-oncogenes play important roles in histogenesis.[7] Although the two-stage model does not refer explicitly to oncogene activation, gene amplification, and promoting effects, which are the modern biological concepts for carcinogenesis, it can include them.

Within the context of the two-stage model, initiators are mutagens, delivered in low doses that increase the chance that an NC will become an IC by mutating one homologue of the "cancer gene." There is a low probability that an initiator will transform an NC into an MC. The model predicts that there is no such thing as a "pure initiator," and that if an initiator is applied in sufficient dose, malignant tumors will result. A promoter is an agent that increases the proliferation of IC, thus increasing the chance that an IC will become an MC. This could occur by spontaneous mutation, mutation by further exposure to a mutagen, or by mitotic recombination. A "complete carcinogen" is a mutagen that results in malignant tumors when it transforms at least one NC to an MC. The distinction between an initiator and a complete carcinogen is thus quantitative, not qualitative. A sufficient amount of an initiator should produce malignant tumors, while a smaller dose of a complete carcinogen should act as an initiator in an initiation–promotion exposure situation, although it will produce few if any tumors.

Benign tumors that arise after initiation and promotion are "intermediate lesions," i.e., a proliferation of IC, heterozygous at the cancer gene locus. Therefore, benign tumors are not necessarily a step to malignancy but an indication that production of IC is abnormally high. Another prediction of the two-stage model is that if a second application of initiator is applied to this tissue, a high proportion of malignant tumors will result. The probability is very small that a mutagen delivered at low dose will cause two events in the same cell at the same gene locus. However, after application of an initiator followed by a promoter, the IC that were produced have multiplied to the point where a second application of a mutagen delivered in low dose has a high probability of changing an IC to an MC. This so-called initiation–promotion–initiation (IPI) protocol is a test of the two-stage model.[6,8] Using urethane as an initiator and 12-O-tetradecanoyl-phorbol-13-acetate (TPA) as a promoter, a significant increase in the yield of malignant tumors in mouse skin was observed after the second application of the initiator.[9]

Within the context of the two-stage carcinogenesis model, therefore, if low-level environmental radon acts as a mutagen and cigarette smoke, say, acts primarily as a promoter, then certain observations would be expected. The radon alone should yield very few tumors, but most of them will be malignant. Radon exposure followed by smoke exposure should yield a larger number of tumors, most of which should be benign. Radon, cigarette smoke, and radon exposures, if repeated, should yield still more tumors, most of which should be malignant. Smoke alone should yield the fewest number of tumors (very likely none). Exposure to smoke and then to radon should have a yield similar to that of radon alone. And exposure to radon–radon, with no intervening smoke, should yield a much smaller number of tumors than IPI.

Fractionation of the dose over time or reduced exposure rate of a mutagen might increase malignant tumor yield by reducing the number of cells killed, or by allowing IC that have been produced to multiply. The first mechanism may be important in high-dose radon carcinogenesis. In low-dose radon carcinogenesis, the second mechanism would be important within the context of the two-stage model. A single moderate or low dose of radon has a certain probability of transforming an NC into an MC directly. Fractionating the same total dose into a series of smaller doses that are delivered over time allows time for IC produced in previous exposures to multiply, thus increasing the likelihood that the next dose will produce an MC.

7.3 HUMAN DATA

Humans exposed to radon and its progeny include underground-mining populations (the epidemiological basis for risk predictions); populations exposed to high natural radioactivity, such as those in India and Brazil,[10] the People's Republic of China, Austria, Finland,[11] and Canada[12]; radon spa workers and patients intentionally exposed to high concentrations of radon in air and water[13,14]; and the general public. In general, despite the potentially large number of nonmining exposures, a causal relationship between lung cancer and radon has not yet been unequivocally established.

Historically, the epidemiological evidence for occupational exposures to radon dates back to the 16th century, when it was noted that Central European miners in Joachimstal and Schneeberg were suffering from a widespread fatal lung disease commonly referred to as "bergkrankheit."[15] Much of this disease was bronchogenic carcinoma, although it was not recognized as cancer until 1879. The etiological role of radon progeny was not generally accepted until the 1960s.[16]

Based on patient exposures and the Japanese atomic bomb survivors, as well as exposures in underground mines, it is now known that cancer is the main health effect associated with irradiation of the respiratory tract. Cancers may develop in all tissues of the respiratory tract, but site sensitivity for carcinoma induction may vary strongly with histological cell type and with the patient's genetic background, medical care, and lifestyle.

Tumors of the bronchi and bronchioles, which are also traditionally referred to as lung tumors, account for the majority of the tumors found in the human respiratory tract. Some Asiatic populations have extremely high levels of nasopharyngeal cancers; levels of laryngeal cancers are high in people of Finland, France, and the Mediterranean countries.[17] The vast majority of cancers originate from the pseudostratified epithelium; some rare tumors originate from bronchial glands. Many cell types are implicated in the histogenesis of bronchial cancers,[18] but secretory cells are considered the most likely progenitors for the entire pulmonary epithelium.[19] They displace basal cells, which were historically considered the progenitor cells.

The cell type found in bronchogenic carcinoma among radon-exposed miners does not define the lung cancer etiology. Small-cell or oat-cell carcinoma is the earliest to appear, but all forms are increased by radon-progeny exposure. In the underground miners with the longest latent intervals, epidermoid carcinoma appeared dominant. The tumor cell type has been shown to vary with many parameters, including smoking history, age at first underground exposure, and latent interval. The percentage of small-cell carcinoma decreased and the percentage of epidermoid carcinoma increased with latent interval.[20] In view of the numerous cell types involved, documenting tumor etiology by type seems unlikely, contrary to earlier reports.[21–23]

Lung cancer deaths in excess of those expected have been observed in several epidemiological studies of underground miners. These include U.S., Canadian, and Czechoslovakian uranium miners, Swedish and British iron miners, Swedish lead and zinc miners, and Newfoundland fluorspar miners. Although other potential carcinogens, such as diesel smoke, traces of arsenic or nickel, and iron ore, are found in these mines, the lung cancer response appears to be predictable based on radon progeny exposure.[2,24–32]

In most of the large epidemiological studies conducted to date, miners with cumulative radon decay product exposures somewhat below about 100 WLM had excess lung cancer mortality. The four major studies where a dose response can be inferred include: 3362 U.S. underground miners followed since 1950, whose exposures range from 60 to 7000 WLM with an average of 800 WLM (0.21–24 J h m^{-3}; average, 2.8 J h m^{-3}); 15,984 Canadian uranium miners with upper estimate exposures ranging from 5 to 510 WLM with an average of 74 WLM (0.02–1.8 J h m^{-3}; average, 0.26 J h m^{-3}); 2400 Czechoslovakian uranium miners followed since 1948 with an exposure range from 72 to 716 WLM with an average of 200 WLM (0.25 to 2.5 J h m^{-3}; average, 0.70 J h m^{-3}); and 1415 Swedish iron miners born between 1880 and 1919 who were alive in 1930 and exposed to from 27 to 218 WLM with an average of 80 WLM (0.09 to 0.76 J h m^{-3}; average, 0.28 J h m^{-3}).[33] To date, from 3 to 8% of the miners studied have developed lung cancer, mostly bronchogenic, that is attributable to radon-progeny exposures (i.e., above that expected from smoking or other causes). A recent review includes U.S., Canadian, and Czechoslovakian uranium miners along with supporting evidence from French uranium miners.[32]

All of the studies reported so far suffer from deficiencies; exposure conditions are not well known and, in many instances, follow-up time for miners is too short to reveal lifetime risk. None of the study groups has yet reached full closure, i.e., not all the miners have died, with or without lung cancer. Because of the deficiencies in the data, the reported excess lung cancer rates vary by a factor of 30, although the most probable ranges are smaller.[29] In the lowest-exposure category for U.S. miners (60 WLM, midpoint [0.21 J h m^{-3}]), a deficit (not statistically significant) rather than an excess in lung cancer deaths is observed.[26]

HEALTH EFFECTS

The mean excess lung cancer rate per million persons per year per WLM exposure calculated by the National Council on Radiation Protection and Measurement (NCRP)[29] was 12 ± 2 (3400 ± 600 $J^{-1} h^{-1} m^3$). A rounded-off value of 10 (2900 $J^{-1} h^{-1} m^3$) was chosen in estimating the lifetime risk from radon progeny exposures. The mean value for the excess rate that was estimated by the ICRP[32] was 8 (2300 $J^{-1} h^{-1} m^3$), with a probable range of 3 to 15 (900 to 4300 $J^{-1} h^{-1} m^3$). The ICRP's estimated relative risk of lung cancer is 1% WLM^{-1} (300% $J^{-1} h^{-1} m^3$) exposure, with a probable range of 0.5 to 1.5% WLM^{-1} (140 to 430% $J^{-1} h^{-1} m^3$). The most recent update of the Colorado plateau miner data places the relative risk coefficient at 0.9 to 1.4% WLM^{-1} (260 to 400% $J^{-1} h^{-1} m^3$) exposure.[2] A relative risk of 1% WLM^{-1} (300% $J^{-1} h^{-1} m^3$) implies a doubling exposure of 100 WLM (0.35 $J h m^{-3}$); i.e., the lung cancer rate for that particular population doubles at 100 WLM (0.35 $J h m^{-3}$) cumulative exposure.

7.3.1 Statistical Projection Models

The epidemiological evidence for health risks in a study population is based on medical data concerning the disease; time patterns of occurrence of the disease; the distribution of the disease, particularly among age, sex, and race groups; cofactors that influence the disease distribution pattern, such as cigarette smoking; the statistical models used for data analysis; and demographic data for exposed and control populations. Ideally, control or reference populations should be identical in every respect to the exposed populations except for the exposures of interest.

The two general types of epidemiological studies are cohort and case-control. Cohort studies segregate subjects on the basis of radon nonexposure and exposure. Retrospective or case-history study cohorts are defined after disease occurs, and underground miners generally fall in this category. Prospective or forward study cohorts are defined before disease occurs. An example of this type would be establishing a cohort from prospective employees before opening a new mine. Case-control studies segregate subjects on the basis of lung cancer nondisease and disease. An investigation of exposure differences then allows a correlation with the disease.

Much of the epidemiological analysis of the past has utilized the modified life-table approach. It is straightforward, since person-years at risk are simply divided into a number of strata, and age-specific, calendar-year-specific mortality rates from some reference or control population (often, the U.S. population) are applied to each.[2] The expected mortality is then compared to the observed mortality, using the following ratio:

$$\text{SMR}_j = \sum O_{ij} / \sum E_{ij}$$

where SMR_j = standardized mortality ratio for cause j, O_{ij} = the observed number of deaths for cause j in stratum i, and E_{ij} = the expected number of deaths for cause j in stratum i from reference population rates.

If the total number of observed deaths in each stratum is large, and if the reference population is the appropriate comparison group, the modified life-table approach would be the analytical method of choice, as no further modeling would be needed. However, after stratification by age, race, sex, calendar year, other confounders such as smoking and health status, and, finally, the exposure of interest (such as radon exposure levels and rates), there are seldom enough observed deaths in each stratum for reliable mortality rates.

An alternative to using the modified life-table approach is some form of statistical modeling. Modeling to estimate health risk is necessary when conclusions must be drawn about risk in regions of the exposure–response relationship for which data are too sparse (or are nonexistent) to estimate risk directly. The models also permit adjustments of the risk for interactions and confounders such as smoking, age, sex, etc. This is particularly important when adult male underground miner data are used to extrapolate risks to general populations. The differences between occupational and environmental radon exposures are obvious: environmental radon exposure levels and rates of exposure are generally much lower than for occupational radon exposure. Also, copollutant exposures are different (mines versus indoor environments), health status may be different (occupational workers as a group are healthier; this phenomenon is called the "healthy-worker effect"), different age groups and male:female sex ratios exist (females in populations versus adult males in underground mines), etc.

A number of different statistical modeling approaches have been used for examining exposure–response relationships. The two most popular are the absolute-risk and the relative-risk models, or some variation of either type.[26,34,35,29,32] Regardless of age at exposure, lung cancer rarely appears before age 40, and never before a minimum latent interval of five to ten years following exposure. Most statistical models, therefore, are modified to include these basic features and an average annual rate of appearance of lung cancer.

The absolute-risk model is based on the assumption that the excess risk is dose-dependent and additive to the baseline (spontaneous) background lung cancer risk. In contrast, the relative-risk model is based on the assumption that the excess risk is dose-dependent and multiplies the baseline lung cancer risk. Typical examples of lung cancer appearance rates are shown in Figure 7.1. A single exposure is assumed to occur at age 20. The absolute-risk concept shows an excess lung cancer rate starting at age 40, following a 20-year latent period, and continuing to be expressed until age 70. An expression period of 30 years is assumed by ICRP,[36] 40 years by the United Nations Scientific Committee on the Effects of Atomic Radiation (UNSCEAR),[27] and a lifetime plateau of risk by the Biological Effects of Ionizing Radiation (BEIR)-III Committee.[26] An exponential decrease in the rate of risk expression due to repair, cell death, or unspecified mechanism is assumed by NCRP[29] in the data shown by the dashed-line curve. Examples of groups using the absolute-risk concept are BEIR-III,[26] ICRP,[36] and NCRP.[29] Examples of groups using the relative-risk concept are the Environmental Protection Agency (EPA),[37] National Institute for Occupational Safety and Health (NIOSH),[2] and ICRP[32] in their modeling of the risk for environmental exposures.

Not enough is known concerning the prediction accuracy of the various projection models when applied to specific mining and non-

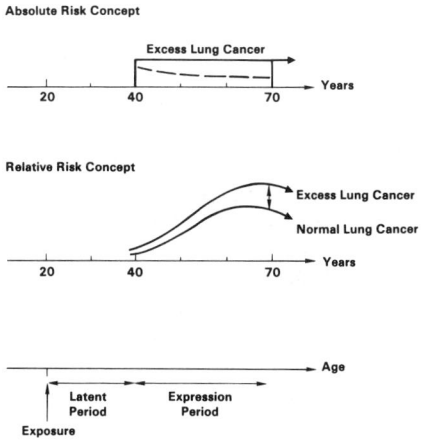

FIG. 7.1. Examples of lung cancer appearance rates.

mining groups. It is possible that neither absolute- nor relative-risk models will describe the lung cancer risk. The various mining cohorts under study will not go to closure for about 20 more years, and until then, the time course of lung cancer development for radon exposures in miners (or in general populations) will not be accurately known.

7.3.2 Lifetime Lung Cancer Risk Coefficient

The estimated mean excess lung cancer rates per million miners per year per unit exposure to radon decay products have been converted by NCRP and ICRP into estimated lifetime lung cancer risk coefficients for exposures of populations (Table 7.1). Both groups conservatively assumed a linear nonthreshold dependence of lung cancer on dose from radon-progeny exposure. The NCRP used the modified absolute-risk concept, delayed the appearance of cancer until age 40, or after a minimum latency interval of 5 years, and decreased the rate of risk expression by assuming a risk loss half-time of 20 years. Differences in lifetime risk between smokers and nonsmokers were ignored.

The ICRP used a modified proportional-hazard model (the relative-risk concept) and the similarity between the observed age distribution of radiation-induced lung cancer rates and the expected rates without radiation exposure to correct for the predominantly promoting influence of smoking. They also corrected for the exposure of miners to long-lived radioactivity in dust and from external gamma radiation.

TABLE 7.1. Estimated Lifetime Lung Cancer Risk Coefficients per Working Level Month per Year Environmental Exposures

Risk coefficient[a]	Reference
91×10^{-4}	National Council on Radiation Protection and Measurements (NCRP), Evaluation of Occupational and Environmental Exposures to Radon and Radon Daughters in the United States, NCRP Report No. 78, National Council on Radiation Protection and Measurements, Bethesda, MD (1984).[29]
100×10^{-4}[b]	International Commission on Radiological Protection (ICRP), Exposure and Lung Cancer Risk of the General Public from Inhaled Radon Daughters, Report of a Task Group of the ICRP (Draft February 1985; in press).[32]

[a] Multiply values by 285 to convert to SI units (J h m^{-3} y^{-1}).
[b] Range 50×10^{-4}–150×10^{-4}, with nonsmokers near the lower end of the range.

These effects are estimated at 10% of the total excess lung cancer risk. In view of the very close agreement in the lifetime lung cancer risk coefficients accessed by the two major radiological protection agencies, a rounded value of 10^{-2} WLM^{-1} y (2.9 J^{-1} h^{-1} m^3 y^{-1}; i.e., 100 per 10,000 persons exposed to 1 WLM y^{-1} for life) is suggested for environmental radon exposures. The estimated uncertainty is ±50%, with nonsmokers near the lower end of the range.[32]

7.3.3 Lung Cancer Modifying Factors

Factors that are known to be associated with an increased incidence of respiratory tract cancers are lifestyle, genetic factors, health status, time-related factors, and, possibly, some unknown factors. Smoking and alcohol consumption, for example, appear to strongly influence the incidence of cancers of the mouth, pharynx, larynx, and esophagus. Smoking, in both sexes, seems to be the most important of all possible modifiers of lung cancer, followed, perhaps, by general air pollution.[38]

7.3.3.1 Smoking Effect

The influence of smoking depends on several factors, including cigarette smoking rate, yields of tars, inhalation habits during smoking, age at start of smoking, and duration of smoking.[39-41] Since smoking habits differ around the world, differences can be expected in the smoking-related cancer incidence. Although the data are not entirely consistent, most indicate that tobacco smoke acts mainly as a promoter. Cessation of smoking is rapidly followed by a decreased risk.

The effect of smoking in radon-exposed miners is still a matter of debate, partly because smoking habits in miners may differ from smoking habits in control populations. The U.S. uranium miner data indicate that smokers have a shorter induction latency period and a higher incidence of lung cancer than nonsmoking uranium miners.[42,35,2] However, follow-up is still short in comparison to that for other mining groups, so the ultimate conclusion may change. Data from Swedish base-metal miners are quite different. These miners also show a shorter induction latency period if they smoke, but data differ among subgroups. As an example, the effect of cigarette smoke is additive in one study,[43] had no particular effect in another,[44] and indicated antagonism in yet another.[45] The role of other potential carcinogens in the mines on radon-related lung cancers is unknown, but circumstantial evidence appears to rule out diesel exhaust and other agents.[46] The ICRP[32] includes a 10% correction to compensate for this.

The main difference between the U.S. and the Swedish studies is in the length of follow-up and, thus, the portion of the life span over which data on deaths were collected. We can partially resolve these disparities by postulating that radiation exposures induced approximately the same finite numbers of cancers in both smokers and nonsmokers.[47,48] However, because cancer appeared earlier in smokers, one sees a larger number of lung cancers in smokers during short periods of follow-up and a larger number in nonsmokers if one looks very late in the period. In addition, there is, perhaps, an approximately equal number of lung cancers when follow-up is over the life span of the two groups. There is less certainty about the relationship of smoking, radon-progeny exposures, and lung cancer at low (environmental) radon-progeny exposure levels.

A recent report on the joint effect of cumulative cigarette smoking and cumulative radon-progeny exposures considers the response to be intermediate between additive and multiplicative and thus, strictly speaking, synergistic.[2] The evidence is not clear whether smoking acts at an early, intermediate stage or at both the early and late stages in the carcinogenic process. The correlation of age at start of smoking with latency is thus obscured. The rapid decline in the risk of lung cancer when smoking ceases indicates an intermediate effect or a combination of early- and late-stage effects.

7.3.3.2 Sex and Age Effect

Lung cancer mortality data regarding women and children are sparse to nonexistent. Presently, women in occidental countries are much less prone than men to develop spontaneous cancer in the respiratory tract. However, it may be expected that sex differences in respiratory tract tumors will decrease because of changes in smoking habits; eventually, women will reach parity with men, probably by the year 2000.[49] After smoking habits were taken into account,[50] there appeared to be no difference in the absolute risk of lung cancer between Hiroshima and Nagasaki male and female atomic-bomb survivors. The ICRP[32] considers the age-specific lung cancer rate per unit of radon exposure equal for males and females.

There is evidence that the age-specific incidence of radioinduced lung cancers follows that of spontaneous cancers and that the relative risk coefficient is age-independent in adults.[32] The induction latency period does appear to be age-dependent, declining from about 45 years

HEALTH EFFECTS

to about 10 years in uranium miners who start mining at the ages of 20 and 60, respectively.[24] Children may have a higher risk for lung cancer than adults. Relative risk was largest among Hiroshima and Nagasaki survivors 0 to 19 years old at the time of bombing (follow-up to 1982), among whom the spontaneous rate is still very low.[50] It is uncertain whether this pattern will continue until extinction of the cohort. On the basis of radon-progeny lung dose alone, 10-year-old children would be expected to be 1.5 to 2 times as sensitive as adults.[29,51] Experiments in animals also suggest that children may be twice as sensitive as adults.[52]

Risk data derived from adult male miners represent only a subgroup of a population for which the healthy-worker effect has been noted.[53] The NCRP[29] extrapolates these data to women and to children solely on the basis of differences in their radon-progeny inhalation-exposure-to-lung dose-conversion factors. The resulting risk coefficients, when applied to a typical environmental level of radon progeny, account for about one-fifth of spontaneous lung cancers. This is in excellent agreement with the current assumption that 83% of the lung cancer deaths in the U.S. are due to smoking,[54] provided that significant lung cancer contributions from environmental carcinogens other than radon are ruled out.

7.3.3.3 Time-Related Factors

One of the most difficult aspects of producing a valid quantitative risk assessment is dealing with the effects of various time-related factors on the exposure–risk relationship. These factors include exposure rate, calendar time, age at first exposure, and time since cessation of exposures. The multistage model of carcinogenesis, already discussed, is an attempt at explaining these temporal patterns.

Cumulative exposure is actually the product of duration of exposure and intensity or rate of exposure. The implicit assumption made when using cumulative exposure in assessing risk is that, all else being equal, high exposure rates for short periods of time are equivalent etiologically to low exposure rates for long periods of time. Some investigators found no statistically significant effect of exposure rate in U.S. uranium miner data.[25,35] However, these investigators used duration of employment rather than actual time spent underground in the mines as a measure of the exposure period. A subsequent analysis of these data[2] shows a statistically significant negative exposure-rate effect that is qualitatively similar to that observed in animals exposed to radon. That is to say, lung cancer incidence is greater at lower rates of exposure for comparable cumulative exposure. Also, the exposure-rate effect is stronger in the lower cumulative exposure ranges. An order of magnitude decrease in exposure rate below 800 WLM (2.8 J h m^{-3}) exposure increased lung cancer rates by about 60%. There is as yet no explanation for the negative exposure-rate effect (other than that postulated above for the two-stage model of carcinogenesis), nor do we know the lowest rate at which this effect persists. More to the point, it is uncertain whether the exposure-rate effect continues below the rates received by miners. Thus, the environmental cancer risk coefficient derived from epidemiological data may be underestimated.

Mortality patterns change over time because of lifestyle changes (e.g., use of tobacco and alcohol), medical care, and other exposure-risk factors. A properly matched calendar-time control population to an exposed cohort is of great importance in an epidemiological analysis. A recent birth decade match for the U.S. uranium miner cohort revealed that miners born in later decades are at greater risk of lung cancer per unit of exposure than were miners at equivalent ages who were born earlier. This was thought to be associated with the gradual lowering of radon exposure rates in mines.[2]

As previously mentioned, the tumor-induction latency period in miners decreases with increase in age at first exposure. Specifically, the shortest periods are observed in men who began mining at a late age, those who smoked heavily, and those who had the most intense exposures. Although radon-induced lung cancer appears at the ages expected for spontaneous cancers (rarely before age 40), many appear between ages 40 and 55 in miners who smoke. There is no evidence for a fixed induction latency period; children, for example, do not show lung cancers until later in life. Also, as previously mentioned, children who have had comparable exposures may be at somewhat elevated risk in comparison with adults, although this is uncertain.

When considering chronic exposures of varying intensity (the usual occupational situation), the relative risk increases with age at initial exposure if a late stage in the carcinogenesis process is affected.[55] Recent evidence is that U.S. miners initially exposed at later ages are at greater risk of lung cancer than those exposed at younger ages, all else being equal.[2]

It has been predicted that when exposure begins some time after infancy, excess relative risk increases, peaks, then decreases with time after exposures if the first stage in the carcinogenesis process is affected.[56] When the next-to-last stage is affected, relative risk decreases strictly with time after the last exposure. The recent update of the U.S. uranium miner cohort study showed a strong negative effect of time since the last exposure. This implies that a miner's chances of surviving lung cancer increase dramatically with each year after he leaves the mines, and that radon acts at a late stage in the carcinogenic process.

An exponential factor is used in the NCRP[29] risk projection model to account for the observed decrease in the rate of risk expression following exposure. Since lung tumors do not usually appear before age 40, exposure at an earlier age allows more time for decay of the risk. (Alternatively, the risk modeling could have proceeded on the basis that vulnerability is greater if exposure occurs at a later age.[57]) There is no direct evidence for the disappearance of cells transformed by radon, although a longer latency time in young miners is consistent with this assumption. The observation has been made[58] that promotion of radon-induced tumors in rats significantly decreased with time elapsed between irradiation and application of a promoter. This is suggestive of the disappearance of cells changed by radon exposure.

7.4 ANIMAL STUDIES

7.4.1 Introduction

Human epidemiological data provide the most important basis for assessing risks of radon exposures. However, additional insight into the nature of exposure–response relationships is provided by animal experimentation and dosimetric determinations. The broader advantages of the animal studies are: (1) experimental verification of suspected cause-and-effect relationships in human epidemiological studies; (2) a quantitative basis for using epidemiological data to predict health effects attributable to background radon exposures; and (3) a quantitative basis for deriving exposure limits for radionuclides other than radon progeny that are deposited in the lung.

Animals can serve as surrogates in studying components of the radon exposure problem. They can be exposed to well-defined levels of radon and associated pollutants, can be sacrificed for the study of developing lesions, or can be kept for their life span to determine late effects, such as cancer. They can also be exposed to radon levels spanning the range from environmental radon levels to the highest levels found in mines, thus allowing the determination of the shape of the radon exposure–response relationship. And, because radon and radon-progeny exposures comprise our major database in man for lung cancer from inhaled radionuclides, as do radium exposures for bone cancer, the data provide a basis for extrapolating radon-related health effects to other radionuclides deposited in the lungs of man. There are, of course, limitations on the accuracy of such derived data, such as those attributable to differences in lung volumes affected by the radiation and any species differences in the disposition of inhaled radionuclides.

Animal studies have been conducted for more than 50 years to examine the pollutants and their levels in underground mines that were responsible for the respiratory effects observed among miners. This work has emphasized respiratory cancer and the interaction of radon with other agents, such as cigarette smoke. Many of the initial studies, however, were concerned with early effects or short-term pathological changes.[59-61] Exposures were based primarily on radon gas concentrations, giving little or no information on the radon-progeny concentrations, which, as previously discussed, contribute the greatest dose to the lung.

Following the growing concern in the United States and France in the 1950s that the early European mining experience might apply to current miners, systematic studies were begun with animals to identify the agents responsible for the increased incidence of respiratory disease and to quantify the relationships. Two U.S. research centers, the University of Rochester (UR) and the Pacific Northwest Laboratory (PNL), and the Compagnie Générale des Matières Nucléaires (CO-GEMA) laboratory in France, have contributed the modern data on radon and other mine-associated exposures in animals. A schematic overview of these studies is shown in Figure 7.2. The outer ring depicts the pollutants in the air of underground uranium mines. (Except for uranium-ore dust, these are typical of other types of mines if, for example, the source of radon is from groundwater.) Inhalation exposure to these pollutants produces the major biological effects shown in the middle ring. The inner circle portrays the surrogate animals for man: various species of rodents and the beagle dog. Approximately 2000 mice, 100 rats, and 80 dogs were employed in the early, completed studies at UR; 800 hamsters, 5000 rats, and 100 dogs in the ongoing studies at PNL; and 10,000 rats in the ongoing studies at COGEMA. Exposures were primarily to radon and radon progeny attached to laboratory room air particles or to ambient aerosols at UR and CO-GEMA, with some ancillary exposures to cigarette smoke, cerium hydroxide, and uranium-ore dust in the French studies. The PNL exposures included radon, radon progeny attached primarily to ura-

HEALTH EFFECTS 231

FIG. 7.2 Overview of animal studies related to uranium miner health problems.

nium-ore dust, cigarette smoke, and diesel-engine exhaust (to simulate the diesel-powered equipment used in the mines).

Updated biological effects data resulting from chronic radon inhalation exposures of mice, hamsters, rats, and beagle dogs are presented here. Emphasis is placed on the carcinogenic effects of radon and radon decay products, including the influences of radon-progeny exposure rate, unattached fraction and disequilibrium, and coexposures to other pollutants. Plausible values for the radon (radon progeny) lifetime lung cancer risk coefficients are also provided.

Due mainly to inadequate experimental design, work prior to the 1970s had not shown that it was possible to produce pulmonary carcinomas in animals, in a systematic way, from controlled exposures to radon and its progeny.[46,62] More recent animal data appear in ICRP[1] and NCRP[29] reports, as well as in research reports from Senes Consultants, Ltd.[63,64] Portions of the analyses of the animal data presented here, particularly the comparison of human and animal radon exposure data, are adapted from the research reports.

7.4.2 Radon Inhalation Studies at the University of Rochester

7.4.2.1 Introduction

Beginning in the 1950s, investigators at UR examined the biological and physical behavior of radon and radon decay products, as well as their dosimetry in the respiratory tract.[65-67] Rats and dogs were exposed to several levels of radon alone and to radon along with radon decay products attached to room-dust aerosols.[68] It was demonstrated that the degree of attachment of radon progeny to carrier dust particles was a primary factor influencing the alpha-radiation dose to the airway epithelium. This dose was due primarily (>95%) to the short-lived progeny ^{219}Po and ^{214}Po, rather than to the parent ^{222}Rn. These findings supplemented earlier data on the relative levels of radioactivity found in the nasal passages, in the trachea and major bronchi, and in other portions of rat lungs after exposure to radon and radon progeny.[69] The respiratory tracts of rats that inhaled radon plus radon decay products contained 125 times more activity than those of animals that inhaled radon alone.

A pioneering series of experiments were initiated in the mid-1950s to determine the biological effects of inhaled radon and radon progeny in mice, rats, and beagle dogs.[67,70-72] The essentially negative biological results of these studies suggested that alpha radiation was inefficient in producing tumors in the respiratory tract.

7.4.2.2 Experimental Results

Adult mice were given chronic radon progeny inhalation exposures for periods from eight weeks to their entire life span at 1800 WLM wk^{-1} (6.3 J h m^{-3} wk^{-1}). Average alpha-radiation dose rates in sacrificed animals were estimated at 0.05, 0.18, 0.02, 0.60, and 2.8 Gy wk^{-1} to whole body, kidney, liver, gastrointestinal (GI) tract/stomach and contents, and lungs/trachea/bronchi, respectively. Dose to bronchial tissue was five to ten times that to whole lung or as much as 2.8 Gy wk^{-1}. Average dose to whole lung was about 2 mGy WLM^{-1} (0.6 Gy J^{-1} h^{-1} m^3). It was noted in associated experiments that life spans of male and female mice were shortened by 50% by life-span exposures to 72,000 WLM (252 J h m^{-3}) radon decay products; estimated mean doses to lung were 110 Gy. Destructive hyperplastic and metaplastic lesions, but not carcinomas, were noted in the tracheobronchial tree.

HEALTH EFFECTS

Other groups of mice, similarly exposed at lower levels, had pathological changes similar to but not as extensive as those in the first life-span exposure study. Changes were more marked in the trachea than in the large bronchi. Repair of lesions was rapid, and by eight weeks after exposures, tissues appeared to be normal. The epithelial lining of terminal bronchioles, however, became flattened or disappeared with increased time after exposure. Lesions were considered variants of those that occur spontaneously in older animals. Adenomas or benign tumors showed qualitative changes suggestive of malignancy.

Male standard and Sprague-Dawley rats were given chronic radon-progeny inhalation exposures for 24 weeks at 1100 WLM wk^{-1} (3.8 J h m^{-3} wk^{-1}). Cumulative exposures were approximately 26,000 WLM (91 J h m^{-3}); estimated mean lung doses were 44 Gy. Histopathological changes were not marked, nor were they greatly different from changes observed in control animals. None of the histopathological changes appeared to be dose related or to progress with time. The virtual absence of findings precluded development of exposure–response relationships.

Dogs were exposed for 1 to 50 days, at 200 WLM d^{-1} (0.7 J h m^{-3} d^{-1}), to provide cumulative radon-progeny exposures ranging from 200 to 10,000 WLM (0.7–35 J h m^{-3}). They were sacrificed at 0, 1, 2, and 3 y after exposure; none remained alive after 3 y. Results were as follows:

- Average lung dose per unit exposure was 1.7 mGy WLM^{-1} (0.5 Gy J^{-1} h^{-1} m^3) with a range of 0.8 to 7.9 mGy WLM^{-1} (0.23–2.3 Gy J^{-1} h^{-1} m^3). Doses per unit exposure to tracheal epithelium and bifurcations in the lung averaged 50 mGy WLM^{-1} (14 Gy J^{-1} h^{-1} m^3), with a range from 3 to 210 mGy WLM^{-1} (0.9–60 Gy J^{-1} h^{-1} m^3). The dose to tracheobronchial epithelium from ^{218}Po was considered three to four times higher than that from ^{214}Po.
- Pathological lesions observed for 200- to 10,000-WLM (0.7- to 35-J h m^{-3}) exposures were subtle, variable, diffuse, and very small, involving only a small fraction of the lung. No significant differences from control groups were noted immediately following exposures. The number of small inflammatory foci increased with exposure level in dogs sacrificed at 1 and 2 y; however, by the third year of sacrifice, the relationship disappeared at the lower exposure levels and was equivocal at the higher exposure levels. Small patches of thickened alveolar walls, with some metaplasia

of alveolar cells and hyperplasia of bronchial epithelium, were noted. No lesions were noted in the upper bronchial region.

7.4.2.3 Discussion and Conclusions

While injury to the bronchial tree was extensive during irradiation, the lesions did not lead to a significant incidence of bronchial tumors or cancer; after irradiation ceased, the lesions were quickly repaired. In the early experiments, the only late, permanent changes apparent occurred in the alveolar and, possibly, in the bronchiolar region of the lung. They were observed for a wide range of doses and for a period of up to 3 y in the dog, and for up to 1 and 2 y in the rat and mouse. Some of the changes may have been preneoplastic, but the high-level exposures, with consequent life-span shortening, and the early termination of experiments, precluded further development. The UR experiments were most noteworthy in establishing the exposure-to-dose values in whole lungs and portions of lungs and in other organs. The measured average value of about 1.7 mGy WLM^{-1} (0.5 Gy J^{-1} h^{-1} m^3) for the whole lungs of mice and dogs was comparable to measurements in dogs (2–4 mGy WLM^{-1}; 0.6–11 Gy J^{-1} h^{-1} m^3) at PNL.[73] Comparable lung doses have been calculated for hamster lungs[74] and for human lungs.[75–79] The few early and late pathological effects did not allow establishment of carcinogenic exposure–response relationships. The influence of the radon decay product carrier aerosol (laboratory room air containing dusts and oil and water droplets) on the results of these experiments is uncertain, though it probably led to more rapid solubilization of the progeny in blood (stated clearance was 10 min), with a resultant decrease in irritation or fibrosis in comparison with ore-dust/silica aerosols.

7.4.3 Radon Inhalation Studies in France

7.4.3.1 Introduction

Studies were begun in the late 1960s and early 1970s to determine whether radon and its decay products induced tumors in rats and to provide data supporting the epidemiological data on radon-progeny carcinogenesis.[80–87]

Prior to 1972, adult male Sprague-Dawley rats were exposed to ambient air mixed with radon derived from 25% uranium-content ore. After 1972, ambient air was mixed with radon derived from barrels of radium-rich lead sulfate. Maximum radon concentrations were 2.8×10^7 Bq m^{-3} and 4.6×10^7 Bq m^{-3}, respectively, for the two exposure periods. Early exposures contained radon-progeny equilibrium factors ranging from about 1 to 30%, while later exposure factors ranged from about 1 to 100%. Because of decay product plateout, the disequilibrium of the decay products increased when the number of animals in the inhalation chambers increased. Exposure periods ranged from about 1 to 10 months; exposure rates ranged from less than ten to hundreds of WLM wk^{-1} (<0.03 to >0.3 J h m^{-3} wk^{-1}), the majority averaging approximately 200–400 WLM wk^{-1} (0.7–1.4 J h m^{-3} wk^{-1}). Cigarette smoke exposures included inhalation by up to 50 rats of the combustion products of nine cigarettes for 10 min, several times a day, for a total of 350 hours.[81]

7.4.3.2 Experimental Results

In two major early experiments,[80] rats were given inhalation exposures to either stable cerium hydroxide prior to radon-progeny exposures or to 130-mg m^{-3} uranium-ore dust concentrations (given alone on days alternating with radon progeny exposures). The purpose was to determine whether lesions induced by dust exposures altered the carcinogenic properties of radon decay products. Potential alpha-energy exposures varied from 500 to 8500 WLM (1.7–30 J h m^{-3}). While exposure to stable cerium hydroxide shortened the induction latency period by two to three months, uranium-ore-dust exposures appeared to have little influence on the tumorigenic process, although the number of animals used in the latter exposures was considered too small to draw a firm conclusion. The prime conclusion reached in these experiments, perhaps, is that radon decay products alone induced tumors in rats. Other major findings were:

- High potential alpha energy exposures produced interstitial pneumonia and severe fibrosis in the lungs, with death occurring within a few weeks to months following exposures greater than about 6000 WLM (21 J h m^{-3}). No lung cancers were produced.
- Radon-progeny exposures between 2000 and 5000 WLM (7–17 J h m^{-3}), delivered over three to four months, produced the highest incidence of tumors. Animals lived longer than those exposed to higher levels, carcinomas

appeared between the 12th and 24th month from the beginning of exposure, and the tumor induction latency period increased with decreases in cumulative exposure.
- Malignant tumors of several different types were found, often in the same animal. These included epidermoid carcinoma, bronchiolar adenocarcinomas, and bronchioloalveolar carcinomas. A range of intermediary lesions was also noted.
- The exposure/tumor incidence relationship, uncorrected for life-span shortening brought about by high potential alpha energy exposures was not linear over the wide range of exposures; the incidences appeared linear at low exposure levels and gradually diminished from linearity at high cumulative exposures.

Subsequent experiments confirmed the pathology described above and extended the range of exposure to 20 to 50 WLM (0.07–0.17 J h m^{-3}).[84,86,87] Some additional findings were:

- Lung cancers in rats invaded pulmonary lymph nodes, as with human exposures, but metastases to other tissues were rare. Tumor size increased with increase in radon decay product exposure.
- No radioinduced oat-cell carcinomas (commonly found in humans) were observed in rats; however, other histological types of lung carcinomas were similar to those in man. Apart from lung tumors, cutaneous epitheliomas of the upper lip and cancers of the urinary system were noted; the Sprague-Dawley rat used in the experiments is known to be very sensitive to the latter.
- Exposure fractionation increased the incidence of lung cancer. Rats exposed to approximately 3000 WLM (10 J h m^{-3}) potential alpha energy at exposure rates of 50 or 300 WLM wk^{-1} (0.17–1 J h m^{-3} wk^{-1}) showed a nearly fourfold increase in cancer incidence (over controls) with exposure protraction (J. Chameaud, personal communication, 1986).
- Tumor incidence was not affected by increased age of adult rats exposed to radon progeny; however, tumor induction latency periods shortened as the age at first exposure increased.
- Exposures to both inhaled radon progeny and cigarette smoke (whole-body) produced synergistic tumor response when the smoke exposures followed the completed radon-

progeny exposures. No change in tumor response over radon-progeny exposures alone were noted when cumulative smoke exposures preceded radon-progeny exposures. The effect of cigarette smoke was attributed to its promoting action[83,81,84]; histological types of cancers were not altered by cigarette smoke exposures. Tumors in the radon progeny- and smoke-exposed animals were larger and more invasive than those in radon-progeny-only exposures, perhaps indicative of a shorter latency period of smoking-related tumors.

7.4.3.3 Discussion and Conclusions

More than 800 lung cancers have been produced in about 10,000 rats exposed to radon progeny along with ambient aerosols and to mixtures with other pollutants in the French studies. Between approximately 20 and 5000 WLM (0.07 and 17 J h m^{-3}) the derived range in mean lifetime risk coefficient, uncorrected for life-span differences from control animals, is about 1.5×10^{-4}–7.5×10^{-4} WLM^{-1} (400×10^{-4}–2000×10^{-4} J^{-1} h^{-1} m^3) exposure (i.e., 1.5–7.5 per 10,000 rats exposed per WLM for life). The risk decreases at higher WLM exposures because of life-span shortening and gradually appears to plateau at about 6×10^{-4}–8×10^{-4} WLM^{-1} (1700×10^{-4}–2300×10^{-4} J^{-1} h^{-1} m^3) between 20- and 50-WLM (0.07–0.17-J h m^{-3}) exposures. No evidence of a threshold below 20 WLM (0.07 J h m^{-3}) was apparent.[87]

7.4.4 Radon Inhalation Studies at the Pacific Northwest Laboratory

7.4.4.1 Introduction

As in the French studies, exposures of dogs and rodents to uranium-mine-air contaminants began at PNL in the late 1960s and early 1970s to identify agents, and their levels, responsible for producing the lesions observed in the respiratory tracts of uranium miners.

Early experiments concentrated on life-span inhalation studies, with adult Syrian Golden hamsters and adult beagle dogs exposed to mixed aerosols of radon, radon progeny, and carnotite uranium-ore dust, along with diesel-engine exhaust (hamster studies) and cigarette smoke (dog studies).[88–90] To provide data missing from the early study, follow-up studies included earlier exposure of beagle dogs to uranium-ore dust alone and exposures of adult male Wistar rats to mixtures of radon, radon progeny, and uranium-ore dust.[91–95,64,96] The exposures of rats were truncated, rather than life span, in order to provide data for the development of exposure–response relationships. The rat studies were also designed to study the roles of carnotite-ore-dust concentration and radon-progeny exposure rate, unattached fraction, and disequilibrium in the production of lung lesions.

7.4.4.2 Exposure Parameters

Uranium-ore-dust concentrations ranged from less than 0.5 to approximately 20 mg m^{-3} in order to study the pathogenic role of the dust and to alter the unattached fractions of radon progeny. This range accommodates all dust conditions in mines; the upper values represent particularly dusty conditions in early mines, while the lower values are more representative of current mine levels. The uranium content varied between 2 and 4% by weight; silica content was on the order of 80%. Diesel-engine exhaust-fume concentrations were about 7 mg m^{-3}; CO levels were held to 50 ppm, NO_2 levels to about 5 ppm, and SO_2 and aliphatic aldehyde levels to less than the 1-ppm detection limit. Cigarette smoke exposures were by mouth- and nose-only and varied, among animals, between 10 and 20 cigarettes per day.

The activity median aerodynamic diameter (AMAD) of the radon decay products was about 0.5 μm, with a geometric standard deviation (GSD) of about 2.0. Equilibrium factors ranged from 0.1 to 0.6 (10–60% equilibrium with radon concentrations); unattached percentages of ^{218}Po ranged from less than 2% to 24%. The unattached percentages of other progeny were at least an order of magnitude lower. Radon decay product exposures ranged from 20 to about 15,000 WLM (0.07 to about 52 J h m^{-3}), representing not only the levels received by former miners but, like the French exposures, accommodating environmental exposure levels as well.

7.4.4.3 Experimental Results

The major biological effects observed in dogs and rodents exposed to uranium-mine-air contaminants were life-span shortening, pulmonary emphysema, pulmonary fibrosis, and respiratory carcinoma. In radon-progeny-exposed animals, lesions observed in organs other than

HEALTH EFFECTS

the respiratory tract were generally considered spontaneous, or only indirectly exposure-related.

Life-span shortening was uncommon in hamsters but commonly observed in dogs and rats. Life-span exposure of hamsters to high radon decay product levels alone (10,000 WLM; 35 J h m^{-3}) or in combination with uranium-ore dust and diesel-engine exhaust caused no significant changes in mortality patterns compared with those of controls. An accumulation of amyloid (an abnormal complex material resembling starch) in various tissues, along with liver and kidney degenerative changes, common in all hamsters, may account for their median life span being only a little greater than 1 y. In contrast, life-span exposures of beagle dogs produced significant life-span shortening. At mean potential alpha energy exposures of 13,000 WLM (45 J h m^{-3}), mean survival times were 4 to 5 y, compared to approximately 15 y for control animals. Mean survival times of controls and smoke-only-exposed dogs were equivalent to one another during the same period. Rat data to date indicate no significant differences in mortality patterns compared with those of controls for exposures up to about 2500 WLM (8.7 J h m^{-3}). Exposures exceeding 5000 WLM (17 J h m^{-3}) produced significant life-span shortening, the effect increasing with exposure. In addition, the mean survival time of tumor-bearing rats (as in the COGEMA data) was always significantly longer than that of non-tumor-bearing rats. This is not an unexpected finding, since the induction latency period of lung tumors is a large fraction of the rat life span, and tumors must grow large enough to detect. The shorter-lived animals may have died too soon for tumors (if any) to be detected. Lethality from neoplasms is more marked in longer-lived species, such as dog and man, which have tumor induction latency periods that are generally much shorter than their life spans.

As already noted, in the radon-progeny-exposed animals, the lesions observed in organs other than the lung are generally considered spontaneous, or only indirectly exposure-related. This is in contrast to the case for most alpha emitters, which translocate from the lung to irradiate other organs. In animals exposed to high concentrations of uranium-ore dust alone (and, presumably, to radon progeny and uranium-ore dust mixtures), sufficient long-lived radioactivity from the precursors of radon can concentrate in the kidneys to impair their function. However, direct evidence of kidney function impairment from exposure to radon progeny alone is lacking. Neutrophilia was observed in dogs exposed to high levels of radon progeny, uranium-ore dust, and cigarette smoke and to cigarette smoke alone. This probably resulted from chronic irritation and pulmonary-cell death.

Pulmonary fibrosis and, to a lesser extent, emphysema are common findings in hamsters, rats, and dogs exposed to radon progeny alone and in mixtures with uranium-ore dust. These effects, however, are not produced to any appreciable extent in groups of animals until exposures to radon decay products exceed several thousand WLM (>17 J h m^{-3}). Lung cancer, on the other hand, is produced at exposure levels as low as 20 WLM (0.07 J h m^{-3}), making neoplastic lesions the most sensitive indicator of radon progeny exposures.

While investigating the causes of non-neoplastic lesions in the radon-progeny-exposed dogs, animals were matched on a mean-radiation-dose-to-lung basis (equivalent to the peripheral lung dose) with plutonium-exposed dogs. The prevalence and severity of both emphysema and fibrosis were comparable. Because the plutonium experiments did not include ore dust, the tentative conclusion was that radiation doses from high concentrations of radon decay products were responsible for the production of these lesions in the radon-progeny-exposed dogs. This conclusion is supported by experiments with uranium-ore dust alone where the severity of pulmonary fibrosis and emphysema was significantly less. Thus, a more rapid development of fibrotic lesions, at least, occurs when radon decay products accompanied uranium-ore-dust inhalation exposures.

Animals exposed to radon decay products alone or in mixtures with other uranium-mine-air pollutants commonly had adenomatous lesions that progressed to squamous metaplasia of alveolar epithelium. Bronchioloalveolar and bronchogenic carcinomas followed this and other preneoplastic changes in the lung. Hyperplasia and squamous metaplasia of nasal epithelium, with eventual development of squamous-cell carcinoma, were also noted in experiments (generally in rodents) when the unattached fraction of radon progeny was high.[96,94] Low or intermediate unattached fractions produced nasal carcinoma in dogs, indicating that either their nasal tissue is more sensitive to carcinogens than that of rodents or that nasal deposition was higher in dogs.[90] There was no indication that high-disequilibrium (low equilibrium factor) radon-progeny exposures without associated high unattached ^{218}Po produced more nasal carcinomas than low-disequilibrium radon-progeny exposures.

Life-span inhalation exposures of adult male and female beagle dogs produced an overall 21% incidence of tumors primary to the lung. The mean potential alpha energy exposure was about 13,000 WLM (45 J h m^{-3}); the mean exposure rate was 71 WLM wk^{-1} (0.25 J h m^{-3} wk^{-1}). Under the conditions of the experiment, concomitant exposure to cigarette smoke had a mitigating effect on radon-progeny-induced

HEALTH EFFECTS

tumors (5% incidence of tumors, compared to 37% in the group not exposed to smoke). This was possibly due to a thickening of the mucous layer as a result of smoking and a stimulatory effect of cigarette smoke on mucociliary clearance, although no empirical evidence was collected during the experiments to substantiate these possibilities.

Life-span inhalation exposures of adult male hamsters produced severe radiation pneumonitis but only four squamous carcinomas (three in the radon-progeny-only group, one in the group exposed to radon progeny and uranium-ore dust) in 306 animals exposed to 10,000 WLM (35 J h m^{-3}) of radon decay products (1.3% incidence). Squamous carcinoma occurred only in association with squamous metaplasia of alveolar epithelium. The latter lesion occurred only in hamsters receiving exposure to radon decay products. Thus, it appears that exposure to radon progeny, development of squamous metaplasia, and development of carcinoma were related. Because so few lung cancers were produced in these high-exposure experiments (possibly because of the animals' short median life span), it was concluded that the Syrian Golden hamster was an inappropriate model for further study of the carcinogenic potential of inhaled mine-air pollutants.

Chronic inhalation exposure, at PNL, of adult male Wistar rats to ambient air or to mixtures with radon progeny and uranium-ore dust have complemented and supplemented the findings at COGEMA. The PNL exposures were at levels as low as 20 WLM (0.07 J h m^{-3}). Rats developed squamous-cell carcinoma, adenocarcinoma, and mixed carcinoma, the incidence depending on both the rate and cumulative potential alpha energy exposure; however, the histopathology study of these exposures is not completed. As in the COGEMA data, tumor incidence at very high exposures was low as a result of life-span shortening.

7.4.4.4 Discussion and Conclusions

An increasing trend (sometimes significant) was observed in lung cancer risk per cumulative potential alpha energy exposure with: (1) a decrease in potential alpha energy exposure rate, (2) an increase in unattached fraction of radon progeny, and (3) an increase in radon-progeny disequilibrium. About 70% of the lung cancers observed between exposures of approximately 300 and 5000 WLM (1 and 17 J h m^{-3}) were bronchogenic carcinomas, and 30% were bronchioloalveolar carcinomas.[95] By sizing associated bronchi and bronchioles, the locations of the tumors were estimated to be about 50% proximal (bronchus-associated) and 50% distal (bronchiole- and alveolus-associ-

contrast to the much greater numbers of human lung cancers in proximal locations.[97,98] The prevalence of tracheobronchial and nasopharyngeal squamous metaplasia and, generally, carcinomas increased with increased levels of unattached radon decay products. An approximate 50% increase in respiratory-tract tumor incidence per unit exposure occurred when the unattached levels of ^{218}Po increased from 2 to 10%.

The data are, as yet, inadequate to draw definitive conclusions regarding the effect of exposure rate and the magnitude of the lifetime risk coefficient below 100-WLM (0.35-J h m^{-3}) exposure. Risk coefficients increased 200 to 300% when exposure rates decreased from 500 to 50 WLM wk^{-1} (1.7 to 0.17 J h m^{-3} wk^{-1}). Some of the rats currently on study have been exposed at about 5 WLM wk^{-1} (0.017 J h m^{-3} wk^{-1}), approximating the average exposure rates of former miners. The exposure-rate effect is clearly evident in Figure 7.3; the highest risk coefficient for a given exposure level in the PNL experiments was obtained at the lowest rate of exposure, and vice versa. The data also indicate an increasing lifetime lung tumor risk coefficient per unit exposure with decreasing cumulative potential alpha energy exposure. It cannot be concluded, at present, that the increase in risk coefficient

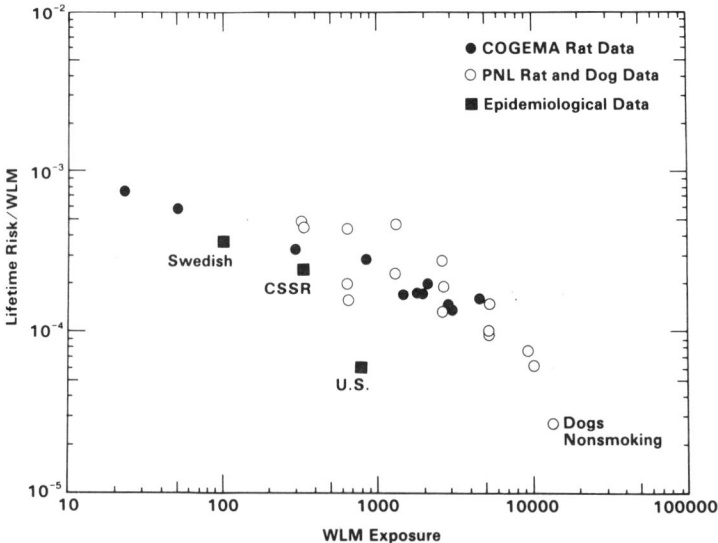

FIG. 7.3 Lifetime lung tumor risk coefficients for radon progeny exposure.

per unit exposure continues with further decrease in cumulative exposure and exposure rate.

There are no firm explanations, at present, for the relative insensitivity of hamsters, compared with rats and dogs, to the induction of lung cancers following radon-progeny inhalation exposures. It may be their generally shorter life span, compared with that of rats, decreases their chance of developing lung cancer. However, rats receiving exposures comparable to those of hamsters develop far more lung cancers, even within a period comparable to the life span of hamsters. It is also uncertain whether the radiation sensitivity of nonsmoking dogs differs from that of rats, since only one datum point can be plotted at high radon-progeny exposure.

7.4.5 Summary of Radon Inhalation Studies in Animals

The biological effects observed in dogs and rodents following the inhalation of radon and its progeny have been primarily respiratory epithelial carcinomas, pulmonary fibrosis, pulmonary emphysema, and life-span shortening. Extrapulmonary lesions are not a significant finding in these exposures, except for certain hematological effects in dogs.

Five variables appear to influence the tumorigenic efficiency of radon progeny in the experimental-animal studies:

- radon progeny cumulative exposure
- radon progeny exposure rate
- radon progeny unattached fraction
- radon progeny disequilibrium
- concomitant exposures to other pollutants.

Tumorigenic efficiency increases with: (1) increase in radon-progeny exposure until life-span shortening reverses the trend; (2) decrease from high radon-progeny exposure rate; and (3) increase in radon-progeny unattached fraction and disequilibrium. Although simultaneous exposure to elevated levels of ore dust or diesel-engine exhaust increased the incidence of emphysema and fibrosis, it did not appear to increase the number of tumors produced by high radon-progeny exposures. In beagle dogs, alternating exposures to smoke and radon progeny at high levels on the same day produced a decrease in lung cancer incidence from that produced by radon-progeny exposures alone. Smoke exposures completed before radon-progeny exposures did not alter the lung cancer incidence in rats; however, smoke exposures at high radon-progeny levels following completed radon-progeny exposures produced a synergistic effect. At current low radon-progeny levels in the mines or in the environment, the influence of exposure to other pollutants is unknown.

Although animal tumor data differ somewhat from human tumor data, the overall incidence data in animals (rats, primarily) are similar to our present estimated lung tumor incidence data in man (Figure 7.3). The derived range in mean lifetime risk coefficient for atmospheres with low percentages of unattached progeny (exposure rates exceeding about 200 WLM wk^{-1} [0.7 J h m^{-3} wk^{-1}], and data uncorrected for life-span differences from control animals) is about 1×10^{-4}–5×10^{-4} WLM^{-1} (300×10^{-4}–1400×10^{-4} J^{-1} h^{-1} m^3) exposure between approximately 100- and 5000-WLM (0.35–17-J h m^{-3}) exposures. Similar cumulative exposure data from underground miners are also estimated to range between about 1×10^{-4} and 5×10^{-4} WLM^{-1} (300×10^{-4} and 1400×10^{-4} J^{-1} h^{-1} m^3). Recent data from rats exposed to 20–50 WLM (0.07 to 0.17 J h m^{-3}), at low exposure rates, placed the lifetime risk coefficient at about 6×10^{-4}–8×10^{-4} WLM^{-1} (1700×10^{-4}–2300×10^{-4} J^{-1} h^{-1} m^3).

Human and animal dosimetric agreement does not necessarily suggest that cancer incidence data in animals can be directly extrapolated to humans because of possible differences in tissue sensitivities and affected sites in the lung. However, the apparently reasonable agreement of both dosimetric and carcinogenic data, at least at the higher exposure rates and for exposure levels exceeding about 100 WLM (0.35 J h m^{-3}), suggests that the animal models (particularly rats) are reasonable surrogates for man.

The similarities in the human and animal data are presently greater than the differences between them. The animal experiments continue to yield valuable information that helps to explain the effects on man of inhaled radon and its progeny and of associated pollutants, especially in mine environments.

7.5 COMPARISON OF HUMAN AND ANIMAL RADON-EXPOSURE DATA

The epidemiological data derived from many types of underground mines show a relatively consistent relationship between lung cancer

HEALTH EFFECTS 245

incidence (nearly the same as death rate from lung cancer) and exposure to radon decay products.[24,25,99,26,27,43,36] This underlying consistency is probably related to the relatively narrow range of bronchial dose per unit potential alpha energy exposure under various exposure conditions in the mines. The few differences result partly because the radon-progeny exposures of miners are imperfectly known and partly because of the influence of both exposure rate and the presence of other mine-air pollutants. The unattached fraction and disequilibrium differences are probably second-order influences except, perhaps, in clean indoor environments.

The comparison of animal data with human data that follows is not meant to constitute proof in humans of effects (or absence of effects) that are demonstrated in animals. Rather, explanation of provisional human data and potential findings in humans are suggested. The hamster data are excluded from this comparison in view of the known low carcinogenic sensitivity of hamsters to inhaled radionuclides.

In rats, tumor production per unit exposure at very high exposures (uncorrected for competing causes of death) was lower than at low to moderate exposures (Figure 7.3). However, the recent analysis of the COGEMA radon data suggests that the age-specific prevalence of pulmonary tumors continues to increase with exposure (even at very high exposures) when the data are adjusted for life shortening.[100] Miners exposed to the highest radon progeny levels in underground mines had the lowest attributable lung cancer rates per unit exposure.[26]

In both the human and rat studies, tumor production appeared to increase with decreasing exposure rate.[84,91,87,98,100] However, an analysis of the human data suggests that the exposure-rate dependence may taper off at environmental and occupational rates and levels of exposure.[101]

In a small group of Swedish zinc and lead miners, a lower lifetime incidence of lung cancer was observed in those who smoked and were exposed to radon progeny than in the nonsmokers similarly exposed.[45] This is tentatively ascribed to the protective effect of either increased mucus production from smoking or of the thickened mucosa resulting from smoker's bronchitis. A similar result was observed in dogs.[90] In rats, cigarette smoke was found to be cocarcinogenic with radon decay products when exposure to smoke followed completion of exposure to the decay products.[81] This effect was not observed, however, when smoking preceded the radon-progeny exposure.[84] Such disparities may partially explain discrepancies in interpreting tentative epidemiological data.

Emphysema and fibrosis have been attributed to radon-progeny exposure in animals—hamsters, rats, and dogs[88,96]—and in underground miners.[102] Simultaneous exposure of animals to ore dust or diesel-engine exhaust generally increased the incidence of emphysema and fibrosis but did not appear to increase the number of tumors produced when exposure to radon progeny exceeded 300 WLM (1 J h m^{-3}) cumulative potential alpha energy.[88,91,84,103]

For equivalent cumulative potential alpha energy exposures, the older the animal at the start of exposure, the shorter the induction latency period and, in humans (the data suggest), the higher the associated risk.[84,26] The highest risk coefficient calculated in humans, about 50×10^{-6} lung cancers per year per WLM ($14,000 \times 10^{-6}$ lung cancers per year J h m^{-3}), is for persons first exposed when over 40 years of age.[26]

The predictions of the various dosimetric models (based primarily on basal cell dose) appear to be borne out in the various species. The tumors induced in experiments with animals are commonly more distal than those in humans. Modeling of Syrian Golden hamster lungs[74] showed that peripheral basal and Clara cells may receive doses approximately equal to or greater than those received by basal cells in the central airways. Human tumors appear more frequently in the upper generations of the bronchial tree. Some absorbed-dose calculations show that basal cells in human upper airways receive the highest dose from radon decay products.[75,76,104] The qualitative agreement of the human and animal dosimetry for the newly defined secretory progenitor cells has yet to be demonstrated.

Lifetime lung tumor risk coefficients are similar in animals (rats, primarily) to those in humans (Figure 7.3). Data are inadequate and too few in number to provide lifetime lung tumor risk coefficients versus cumulative potential alpha energy exposure of mice, hamsters, and dogs. The coefficients based on rat data, uncorrected for life-span shortening, appear to range between approximately 1×10^{-4} and 5×10^{-4} WLM^{-1} (300×10^{-4}–1400×10^{-4} J^{-1} h^{-1} m^3) for all lung tumors (benign and malignant) at cumulative exposures between 100 and 5000 WLM (0.35–17 J h m^{-3}). At exposures considerably less than those where mean life span is significantly shortened (<500 WLM; <1.7 J h m^{-3}), the lifetime risk coefficient in the earlier PNL experiments ranged between 2×10^{-4} and 4×10^{-4} WLM^{-1} (600×10^{-4}–1100×10^{-4} J^{-1} h^{-1} m^3) for all lung tumors.[95] Recent rat data at 20- to 50-WLM (0.07–0.17-J h m^{-3}) exposures, at low exposure rates, indicate that the lifetime risk coefficient ranges between 6×10^{-4} and 8×10^{-4} WLM^{-1} (1700×10^{-4}–2300×10^{-4} J^{-1} h^{-1} m^3).[86] A summary of estimated human lifetime lung cancer risk coefficients appears in Table 7.2.

Except for the greater prevalence of solid alveolar tumors and bronchioloalveolar carcinomas observed in the animals and, possibly, the observation of the oat-cell (K-cell) type of tumors in man, the tumor data are not dissimilar. Although K-cell carcinomas were not found in the animal experiments, they are believed to be involved in pretumoral lesions, with species-dependent phenotypic expression.[105] It has been

TABLE 7.2. Summary of Lifetime Lung Cancer Risk Coefficients per Working Level Month Exposure of Humans[a]

Risk coefficient	Reference
2×10^{-4}–4.5×10^{-4}	United Nations Scientific Committee on the Effects of Atomic Radiation, Sources and Effects of Ionizing Radiation, United Nations, New York (1977).[27]
2×10^{-4}	W. Jacobi, in: *Personal Dosimetry and Area Monitoring Suitable for Radon and Daughter Products*, Proceedings of NEA Specialist Meeting, pp. 33–48, Nuclear Energy Agency, OECD, Paris (1977).[106]
2×10^{-4}	K. D. Cliff, B. L. Davis, and J. A. Riessland, Little danger from radon, *Nature 279*, 12 (1979).[b(107)]
2×10^{-4}–14×10^{-4c}	National Academy of Sciences, *The Effects on Populations of Exposure to Low Levels of Ionizing Radiation* (BEIR-III Report), National Academy Press, Washington, DC (1980).[26]
1.5×10^{-4}–4.5×10^{-4}	International commission on Radiological Protection (ICRP), *Limits for Inhalation of Radon Daughters by Workers*, ICRP Publication 32, Pergamon Press, New York (1981).[36]
$\leq 1 \times 10^{-4}$	R. D. Evans, J. H. Harley, W. Jacobi, A. S. McLean, W. A. Mills, and C. G. Stewart, Estimate of risk from environmental exposure to radon-222 and its decay products, *Nature 290*, 98–100 (1981).[b(28)]
1×10^{-4}–2×10^{-4}	National Council on Radiation Protection and Measurements (NCRP), Evaluation of Occupational and Environmental Exposures to Radon and Radon Daughters in the United States, NCRP Report No. 78, National Council on Radiation Protection and Measurements, Bethesda, MD (1984).[b(29)]

[a] Unless otherwise specified, values pertain to occupational exposures of underground miners; multiply values by 285 to convert to SI units (J h m^{-3}).
[b] Environmental exposures.
[c] Values were converted from published annualized coefficients, assuming 30 years for cancer expression.

postulated that in rats, K-cells convert to mucus-secreting cells, which may eventually become adenocarcinomas.[85]

Data observed in animals that are not unequivocally demonstrated in human exposures to radon progeny are: (1) the increase in tumor production with increase in radon-progeny unattached fraction and disequilibrium, and (2) the importance of the temporal sequence of exposures to cigarette smoke and radon progeny. (The absence of these findings, however, does not constitute proof that similar data might not be obtainable in human exposures but may rather reflect the paucity of this type of human data.)

8

Mitigation

JUDITH E. COOK AND DANIEL J. EGAN, Jr.

8.1 INTRODUCTION

The application of radon reduction strategies and approaches—known as radon mitigation—is a new, specialized field within the home-building industry, as well as the subject of much ongoing research and development activity. Because it is virtually impossible to completely free the indoor environment of radon, the goal of radon mitigation is to *reduce* radon in the indoor environment as much as possible. Using currently developed methods, it is possible to get substantial indoor radon reductions, often to the 4 pCi L^{-1} (0.02 WL, or 148 Bq m^{-3}) level recently suggested as guidance by the U.S. Environmental Protection Agency.

The major source of radon in most structures is radon-containing soil gas. It is believed that the basic mechanism that brings soil gas into a house is the pressure difference between the indoor and the outdoor environment. Pressure inside closed houses is generally slightly lower than the outdoor pressure. This pressure difference is increased in winter, as a "stack effect" (as in "smoke stack") is created by the continual rising of heated air. At the lower levels of the house—including places where the house contacts the soil—pressure is lowered, creating a "sucking" action that draws the radon-containing soil gas into the house. At the higher levels the heated air "exfiltrates" around the upper stories

---

JUDITH E. COOK • Office of Research and Development, U.S. Environmental Protection Agency, Research Triangle Park, North Carolina.   DANIEL J. EGAN, Jr. • Office of Radiation Programs (ANR-460), U.S. Environmental Protection Agency, Washington, D.C.

and the roof. In addition to the stack effect, wind effects, use of appliances that consume indoor air, and imbalanced air flow through the house contribute to house "depressurization." Because the degree of depressurization varies with weather and household activities, radon concentrations will vary within a structure over time. If indoor air cannot move freely from one area to another, there can be spatial variations in radon concentrations as well.[1]

This chapter emphasizes techniques for mitigating the naturally occurring radon that enters houses. It also emphasizes techniques for mitigating radon in existing houses as opposed to new houses. These emphases reflect the focus of research to date. The techniques have not been applied to office or other public-use buildings, because it is generally believed that public-use buildings are safer. People usually spend less time in them; they are better ventilated; and they are multistory. Further surveys of building codes, commercial heating/ventilating/cooling systems performance, and radon levels are needed to confirm this supposition.

8.1.1 Overview of Mitigation Techniques[1]

There are four possible means of reducing radon in indoor air: the indoor concentration can be ventilated, the entry of radon can be reduced, the source of the radon can be removed, or the radon itself (and its decay products) can be removed.

8.1.1.1 Ventilating Indoor Concentrations

Ventilation simply means increasing the flow of outdoor air into the house, which dilutes and replaces the radon-laden indoor air. Ventilation is a simple method to use, because all that is required is that windows and vents on all sides of an area be opened equally. In addition, opening windows and vents neutralizes indoor depressurization, which reduces substantially the pressure-driven flow of radon into the house. Unfortunately, ventilation is not practical in extreme weather or in areas where houses are susceptible to unauthorized entry.

8.1.1.2 Reducing Radon Entry

Another means of reducing radon in indoor air is to reduce its entry. All cracks and openings in a house's structure are pathways for

MITIGATION

the entry of radon-containing soil gas, so it follows that sealing them will reduce radon entry. More effective reduction of radon soil-gas entry can likely be achieved by using mechanical systems or natural phenomena to draw, or force, the soil gas away from the house's lowest level. Radon can be removed from incoming water either by using aeration or granular activated carbon. Radon removed by aeration can be vented to the outside. Carbon adsorbs radon and its decay products.

8.1.1.3 Removing the Radon Source

Removing the radon source is a special case involving the waste products of uranium production, mill tailings, which were used in the construction of some houses in the past. The process of separating uranium from the waste had the effect of concentrating the radium content of the mill tailings, and they became significant sources of indoor radon in houses where they had been used as "gravel" under the slabs. In these cases, the slabs of the houses were torn out, the mill tailings excavated, and new slabs poured.

8.1.1.4 Removing Radon and Radon Decay Products from Indoor Air

In theory, it should be possible to pass indoor air through some type of filter to which the radon and its decay products would adhere, thereby removing them from the air. There are several types of air cleaners now on the market for particle removal, and the radon decay products are in particle form. However, the issue of whether these devices can actually reduce the risk of lung cancer, as discussed in Chapters 5 and 6, is complex.

8.1.2 Short-Term Mitigation Actions

When extremely high concentrations of radon are found in houses, local radiation officials may urge temporary relocation of the occupants. In these instances, further exposure, even for another week, may be judged untenable.

Fortunately, most elevated indoor radon concentrations are not in this range, and homeowners can use a number of fairly simple interim measures to reduce their exposure, while they contemplate permanent radon mitigation measures. Occupants should stop smoking, especially in the house, and visitors should be discouraged from smoking. Occupants should reduce the amount of time they spend in parts of the house where radon concentrations are highest, for example, the basement. If possible, windows should be opened on all sides of the house to increase ventilation and reduce depressurization. Fans can be used to increase air flow through the house, especially through the basement (they should always blow *into* the house). If the house is built over a crawl space, all crawl space vents should be fully opened and remain so throughout the year.[2] Of course, exposed pipes would have to be protected from freezing. If there are obvious radon entry points that can be closed easily, these should be closed at once. For example, cover a dirt sump in a basement or close off from the rest of the house and discontinue use of a dirt basement. Avoid further depressurizing the house by opening windows when depressurizing appliances, such as furnaces, clothes dryers, woodstoves, and space heaters, are in use, and discontinue the use of ceiling fans.[1]

8.1.3 Limitations of Radon Mitigation[1]

Radon mitigation is a developing field and not yet as exact a science as, for instance, designing a heating system. Much has been learned recently about designing mitigation systems, but there are still a number of areas where trial and error are necessary. For instance, many design and construction characteristics of a house that influence the performance of radon mitigation systems are hidden from view. In these cases, some modifications to the mitigation systems may be necessary after they are installed and tested.

8.2 EVALUATION OF SOURCES AND ENTRY MECHANISMS

8.2.1 House Construction Types

Aside from the uranium/radium content and the permeability of the soil where the house is built, house type is one of the major parameters that influences the degree of radon entry. Most houses in this country are variations and combinations of three basic house types: basement houses, houses on crawl space, and houses built on concrete slabs. Each presents a different mitigation problem.

Basement houses have an excavated room, or rooms, below ground level that serve the dual function of being the house's foundation and living or storage space for the occupants. The excavated space can be constructed with foundation walls of different materials, e.g., concrete blocks or cinder blocks, poured concrete, and sometimes field stone or treated wood. They can have either a concrete or a dirt floor. Being below grade (below ground level), a basement offers a wide floor and wall surface area with many (sometimes concealed) cracks, gaps, and openings through which radon soil gas can be drawn into the house. In addition, if the basement has block walls, radon soil gas can be drawn through their porous surface into the house.

A crawl space house is built on a low foundation, partially above and partially below grade. All living areas are above grade, and the crawl space below them generally accommodates heating/air-conditioning ducts and pipes in a space sufficient for a person to crawl about to service them. Crawl spaces usually offer the simplest means of radon mitigation, since they can effectively serve as a ventilated, neutral-pressure buffer between house and soil. To ventilate a crawl space year-round may require that the sub-flooring and the water pipes be insulated. Crawl spaces can be less amenable to mitigation if they open into the living space of the home, that is, if they are actually "mini-basements."

A slab house uses a concrete slab as the base of the house with living spaces constructed directly over it. Some houses are slab-on-grade, while others are built on slabs below grade. As with basement houses, these slabs offer a wide surface area with many (sometimes concealed) cracks, gaps, and openings through which radon soil gas can be drawn into the house.

8.2.2 Possible Entry Points

Any opening that somehow comes in contact with the soil surrounding and below a house can be an entry point for radon. For example, radon-containing soil gas can be drawn into a house through the sump pits that exist in many basement homes. When water rises high enough in the ground to enter the house's foundation, the sump pump in the pit automatically begins to function, pumping water out of the network of drainpipes that encircle and protect the foundation. However, when the drainpipes are dry, radon soil gas can be drawn through their tiny perforations into the drain system, and through the sump pit into the basement.[1]

Common entry points for radon, shown in Figure 5.4, are cracks in basement walls and floors, seams where basement walls and floors meet, seams in concrete floors poured intentionally as expansion joints, floor drains connected to the soil with no trap to prevent gas entry, holes through basement walls or floors for pipes and the like, the porous surface of concrete blocks, and concealed openings in structures associated with masonry chimneys or fireplaces.[1]

Another source of radon in some houses has been water from private wells (or possibly from small municipal well systems). If the radon in the water is in sufficiently high concentrations, its release into the indoor air through showering, clothes washing, etc., can contribute to airborne indoor radon.[3] (See Chapter 5.)

8.2.3 Possible Depressurization Mechanisms[1]

Since the basic mechanism that brings radon soil gas into a house is depressurization, activities that increase depressurization must be minimized, especially in winter. Homeowners can unwittingly increase the depressurization of the house by using appliances that "consume" indoor air. A fire in a fireplace, for instance, consumes air.

8.2.4 Water and Building Materials

Radon can enter the house in dissolved form in the water supply and be released into the indoor air by such activities as bathing and clothes washing. The concentration in water must be very high to influence the indoor air concentration. A commonly used rule of thumb is that 10,000 pCi L^{-1} (370,000 Bq m^{-3}) of radon in water will produce 1 pCi L^{-1} (37 Bq m^{-3}) when released into the air. Researchers have, however, found houses where the major cause of elevated indoor radon concentrations was radon in the drinking water supply. In these cases, systems were installed to remove the radon before it entered the house. Radon in water is a problem mainly for homeowners with private wells, and occasionally for small municipal groundwater supplies. Radon that

MITIGATION

may be dissolved in larger municipal supplies is usually released into the air before it reaches the consumer.[3]

Radon can also be introduced into the house when radium is present in building materials. The solution, of course, is to avoid the use of such materials. Granite can be a source of radon, but it is important to note that the incidence of naturally occurring radium in building materials is almost always a minor problem as compared to the incidence of radon entering a house in soil gas. The most widely publicized example of radon contamination from building materials was in uranium mining areas where builders used the waste products of uranium mining, mill tailings, as a substitute for gravel under the concrete slabs of some houses.

8.3 OPTIONS FOR RADON REDUCTION

8.3.1 Ventilation—Diluting and Replacing Radon-Laden Indoor Air

Some degree of house ventilation occurs continually, even in closed houses, as the lower pressure inside draws outside air in through any available pathway. This continual replacement of indoor air with outdoor air, however small, is referred to as "air change," and the rate at which this replacement occurs is measured in "air changes per hour" (abbreviated ach). This is a measure of how long it takes to completely replace all the air in a house with outside air. In the average American house, one complete air change occurs approximately every one to two hours (i.e., an air change rate of 1 to 0.5 per hour). Newer, more energy-efficient houses could have as little as 0.1 air change per hour. Older, draftier houses could have as many as 2 air changes per hour.[1]

One of the purposes of the ventilation techniques discussed in this section is to increase the number of air changes per hour, which increases the dilution and replacement of radon-laden indoor air. The other purpose—which may turn out to be primary with further research—is to neutralize the pressure difference between indoors and outdoors. Since the exchange rate is much less in energy-efficient houses, they are likely to benefit more from increased ventilation than are houses that already have higher exchange rates. In other words, it may be realistic to increase a ventilation rate of 0.25 ach to 0.50 ach, while it may not be realistic to increase a ventilation rate of 1.5 ach to 3 ach (the house could be uncomfortably cold), although each of these increases would give the same percent reduction in radon concentration. Based on dilution considerations and excluding the effects of pressure neutralization, each doubling of the ventilation rate would reduce radon concentrations by a factor of two. For example, an initial indoor radon concentration of 20 pCi L^{-1} (0.1 WL, or 740 Bq m^{-3}) in a house with a ventilation rate of 0.25 ach can be reduced to about 2.5 pCi L^{-1} (0.01 WL, or 92.5 Bq m^{-3}) by increasing the ventilation rate to 2 ach. Actual reductions would probably be even higher as a result of the pressure neutralizing effect of opening windows and vents.[1]

Clearly, the limiting factors in increasing ventilation rates are human comfort and energy expense. Generally, temperatures between 68° and 78°F (20° and 25°C) and relative humidities between 30% and 70% are comfortable to most people.[4] National data on temperatures[5] indicate that, on the average, there are up to four months each year when ventilation can be used without discomfort to the occupants of the house and without increasing heating or cooling costs. Beyond these four months, occupants may have to live with some discomfort and some increased heating/cooling expense.[1]

Since the most important area to ventilate is the area nearest the soil—where the radon-containing soil gas is entering the house—one option for applying ventilation, if it is feasible, is to close off and not use a basement that is being ventilated during extreme weather. The human comfort problem is thereby avoided, as is the energy increase in some measure, although leakage from the basement into the living areas will still increase energy costs by about 20%. The only requirements in this instance are that pipes in the basement be protected from freezing and that the basement be abandoned[1]—not always an attractive option.

Three ventilation alternatives are discussed in the following sections, with information on their expense and their ability to reduce radon levels.

8.3.1.1 Natural Ventilation

The easiest form of ventilation to use is natural ventilation. All that is required is that windows, or doors, and crawl space vents be opened equally on all sides of an area. Opening windows and vents on more than one side of the house is important in order to ensure that the pressure indoors and out remains neutral. If, for example, windows were opened only on the downwind side of a house, it would depressurize further, increasing the flow of soil-gas-borne radon into the house.[1]

Because natural ventilation is driven by winds and pressure and temperature differences between the indoor environment and the outdoor environment, it cannot be well controlled. Therefore, one can have only moderate confidence that natural ventilation will constantly keep radon concentrations in an acceptable range.[1] Natural ventilation would not be effective for radon concentrations above about 40 pCi L^{-1} (0.2 WL, or 1480 Bq m^{-3})[3] if dilution were the only mechanism at work. However, the pressure-neutralizing effect of opening windows and vents (mentioned in Section 8.3.1) could likely make natural ventilation effective on even higher concentrations.[1] If natural ventilation is used year-round in most of the country, it will increase heating/cooling costs up to three times normal in a house with an initial exchange rate of 0.25 ach.[1]

8.3.1.2 Forced Air Ventilation

Forced air ventilation, too, is relatively simple in that fans are used to *force* air through an area, rather than relying on prevailing winds to do this.[1] The advantage of forced air ventilation is that it enables the *control* of air flow through an area. Thus, one can have increased confidence that it will keep radon concentrations in an acceptable range.[1] Forced air ventilation would not be effective for radon concentrations above 40 pCi L^{-1} (0.2 WL, or 1480 Bq m^{-3}),[3] if dilution were the only mechanism at work. However, the pressure-neutralizing effect of opening windows and vents (mentioned in Section 8.3.1) could likely make forced air ventilation effective on even higher radon concentrations.[1]

If forced air ventilation is used year-round in most of the country, it will increase heating/cooling costs up to three times normal in a house with an initial exchange rate of 0.25 ach. In addition, operating fans year-round would cost about $100. This estimate does not include more elaborate installations with new wiring, duct work, dampers, filters, and the like.[1]

For both natural and forced air ventilation, balanced air flow is of utmost importance. Mistakes can mean the difference between reduced radon concentrations, no reduction at all, or—in some cases—increased entry of radon soil gas into the house. Opening only upstairs windows or using an attic exhaust fan could create negative pressure on the basement. Hence, the primary area to ventilate is always the basement, or lowest level, and fans are always placed so that they blow *into* an area.[1]

8.3.1.3 Forced Air Ventilation with Heat Recovery

The use of heat recovery ventilators enables the use of forced air ventilation without the complete loss of all heated, or cooled, air as the air exchange rate is increased. Heat recovery ventilators, also called air-to-air heat exchangers, use a heat transfer surface to warm—or cool—incoming air. The heat transfer surface is heated or cooled by the air being exhausted from the house.[1]

The heat recovery ventilator offers reasonable potential for treating houses with radon concentrations up to about 40 pCi L^{-1} (0.2 WL, or 1480 Bq m^{-3}).[3] By making use of the heat or cooling in the outgoing air, heat recovery ventilators reduce considerably the amount of extra energy needed to heat or cool a ventilated area. Typically, whole-house heating/cooling costs are only about one and one-half times normal, or less. These devices range in cost from $400 to $1500, and are capable of energy recoveries up to 70%.[1]

8.3.2 Preventing Radon Entry

8.3.2.1 Reducing Entry Points[1]

It is possible to reduce radon entry to some degree simply by sealing all entry points that can be found. In most cases, however, sealing by itself will probably not be sufficient, because radon entry routes are too numerous and many of them are concealed. Sealing is usually recommended as a first step in any radon mitigation strategy, because it is something that most homeowners can do themselves. It can't hurt, and might help, especially if there are some big holes. It may ultimately be needed anyway to make other mitigation systems, like sub-slab ventilation, work effectively.

The first sealing step would be to seal the largest and most obvious radon entry routes, including dirt basements, sump pits, and floor drains connecting to the soil without traps. The best solution for dirt basements is to excavate the fill dirt in the area and replace it with concrete. Sump pits can be capped with an impermeable covering, like sheet metal, sealed at all joints, and a fan used to draw radon-laden air from under the cap and exhaust it to the outside. For floor drains to soil, traps can be added, or, if necessary, removable stoppers used to prevent radon soil gas entry. Holes in the top row of concrete blocks and large holes in walls and floors should also be among the first openings sealed.

Concrete blocks have hollow spaces inside that connect and form a network inside block walls. Radon can be drawn in through the network of spaces and enter the house from the openings in the top row of blocks. To seal these, one may stuff crumpled newspaper into the hollow spaces and concrete over it if the top blocks are easily accessible. If they are not easily accessible, then a urethane foam can be extruded into the hollow spaces through a hose and nozzle assembly.

In addition to large, obvious openings, a conscientious homeowner could also seal smaller openings, although the impact will be much less. These include openings where pipes and ducts enter the basement, mortar joint cracks between blocks, gaps between block and brickwork surrounding basement fireplaces, and pores in the surface of concrete blocks.

Cracks and utility openings can be sealed by first enlarging them and then filling them with caulk, grout, or sealant. Joints between the wall and the slab can be enlarged, filled with sealant, and then covered with mortar.

Epoxy sealants or waterproof paints are used to reduce the flow of radon through porous walls, especially block walls. Meticulous surface preparation is required to ensure that these coatings will adhere to the surface.

The effect of sealing on the radon concentration in specific houses is unpredictable, because of wide variations in the strength of the source material in the soil, because each house is different, and because the unseen cracks and openings in a house's foundation may be letting in more radon than those that can be found and sealed. Also, houses continue to settle over time and this settling can create new pathways for radon entry. In addition, sealing jobs cannot be expected to provide a constant barrier over time; house settling and other wear and tear can reopen cracks and gaps. Thus, one can have only low to moderate confidence that sealing will effectively control indoor radon concentrations. Sealing major sources, like dirt basements and sump pits, will have a more marked effect on indoor concentrations than will the sealing of small cracks and openings.

Sealing major exposed radon sources within the house structure can range in cost from as low as $100 to several thousand dollars, as, for example, when concreting a dirt basement. Most sealing of cracks and small openings can be accomplished for under $100.

8.3.2.2 Venting Radon from Soil Surrounding the House

There are three techniques by which radon soil gas can be vented from the soil surrounding a house. The basic mechanism in all three is to draw a suction greater than the suction created by the depressurization of the house. This reverses the predominant flow of gas so that it flows away from the house. To understand these techniques, it is first necessary to explain briefly a few basics of house construction.

Houses of either the basement or crawl space type begin below grade with footings of poured concrete. Trenches somewhat bigger than the planned walls are dug, and concrete is poured into them to provide a firm "footing" for the foundation walls that they will support. Block foundation walls have hollow spaces inside. Each successive course of blocks is laid so that the center of the top block covers the ends of the two blocks below. This "ties" the wall of blocks together, and also creates a network of hollow spaces inside the wall that connects both vertically and horizontally. Foundation walls can also be made of field stone or timbers, but in all these cases the foundation walls are usually built over footings of concrete. Mortar is used to attach the bottom of the foundation walls to the footings.[1]

It is believed that a significant amount of radon soil gas may enter the house in the area of the footings, that is, around the mortar that attaches the bottom of the foundation walls to the footings, around the mortar that holds the blocks or stones together, and through the porous exterior surface of blocks in the foundation walls.[1]

In many basement houses, construction is also characterized by a concrete slab which forms the floor of the excavated room. Basement houses and slab-below-grade and slab-on-grade houses have this slab-over-soil construction in common, and in this case it is believed that radon soil gas enters through utility perforations, cracks, spaces, and joints in the floor, as well as through sumps and floor drains. Slabs are usually poured over aggregate, most often crushed rock, to give them a firm base.[1]

The sections that follow describe drain-tile soil ventilation, sub-slab ventilation, and wall ventilation, all of which are designed around the unique construction characteristics of different types of houses to draw radon away from the house's foundation.

8.3.2.2a *Drain-Tile Soil Ventilation.*

Drain-tile soil ventilation is a good radon reduction option for a house with a drain-tile system completely encircling it. As described in Section 8.2.2, drain-tile systems encircle the foundations of many houses to protect them from water.

If such a system loops the entire house, is completely intact with no tiles crushed, silted in, or missing, and has the tiles attached to each other, rather than simply touching each other, it offers a relatively simple and cheap ready-made means of drawing suction on the soil surrounding the house's foundation. This option is the most esthetically pleasing of the soil ventilation alternatives, and can be very effective if the full loop gives good distribution of the suction.[1] Of course, the only way one can be sure that a drain-tile system meets the above criteria is to have a completely new system installed, and this would be quite expensive. Adding drain-tile soil ventilation to an existing drain-tile system is fairly inexpensive, however, so it is generally cost-effective to try this method.[1]

Drain-tile soil ventilation uses a fan to draw suction on the network of drain tiles. This suction draws radon soil gas into the tile network, thereby preventing it from entering the house in the vicinity of the footings. Since the drain-tile network encircles the house at the base of the foundation, the suction in the system can also draw soil gas from under the house's slab (if it has one) as well.[1] Adding such a system to an existing drain-tile network is fairly simple. In the case of a drain-tile system that connects to a sump pump in the basement, the entire sump pit area is capped with an impermeable material and sealed, and a fan is used to draw suction on the sump pit and the drain-tile system attached to it.[1]

In the case of a drain-tile system that drains to a location remote from the house—to an above-grade discharge or a dry well—the drain line is located and cut, and a fan, trap, and riser assembly is added (see Figure 8.1). Thus, water can still drain from the drain tiles, and the fan can draw suction on the drain tiles without drawing in air from the discharge area.[1]

In both of the above methods, the fan is located outside and must be enclosed in a protective housing to protect it from weather and debris and to protect animals from harm. Since the radon concentration in the air exiting the fan can be very high, homeowners are cautioned to locate it either in an area remote from the house, or at a safe height. The fan should also be inaccessible to children.[1]

Drain-tile soil ventilation can be used for reducing any radon concentration, although for concentrations above 200 pCi L^{-1} (1.0 WL, or 7400 Bq m^{-3}),[3] it may not be able to get below 4 pCi L^{-1} (148 Bq m^{-3}). Since drain-tile soil ventilation functions by drawing soil gas away from the house's footings, it might not work as effectively if there are interior walls in the basement sitting on footings of their own. In this latter case, the fan might not always draw sufficient suction to keep soil

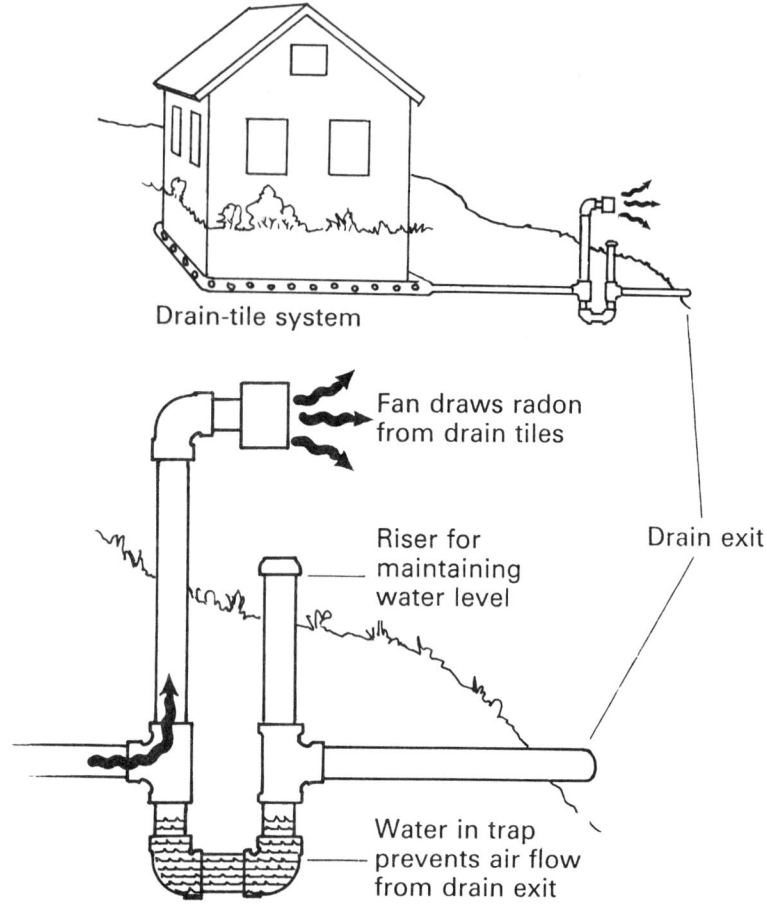

FIG. 8.1. Drain-tile soil ventilation system, draining to remote discharge area.[3]

gas from entering around the footings of these walls. It would be extremely unusual for such interior walls to have drain tiles of their own. Even in this latter case, drain-tile soil ventilation may be made to work adequately by using a higher-powered fan.[1]

If a homeowner were to have a contractor install a new drain-tile system and include in the installation the fan, trap, and riser, the entire system would cost about $1200, assuming that no unusual problems were encountered. A do-it-yourself installation would probably require about $300 in materials.[1] Power to operate the fan in the drain-tile system year-round would cost about $25. Since the suction would draw some warmed or cooled air out of the house, heating or cooling costs could be expected to increase by about $125 per year.[1]

8.3.2.2b *Sub-Slab Ventilation.*[1] For basement houses and slab-on-grade houses, ventilation of the area below the slab may be used to draw accumulated radon soil gas out of the aggregate, or soil, beneath the slab.

Drawing suction on the sub-slab area can be accomplished in several ways: (1) individual pipes can be inserted into the slab and a fan used to draw suction (see Figure 8.2); (2) if drainpipes exist below the slab, these can be used with a fan to draw suction on the sub-slab area; (3) a network of perforated pipes can be laid, the slab poured over them,

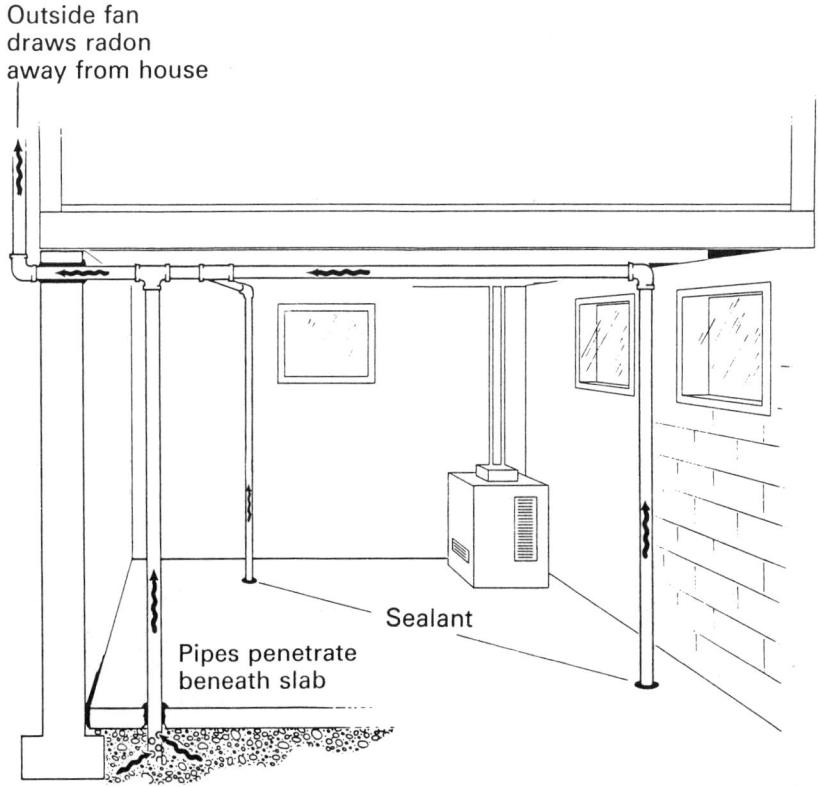

FIG. 8.2. Individual pipe variation of sub-slab ventilation.[3]

and a fan used to draw suction; or (4) an extensive network of perforated pipes can be laid and attached to a stack and natural phenomena possibly used to draw radon out of the sub-slab area.

For existing houses, the most practical solutions are the individual pipe method and the drainpipe method. In the individual pipe method, several pipes are inserted vertically into the aggregate through holes drilled in the slab. The number of pipes needed is dictated by the permeability under the slab and the size of the slab. The inserted pipes are connected to each other by horizontal pipes, usually running around the ceiling and connected to a fan. The fan draws suction on the entire sub-slab area through the inserted pipes and exhausts the radon soil gas outside the house, usually at roof level. Another possibility for existing houses is a variation on drain-tile soil ventilation. The perforated drainpipes that were laid for water drainage under some slabs during construction usually drain into a sump within the house's footings. By using a fan to draw suction on the sump and the drainpipes, the sub-slab area is ventilated.

Both these methods rely on a good layer of aggregate or a permeable soil below the slab to allow the effects of a few suction points or the drainpipes to radiate to the entire slab. Permeability can be tested fairly easily before installation is begun by drilling a hole into the slab, inserting a pipe into it, and attaching a fan to it temporarily. With the fan operating, smoke tracer tests at joints and cracks remote from the fan (see Section 8.4) will give a good indication of how much air can move through the aggregate or soil. If permeability is poor, more suction points may be needed, or a network of perforated pipes will have to be laid.

In the perforated pipe network method, an extensive layer of perforated pipes is laid horizontally, a fan attached to it, and the slab poured over the pipes. Because of the expense of tearing out an existing slab and replacing it, this method is best suited to new construction. It has also been used in existing houses when the slab had to be torn out anyway, because there was contaminated material under it (e.g., uranium mill tailings) or because there were structural problems.

In some houses, an extensive sub-slab piping network may provide adequate ventilation in a passive mode, without a fan. By connecting the piping network to a stack that exhausts at roof level, suction is created through natural thermal effects inside the stack and a reduced pressure at roofline caused by wind movement. If the flow resistance through the aggregate is low, the weak suction created by this stack effect may be sufficient to ventilate the sub-slab. Passive ventilation appears to work only in cold weather.

MITIGATION

Another variation of sub-slab ventilation that is being tested is *forcing air* into the sub-slab area, rather than drawing suction on it, that is, pressurizing rather than depressurizing. This would have the effect of *pushing* radon soil gas away from the slab and foundation.

None of the methods described above will function effectively without sealing openings in the slab. Without sealing, house air could be drawn into the system, overwhelming its suction power. Holes in the slab, large seams (cold joints), openings around utility penetrations, large settling cracks, and large joints where the wall and floor meet must be sealed with mortar. Small openings can be sealed with asphalt, caulk, or similar sealants.

Sub-slab ventilation is one of the most effective radon reduction methods known at the present time and can be used for any radon concentration. Because of the cost, however, homeowners might elect to use less expensive methods for radon concentrations below 40 pCi L^{-1} (0.2 WL, or 1480 Bq m^{-3}). The effectiveness of sub-slab ventilation may be reduced by the existence of block walls in a basement, because it is difficult to draw enough suction to keep radon soil gas from entering through the walls. Closing the hollow spaces in the top course of blocks, as well as other gaps and openings in the walls, will improve sub-slab ventilation of block-wall houses considerably, and using high-powered fans can overcome much of the leakage into a system.

Having a contractor do an uncomplicated installation of an individual pipe sub-slab system could cost between $1000 and $2500. A similar installation of a piping network, including the labor to cut channels into an existing slab to lay the pipe, could cost between $2000 and $7500. A sub-slab ventilation system would place about the same annual energy load on a house as wall ventilation and drain-tile soil ventilation: power to operate a fan year-round could be about $25 and, assuming the fan draws some house air into the system, heating/cooling costs could increase about $125.

8.3.2.2c *Wall Ventilation.*[1] If a basement has block walls, another option for reducing radon soil-gas entry is to draw suction on the walls themselves. The same network of hollow spaces that enables radon to *enter* the house can be used to draw radon *away* from the house.

The basic approach in wall ventilation is to attach a fan to the network of spaces inside each wall, draw radon soil gas out of the walls, and exhaust it to the outside. There are two variations of wall ventilation: (1) the single-point pipe method and (2) the baseboard method (see Figure 8.3). For either of these methods to work effectively, *all* walls must be treated, including any interior walls that penetrate the concrete

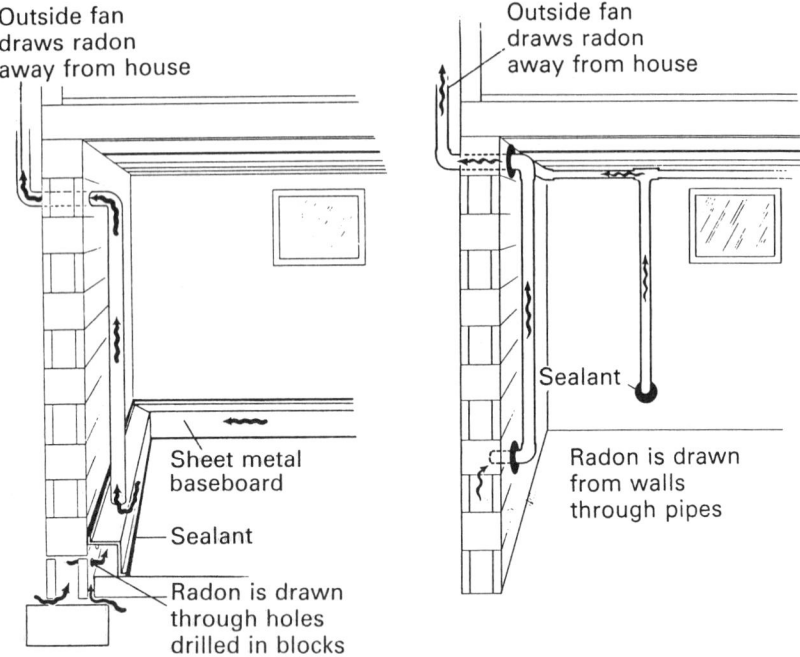

FIG. 8.3 Two variations of wall ventilation: the baseboard method; the single-point pipe method.[3]

floor. Both methods also require that all large openings in the walls be closed. Otherwise, the house air being drawn in through these openings will simply overwhelm the system and it will have very little suction power left for radon soil-gas control.

Closing all large openings in the walls means the same sealing of the hollow spaces in the top row of blocks as was described in Section 8.3.2.1. Other large openings must be closed. These include space around pipes where they enter the basement and any other visible holes and gaps.

With large openings effectively closed, the wall ventilation system can be installed. Usually, the choice between the single-point pipe method and the baseboard method is dictated by the conditions in the basement, by the expense involved, and by the relative importance of the usability and appearance of the area.

The single-point pipe method is the cheapest. It involves drilling one hole into a hollow space in each wall and inserting a pipe into each drilled hole. The pipes usually lead up to connecting pipes encircling the entire perimeter of the inside of the basement. At the end of the pipe network is a fan that draws suction on the walls. Many variations of this method are possible, depending on the unique requirements of the basement. For instance, the pipes could be inserted into the network of hollow spaces from the outside of the house, and the fan could be located inside, or outside, the house.

In the more expensive baseboard method, holes are drilled into the hollow spaces around the entire perimeter of the walls near the floor. Then a two-sided baseboard "duct" is constructed of sheet metal, or some other suitable material, and attached with sealant and screws to the wall above the drilled holes and to the floor in front of them. In this way, all holes—and the joint where the wall meets the floor—are completely covered by the baseboard duct, and the duct system is attached to a fan. Because there are holes all along the perimeter of the basement, a more uniform suction is drawn on the walls than with the single-point pipe method.

If wall openings are not sufficiently closed, the fan used to draw suction or to pressurize the walls cannot work effectively. Among possible openings that may require closing are the hollow spaces in the top row of blocks, the space between brick veneer and the exterior wall, and spaces between fireplace structures and the walls of the basement. A higher-powered fan may help to increase the efficiency of a wall ventilation system. In some cases, wall ventilation may simply not be adequate to keep radon-containing soil gas from entering through openings in the slab.

Wall ventilation is usually added as a supplement to sub-slab ventilation if sub-slab ventilation does not function effectively alone. To have a contractor do an uncomplicated installation of a single-point pipe system in an unfinished basement would probably cost about $2500. A similarly uncomplicated contractor installation of a baseboard system in an unfinished basement would probably cost about $5000. An uncomplicated installation would be one in which the hollow spaces in the top course of blocks are easy to access and close and in which there are few appliances or other obstacles around which the pipes must be installed.

The cost of a do-it-yourself installation, although not generally recommended, could be as little as $100 to $500 for pipes, sheet metal, fans, and miscellaneous supplies, depending on the number of fans required and the size of the basement.

A wall ventilation system would place about the same annual energy load on a house as drain-tile soil ventilation. Operating a fan year-round could cost about $25 and, assuming the fan draws some house air into the system, heating/cooling costs could increase about $125.

8.3.2.3 Reducing Pressure Differentials

Because pressure-driven flows are believed to be the basic mechanism by which radon soil gas enters a structure, it is important to minimize any *additional* depressurization of the house, especially in winter and in the areas where radon soil gas enters—basements or rooms directly over the soil.[1] It appears that the major sources of depressurization inside the house are combustion appliances which *consume* indoor air, further lowering indoor air pressure, and thermal bypasses which facilitate the stack effect. Among these are furnaces, water heaters, clothes dryers, woodstoves, fireplaces, and space heaters. Of these, furnaces and water heaters probably depressurize the house the most, because they generally operate a much greater percentage of the time than do any of the others.[1] The American Society of Heating, Refrigerating, and Air-Conditioning Engineers (ASHRAE) has recommended since 1981 that direct outside supplies of "makeup" air be provided for combustion appliances, because they believe this is necessary for effective and controlled ventilation and acceptable indoor air quality.[6] By supplying each air-consuming appliance with its own air supply through separate ductwork to the outside, further depressurization of the house, and increased flow of radon soil gas into the house, is prevented. In the case of a fireplace, depressurization can be prevented simply by cracking a window while the fireplace is in use.[1]

Another possible source of depressurization inside the house is local exhaust fans, e.g., ceiling fans, that are used intermittently. They are not as important as combustion appliances, but they do draw suction on the interior of the house when they are operating. Thus, it would be advisable to keep use of such fans to a minimum, especially in winter. Also, if portable fans are used to ventilate the house, always ensure that fans blow *into*, not out of, the house.[1]

It is difficult to generalize about the impact on indoor radon concentrations of preventing appliance depressurization. There are too many variables. Among these variables are the number and operating conditions of the appliances, the type of house in which they are installed, the strength of the radon source material, and others. Best estimates are that the average annual radon reduction benefit may be between 0% and 50%. It is impossible to estimate the cost of installing separate ductwork to an unknown array of appliances.[1]

8.3.2.4 Removing Radon from Water

The two methods available for removing radon from drinking water involve the use of aeration and granular activated carbon. Both methods can be used in the home or at the source of the water.

When water containing radon is exposed to air, some of the radon escapes. Thus, aeration is a viable means of removing radon from water. A diffused aeration tank typically can remove more than 95% of the radon, and spray aeration has achieved efficiencies of 93%.[7] Packed tower aeration, which has been shown to be effective in removing volatile compounds from water, also appears to have potential for removing radon from water, although it has yet to be tested at pilot- or full-scale.

Granular activated carbon (GAC) has been used to adsorb noble gases such as radon. Efficiencies as high as 96% have been reported.[7] Because of its short half-life (3.8 d), much of the radon decays on the GAC bed before breakthrough.

An aeration system for an average house would cost about $1000. Annual operating costs would be about $80. A granular activated carbon system for an average house would cost between $500 and $1500 with annual operating costs of about $20 to $40. The disadvantage of the granular activated carbon system is that the occupants of the house are exposed to the radioactive material adsorbed on the GAC.

8.3.3 Air Cleaning

Another approach to reducing the risks of radon is to *remove* radon and its decay products from the indoor air. There are various types of devices on the market that can remove particulates from the air, and these devices can remove radon decay products that have attached to these particulates. However, it is unclear whether these air cleaners can effectively remove radon itself from the air.

This uncertainty exists because there are insufficient data to enable precise description of the ways in which radon and its decay products may cause lung cancer. It is generally believed that the most dangerous situation involves inhaling decay products attached to relatively small particulates, which are more likely to deposit in the deepest, most sensitive parts of the lung. Air cleaners are thought to be more efficient at removing larger particulates than smaller ones. If this is so, then the radon still in the air after it has passed through an air cleaner will generate decay products that will attach to smaller particulates than would have been the case without air cleaning. Thus, although the risks may have been somewhat reduced by removing larger particulates and the decay products attached to them, the risks may also have been somewhat increased because the remaining decay products can become attached to particulates that will make them more dangerous.

The same issue arises even if the air cleaner is very effective at removing particulates of all sizes, because the remaining radon can then generate decay products that will not attach to any particulates, becoming "unattached decay products." Although data are inconclusive, some scientists believe that such unattached decay products may be even more effective at causing lung cancer than decay products attached to particulates. (See Chapter 7).

Thus, while air cleaners that are now available are likely to be effective at reducing the overall concentration of radon decay products in indoor air by reducing particulate concentrations, it appears that they may not be as effective in reducing the corresponding health risks. Additional research is needed to resolve these uncertainties, although the necessary scientific studies may be analytically complex. On the other hand, development of air cleaning systems that simultaneously remove radon itself, if this can be accomplished in a practical fashion, would offer significant benefits.

8.4 EVALUATION AND MAINTENANCE OF RADON MITIGATION SYSTEMS[1]

None of the methods described above can be installed and forgotten. They all must be evaluated periodically to ensure that they are still working, and they all require maintenance.

Evaluation of the radon mitigation methods described in Section 8.3.2.2 after they are installed is usually done by smoke tracer tests. The goal is to determine if the system is drawing sufficient suction to keep radon soil gas out of the house. With the mitigation system operating, a smoke generator, e.g., a smoke tube, is passed over the surface of walls, along the wall-to-floor joint, and over any other likely entry points. Smoke should be consistently drawn into the area being tested. In those places where it is not, there is reason to suspect that radon is still entering the house because of insufficient suction. Other

MITIGATION

diagnostics that would be conducted by the system's installer include flow and pressure measurements in vent pipes and pressure measurements under the slab and in block wall cores.

Maintenance of radon mitigation methods involves inspecting outside fans for damage or icing, periodic oiling of fans, checking seals where fans are attached to pipes, checking seals over basement cracks, gaps, and openings, and checking traps to be sure they are still filled with water. In the case of natural and forced-air ventilation, maintenance would also involve periodic checking to ensure that all windows and vents remain uniformly open on all four sides of the area being ventilated. For heat recovery ventilation, it would be necessary to check periodically to ensure that there is a balanced flow of air into and out of the system.

8.5 DEVELOPING A MITIGATION STRATEGY

In the previous sections an array of mitigation methods together with their relative effectiveness and costs have been presented. These are only the raw material with which a homeowner would develop a mitigation strategy. A complete strategy would likely include the following steps:

1. Screening measurement to determine if radon levels are elevated.
2. Follow-up measurements to determine the extent of the problem.
3. Taking short-term measures to protect occupants while long-term mitigation measures are being decided upon.
4. Contacting local radiological health officials, environmental health officials, or an experienced radon mitigation contractor to seek guidance.
5. Contractor conducting house diagnostics to determine where radon is entering.
6. Contractor installing radon mitigation system.
7. Contractor conducting postmitigation measurements to evaluate the effectiveness of the installed system.
8. Contractor making any necessary modifications to the system.
9. Contractor again conducting postmitigation measurements.

10. Homeowner conducting periodic checks of the system to ensure that it continues to function effectively.

As outlined above, many homeowners will understandably need the assistance of radiation and radon mitigation experts in dealing with a suspected indoor radon problem. Assistance with radon measurements can be obtained from laboratories and businesses who routinely conduct radon measurements. The U.S. Environmental Protection Agency (USEPA) conducts a voluntary Radon Measurement Proficiency Program which allows firms to demonstrate their capabilities in measuring indoor radon. Lists of participating firms in various areas are available from USEPA regional offices.[2]

Assistance in making important decisions about how to proceed is available from the radiation health officials in most states. In some states this assistance is available from the state's environmental protection agency.

Some states are also conducting training courses for building contractors who wish to become proficient in radon mitigation. Information on which contractors have taken such training would also be available from the state radiological health office or environmental protection agency.

9

Risk Assessment and Policy

WILLIAM A. MILLS and DANIEL J. EGAN, Jr.

9.1 RISK ASSESSMENT (William A. Mills)

9.1.1 Introduction

In this chapter, an assessment of the risks of developing lung cancer from exposures to radon will be discussed, and an attempt made to place such risks in perspective by comparisons to the "natural" risk of developing lung cancer and to the risk of developing it from cigarette smoking.

Unlike many environmental problems, most exposures to radon are not brought about by technology or some other human activity. Radon is ubiquitous in our natural environment and its presence in the environment is older than life itself. Even human exposure to indoor radon is not new; early humans living in caves most certainly experienced "elevated" levels of "indoor" radon. Recognition of the possible health consequences associated with exposure to radon is also not new, as evidenced by the studies of underground miners referred to in Chapter 7. From such studies, various estimates of the number of lung cancer deaths per year in the United States that might be attributable to radon exposure have been made. The estimate published by the U.S. Environmental Protection Agency (USEPA) that up to 20,000 lung cancer deaths per year may be due to radon exposure is the number

---

WILLIAM A. MILLS • Oak Ridge Associated Universities, Washington, D.C. DANIEL J. EGAN, Jr. • Office of Radiation Programs (ANR-460), Environmental Protection Agency, Washington, D.C.

most often quoted.[1] The National Council on Radiation Protection and Measurements (NCRP) estimates that about 10,000 lung cancer deaths per year in the U.S. population may be due to radon exposure.[2]

Few data are available on the exposure and resulting risk of populations in indoor environments from which valid estimates of health risk can be derived. Thus, the use of risk coefficients (i.e., risk per unit exposure), derived from studies of underground miners, to estimate risk to members of the general population is subject to many uncertainties and assumptions. Differences in the treatment of these uncertainties and assumptions can result in a wide range of estimates. At present, the number of "true" radon exposure-induced cases included in estimates for large populations cannot be distinguished, primarily because the contribution of smoking to lung cancer risk is predominant and the risk from the "passive" smoke in indoor environments is not well known. Thus, estimates of lung cancer due to radon exposure in large populations are not in addition to those caused by cigarette smoke but include in the estimates those who may already be at considerable risk from other causative agents.

Risk assessment involves assuming a source of exposure, a set of exposure conditions, and a risk factor or coefficient, and then using these assumptions to estimate the risk to individuals and to large population groups, such as the total U.S. population. In this section on risk assessment, the model and estimates published by the NCRP are adopted.[2] (See Chapter 7.)

9.1.2 Source of Radon

Table 9.1 shows the global sources of atmospheric radon-222. These releases are very large compared to the concentrations in any given volume of air. However, because of the large volume of air that is available to dilute this release, the average value of the concentration of radon in air is about 0.1×10^{-12} curie per liter of air (0.1 pCi L^{-1} or 3.7 Bq m^{-3}).[3] Over some land areas the concentration may be ten times higher and over large accumulations of uranium milling wastes 100 times higher. In terms of concentration of radon progeny, an average value of 0.001 WL (21 nJ m^{-3}) is representative for outdoor radon concentration. (See Chapter 1 for definition of units used in this chapter.)

The major pathway for exposure of members of the general public is through exposure indoors, where on the average 70–80% of the

TABLE 9.1. Sources of Global Atmospheric Radon-222[a]

Source	Release (Ci y^{-1})[b]
Emanation from soil	2×10^9
Groundwater (potential)	5×10^8
Emanation from oceans	3×10^7
Phosphate residues	2×10^6
Uranium tailings piles	2×10^4
Coal residues	2×10^4
Natural gas	1×10^4
Coal combustion	9×10^2
Human exhalation	1×10^1

[a] Reference (2), Table 3.1.
[b] All quantities have been rounded to one significant figure.

time is spent. Because closed structures do not allow for extensive mixing of air, the concentrations of radon in buildings tend to be higher than outdoor concentrations. However, recently levels as high as 10 WL or more have been found. These indoor radon concentrations can vary widely from essentially the ambient air outdoor value to values that are a few thousand times higher. On the average, the level of indoor radon progeny is reported by the NCRP to be about 0.004 WL (83 nJ m^{-3}).[3] These higher exposure levels can be much larger than levels presently found in underground uranium mines, resulting in annual exposures that exceed the limits established for occupational conditions. The present U.S. occupational limit for underground miners is 4 WLM y^{-1}.[4]

9.1.3 Derivation of Estimates of Radon Risk

In Chapter 7 the evidence for the association between exposure to radon or radon progeny and an excess incidence of lung cancer is presented and analyzed. Both human epidemiological studies and experimental animal studies are considered in deriving estimates of risk to humans.

In reviewing the data on excess lung cancer attributable to radon exposure, the NCRP selected "a rounded annual rate of 10 lung cancers per million persons per year per WLM based on existing human evidence."[2] This annual rate is derived by dividing the excess number of lung cancers observed by the summation of the products of the number of collective person-years at risk times the cumulative exposure for the exposed population. The lifetime risk per unit of exposure is then estimated using known age variations in risk to be 1×10^{-4}–2×10^{-4} WLM^{-1} or an average value of 1.5×10^{-4} WLM^{-1} (4.3×10^{-2} J^{-1} h^{-1} m^3), which is similar to other estimates.[5,6]

Combining the average indoor air radon level of 0.004 WL with the average risk of 1.5×10^{-4} WLM^{-1} yields the following rough estimate for the average number of lung cancers per year in a continuously exposed population of 240×10^6 people:

$$0.004 \text{ WL} \times 50 \text{ WLM y}^{-1} \text{ WL}^{-1} \times 1.5 \times 10^{-4} \text{ WLM}^{-1}$$
$$\times\ 240 \times 10^6 \text{ people} = 7200 \text{ cases per year}$$

or about 7000 cases per year.

In its review, the NCRP found an excess lung cancer mortality in miners only above cumulative exposures of about 100 WLM (0.35 J h m^{-3}) and a decrease in the risk per unit exposure when cumulative exposures were greater than 1000 WLM (3.5 J h m^{-3}).[2] These findings are very important qualifications in assessing the potential health significance of radon exposures. They suggest that a direct proportionality between exposure and risk, for all levels of cumulative exposures, is not supported by the available data. Using a direct proportionality can overestimate risk at cumulative levels significantly below 100 WLM (0.35 J h m^{-3}) and above 1000 WLM (3.5 J h m^{-3}).

The lung cancer risk model derived by the NCRP is described as a "modified absolute risk model for lung cancer" with the following characteristics:

- Tumor incidence does not manifest itself until age 40, regardless of the age at exposure, and after age 40, a minimum single value for the latent interval of five years applies.
- Tumor rate is not uniform with time but is decreased from the time of exposure by an exponential factor with an effective half-time of 20 years.[2]

The half-time or exponential factor accounts for the loss of damaged lung stem cells through cell repair, cell death, or unspecified mechanisms. Figure 9.1 shows the differences in risk using this factor in the cases of single exposures occurring at age 20 and at age 45. The areas under the curves are proportional to the lifetime risk, showing smaller risk due to incorporating the exponential factor into the risk

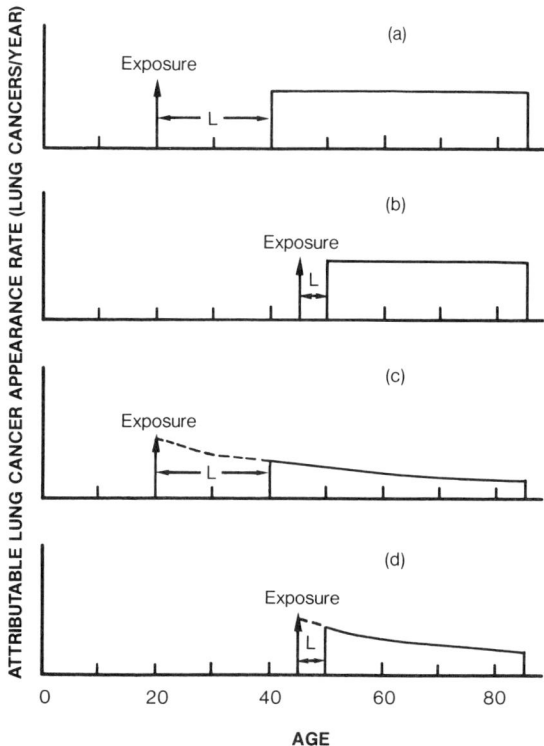

FIG. 9.1. Models for annual appearance of lung cancer attributable to a single exposure of radon progeny at age 20 (a and c) or age 45 (b and d). Models (a) and (b) are not corrected for cell loss and/or repair; in models (c) and (d) a 20-year half-life is introduced to accommodate cell loss and/or repair. L is the latent interval. [From Reference (2), Figure 10.1.]

model. This modified model of NCRP is consistent with the available data.

Using life tables to account for competing risk of death from other causes, lifetime risks of lung cancer have been estimated for various exposure conditions. These estimates are given in Table 9.2 for periods of exposure duration from 1 to 30 years and lifetime. Ages at first exposure begin at age 1 and cover up to age 70. Estimates are given for a radon progeny exposure rate of 1 WLM y^{-1}. Also shown are the estimated total number of lung cancers in a U.S. population of 100,000 persons when exposed for the various exposure periods.

These risk estimates are graphically shown in Figure 9.2 along with

TABLE 9.2. Lifetime Lung Cancer Risk per WLM $y^{-1 a,b}$

Exposure duration	Risk (multiplied by 1000) for exposure at age:					Lung cancers in a population of 10^5 persons[c]
	1	10	20	50	70	
1 year	0.064	0.091	0.13	0.17	0.07	13
5 years	0.34	0.50	0.69	0.84	0.28	66
10 years	0.77	1.1	1.5	1.4	0.38	130
30 years	3.4	4.8	5.5	2.5	0.38	380
Life	9.1	9.1	7.7	2.7	0.38	560

[a] Reference (2), Table 10.2.
[b] For radon progeny measured under environmental rather than underground mining conditions.
[c] For a population with age characteristics equal to that in the whole United States in 1975.

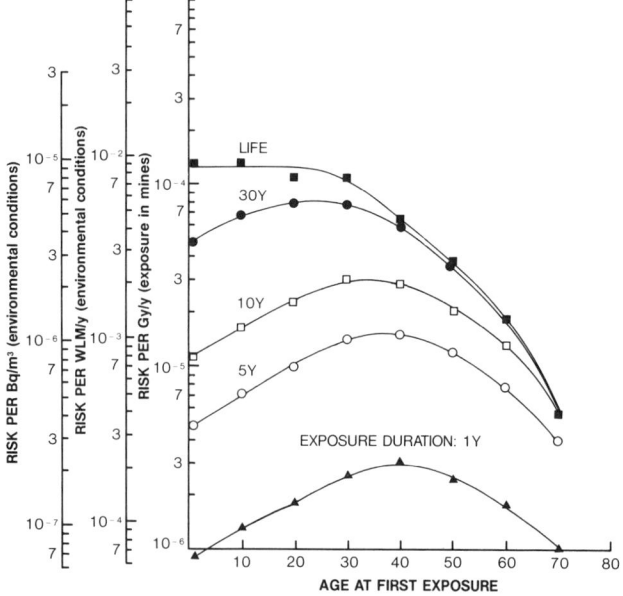

FIG. 9.2. Lifetime risk of lung cancer due to exposure to atmospheric radon (Bq m^{-3}) and concentration of radon progeny (WLM) and in terms of radiation absorbed dose (Gy). The assumed conversion factors are: equilibrium ratio of Rn:RaA:RaB:RaC = 1.0:0.9:0.7:0.7 with an atom ratio of unattached RaA/Rn = 0.07 and a ratio of Gy/WLM = 0.007. Note: 1 Bq m^{-3} = 0.027 pCi L^{-1}; 1 Gy = 100 rad. [From Reference (2), Tables 10.1, 10.2, and 10.3.]

9.1.4 Uncertainties in Risk Assessment

An understanding of the uncertainties is important in assessing the health risk to the individual. These uncertainties influence estimates of the potential for exposure to radon and of the actual radiation dose received. In addition to variations in soil concentrations of radioactivity, including radon, many other "outdoor" environmental variables affect indoor radon levels. Certain soil and climatic conditions affect the transport of radon from soil to indoors. Diurnal and seasonal variations are remarkable and important considerations in interpreting measured concentrations. Inside the house, many variables influence exposure levels. The type of heating and cooling systems and general construction (external and internal) of the house can change radon levels. Furnishings, such as carpeting and drapes, can change the amount of radon progeny deposited on surfaces by removing them from the air to be inhaled. Where high concentrations of radon exist in water supplies, variations in the use of water can result in different levels released into room air. For the individual exposed to a given level of radon there are also a number of biological variables that affect radiation dose to the lung's bronchial epithelial layer, which is the primary site of progeny deposition. These biological variables include breathing rates, lung clearance rates, and overall lung morphology. All the above factors, and many others, affect values and conditions assumed in assessing risk for individual cases and should be evaluated in cases where individual exposures merit attention.

9.1.5 Estimates of Risk from Exposure to Indoor Radon

It is useful to estimate the normally expected exposure to radon experienced by persons in the U.S. population and to estimate the associated risk of this exposure. Also, it is important to examine the range in levels of exposure and the frequency distribution of these levels.

Indoor radon levels much higher than the estimated 0.004 WL (83 nJ m^{-3}) average have been found in some U.S. housing. However, it is not known with any reasonable degree of certainty how many such

TABLE 9.3. Distribution of Population Exposure to Radon[a,b]

Average (WLM y^{-1})	Percent of population
<0.2	69.3
0.2–0.5	23.6
0.5–1.0	5.7
1.0–2.0	1.2
2.0–4.0	0.14
>4.0	—
Total	100.0

[a] Reference (3), Table 9.7.
[b] Assuming an average exposure of 0.2 WLM y^{-1} and a geometric standard deviation of 2.5.

houses exist and what may be the distribution of levels in U.S. housing. Studies attempting to define this frequency distribution suggest that a few percent of single-family dwellings (perhaps as many as one million houses) may have levels that exceed 8 pCi L^{-1} (300 Bq m^{-3}), equal to 0.04 WL (830 nJ m^{-3}) or 2 WLM y^{-1}. A value of 2 WLM y^{-1} is selected as a point of reference because it is the level of annual exposure rate recommended by the NCRP for remedial action.[3] A value of 0.02 WL (420 nJ m^{-3}) has been suggested by USEPA as a goal to be achieved by any remedial action taken to reduce higher levels.[1]

Among estimates of the frequency distribution of radon concentration in housing are those shown in Table 9.3, which were derived from data obtained for a variety of sources and purposes.[3] Other investigators using a similar data base report a lognormal frequency distribution with a geometric mean of 0.9 pCi L^{-1} and a geometric standard deviation of 2.8, with an arithmetic data mean of 1.5 pCi L^{-1} (55 Bq m^{-3}).[7] The two estimates are in reasonable agreement. These distributions are likely to be shifted to somewhat higher values as additional data for U.S. housing are collected. However, this shift would not likely have a significant impact on the overall number of lung cancers that may be attributable to radon exposures in the United States, because the vast majority of the U.S. population would still be exposed to the lower levels of radon.

As noted earlier, the average indoor radon exposure level in the United States is estimated to be about 0.004 WL (83 nJ m^{-3}), which, at an equilibrium ratio of 0.5 between the concentration of radon progeny and radon, is equivalent to an air radon concentration of about 0.8 pCi L^{-1} (30 Bq m^{-3}) or about 0.2 WLM y^{-1}. Using the con-

RISK ASSESSMENT AND POLICY

version that one year of continuous exposure to 1 WL equals 50 WLM per year, an individual would accumulate 14 WLM (0.05 J h m^{-3}) over 70 years. From Table 9.2 the estimated lifetime risk for the one-year-old continuously exposed for 10 years at 0.2 WLM y^{-1} (80 nJ m^{-3}) is (0.00077)(0.2) = 1.5 × 10^{-4} or about one chance in 10,000 of suffering lung cancer. Likewise, the risk from this exposure level for 30 years is less than one chance in 1000. Exposure for a lifetime to 0.2 WLM y^{-1} results in a risk of 0.18% or less than two chances in 1000 of contracting lung cancer. Using different assumptions, USEPA's estimates for the same level of exposure are 0.3 to slightly more than 1%.[1]

9.1.6 Comparative Lung Cancer Risk

To provide a perspective on the risk of lung cancer attributable to exposure to radon, comparisons can be made to the expected risk of lung cancer in the United States. The American Cancer Society reported that an estimated 472,000 cancer deaths would occur in the United States during 1986 and, of this number, 130,000 would be lung cancer deaths (89,000 in males, 41,000 in females).[8] One in five of all deaths is caused by cancer and one in 20 is due to lung cancer, or five percent. The American Cancer Society also estimates that cigarette smoking is responsible for 85% of lung cancers among men and 75% among women, or about 83% overall. The lifetime risk of lung cancer for the nonsmoker is about one percent or less. Even for the nonsmoker, "passive smoking" may be a contribution to this one percent and has been estimated to produce about 5000 lung cancer deaths per year in U.S. nonsmokers aged 35 years or more.[9] On the average, a smoker's risk is about 10 times a nonsmoker's risk, varying from about 4 times for the light smoker (one-half pack per day) to about 20 times for the very heavy smoker (more than two packs per day).

The role of smoking as a confounding factor in assessing radon risk is not clear from analysis of the uranium miner data. The effect of smoking on radon risk assessments may be small or large depending on whether one assumes the risks from smoking and radon exposure to be additive or multiplicative.[10] A recent analysis of the Colorado Plateau uranium miner data suggests that this population has a lung cancer rate about three times as great as cigarette smokers generally have.[11] It is further suggested, however, that this may not be a

TABLE 9.4. Lifetime Risk of Lung Cancer Relative to Nonsmoker's Risk

	Relative risk
Average for U.S. population	5×
Nonsmoker	1
Smokers	
Heavy (>2 cigarette pack per day)	20×
Light (½ cigarette pack per day)	4×
Exposure to radon progeny	
Average outdoor level (0.001 WL)	0.05×
Average indoor level (0.004 WL)	0.2×
Enhanced indoor level (0.04 WL[a])	2×

[a] 0.04 WL, at 0.5 equilibrium for radon and its progeny, is equivalent to a radon concentration of 8 pCi L^{-1} and results in an exposure (continuous) rate of about 2 WLM y^{-1}.

multiplicative effect. No increase was found in the incidence of lung cancer in miners for non-cigarette smokers with less than 300 WLM (1 J h m^{-3}) of cumulative exposure. On the other hand, it has been suggested that a synergistic interaction between cigarette smoking and indoor radon exposure does exist.[12,13] The high incidence of lung cancer in cigarette smokers is attributed to the cumulative alpha-radiation dose to the lung from the radon progeny in indoor air and the inhalation of the natural radioactivity in tobacco smoke, specifically, ^{210}Po and ^{210}Pb. It is suggested that current lung dosimetry models for exposure to indoor radon progeny greatly overestimate lung cancer risks for nonsmokers in the general population.

Table 9.4 shows estimates of lifetime risk of developing lung cancer depending upon smoking histories and the lifetime level of exposure to radon. Values shown are relative to the risk incurred by the nonsmoker, which is about one percent or less and slightly higher in males than in females. The nonsmoker risk estimate includes any contribution to risk from average exposures to radon and passive smoking.

It is very difficult to assess the "true" risk of indoor radon to members of the general public in light of the more proven causative agent, smoking. A satisfactory method of treating the confounding factor of smoking in risk assessment and in establishing levels for health protection has yet to be developed. Until such a method is developed, questions remain as to the actual radon imposed risk of lung cancer in the nonsmoker and as to the overall effectiveness of protection measures. Probably, the radon risk to the nonsmoker is must less than presently estimated and therefore protective measures are likely to be most

RISK ASSESSMENT AND POLICY   283

effective in reducing radon risk to smokers, who are already at high risk.

9.1.7 Summary

1. Many geological, meterological, physical, and biological factors affect the levels of radon in the environment, radiation dose to individuals, and estimates of individual health risk.
2. Levels of indoor radon exposure in U.S. housing can vary from very low, comparable to outdoor environments, to very high, comparable to several times that controlled in underground mines. The frequency distribution of levels appears to be lognormal with a geometric mean of about 0.004 WL (83 nJ m^{-3}); thus, the majority of exposures in U.S. housing is low.
3. Lung cancer is the major health detriment from indoor radon exposure.
4. Human risk of lung cancer from radon exposure is estimated from epidemiological studies of underground miners, primarily uranium miners. Results from these studies suggest a high risk in exposed miners who smoked, but a lesser risk from radon for nonsmokers.
5. Based on NCRP modeling and risk estimates, the number of U.S. lung cancer deaths attributable to radon exposure could be as high as 10,000 per year. These deaths are not necessarily independent of smoking histories, including exposure to "passive" smoke.
6. Lifetime exposure to radon at a rate of 2 WLM y^{-1}, which is the remedial action level recommended by NCRP,[3] corresponds to a lifetime risk of about two percent or less. For comparison, the risk to nonsmokers in the U.S. is about one percent or less, a value which includes average exposures to radon and passive smoke.

9.2 POLICY (Daniel J. Egan, Jr.)

Radon has always existed in man's environment, exposing people to some risk of lung cancer. However, it has become a major environmental concern in many regions of the country only after discoveries in eastern Pennsylvania beginning in December 1984, which are discussed in Chapter 2. This section will examine how this concern emerged, will describe how several governments have responded to the concern, and will review the diverse policy issues raised. While efforts of Pennsylvania, New Jersey, Florida, and the USEPA will be highlighted, many other state and local governments and Federal agencies are also taking actions and developing policies with regard to radon. Much remains to be learned about the effectiveness of various types of policies regarding radon, and the field is certain to be a dynamic and contentious one for years to come.

9.2.1 The Evolution of Current Concern

Elevated levels of radon in homes first became a significant issue for houses built on the residues of mining or milling operations that redistributed the naturally occurring radium in the Earth's crust. In particular, discoveries of elevated levels in homes built on uranium mill tailings in Grand Junction, Colorado, and on reclaimed phosphate mining lands in some areas of Florida led to programs to assess and mitigate the potential hazards in existing and future homes in those areas. However, the focus of these programs was on the effects from the mining and milling residues, rather than elevated levels that might be caused by the undisturbed distribution of minerals.

Not until the 1970s, when energy conservation became a major concern because of skyrocketing energy prices, did the potential effects of naturally occurring radon (other than that from mining operations) attract significant attention. This interest led to several studies to assess the distribution of radon in homes, and the results are discussed in Chapter 5. However, concern about indoor radon was still limited to a small community of scientists until the December 1984 discovery of high radon levels in the Watras house in Boyertown, Pennsylvania (discussed in Chapter 2). This event precipitated a series of further discoveries which showed that the risks from indoor radon, at least in selected areas of the country, could be far worse than anyone thought would be the case.

The Watras discovery quickly led the Commonwealth of Pennsylvania to investigate additional homes in that area. More homes with very high concentrations were found, although none were quite as high as the Watras house itself. These additional discoveries then led Pennsylvania to make free radon detectors available to all residents of the area underlain by a geologic formation known as the Reading Prong. The results of this survey, consisting of about 20,000 homes

TABLE 9.5. Comparison of NCRP-Estimated Distribution of Population Exposure to Radon and Distribution of Screening Measurements in Homes in the Pennsylvania Reading Prong[a]

Radon level (pCi L^{-1})	Percent of population	
	National[b]	Reading Prong
0–2	93	15
2–4	6	24
4–8	1	25
8–16	0.1	13
12–20		10
20–60		9
60–100		2
100–200		0.9
>200		0.5

[a] Screening measurements such as those made in the Pennsylvania Reading Prong may be higher (perhaps by a factor of 2–3) than the average annual exposure levels addressed in Reference (3).
[b] Projected national distribution from Reference (3).

that received and returned the detectors, are summarized in Table 9.5, where they are compared to the distribution of indoor radon generally thought to be representative before the Pennsylvania discoveries. The two sets of numbers are not directly comparable because the 1984 study describes estimates of actual exposure while the Pennsylvania data consist of screening measurements which are likely to overestimate the actual exposure (perhaps by a factor of two to three). The dramatic differences in the distributions are still apparent, however, and they indicate how much greater the risks from indoor radon can be, in a significant sample of homes, from those that were thought to be the case as of the end of 1984.

9.2.2 Government Initiatives to Date

The new realization of the magnitude of the potential dangers of indoor radon has brought with it calls for various types of government action to assess the extent of the indoor air radon problem, and to help provide the technical means to identify unacceptably high risks and to reduce them when they are discovered. There are many ways in which these broadly enunciated needs may be approached, however, and many different levels of government for which various types of action may be appropriate. As a first step toward investigating the policy issues, the recent radon-related initiatives of the governments of Pennsylvania and New Jersey and of the USEPA will be summarized. In addition, recent developments regarding the older program of the state of Florida will be highlighted because of their importance to the evolution of the national concern.

9.2.2.1 Pennsylvania

It started with an urgent phone call to the Commonwealth of Pennsylvania telling of the discovery of shockingly high radon levels, in excess of 2600 pCi L^{-1} (96,000 Bq m^{-3}), in the basement of the Watras home. The Secretary of the Pennsylvania Department of Environmental Resources (DER) soon advised the Watras family to leave their home until something could be done to reduce the radon levels. During the next year, DER staff proceeded to contact people throughout the township and county where the Watras home was located, explaining the potential risks and offering to assess the radon levels in their homes. Most of the people that the DER staff contacted (around 70 percent) turned down the offer, perhaps preferring not to know about (or at least not have the state government find out about) any increased lung cancer risks they might face from elevated radon levels. However, the 30 percent that allowed DER to make measurements made it clear that the Watras home was not an isolated incident. While that house remains the most contaminated home found, a number of other homes in the vicinity were discovered to have radon decay product exposures in the 1–10 WL range. The additional findings convinced the government of the state of Pennsylvania that additional steps were necessary.

These additional steps included several major initiatives. First, free radon measurement devices were made available, through coupons in newspaper advertisements, to all residents whose homes were built on the geologic formation known as the Reading Prong. About 20,000 of the estimated 70,000 eligible families took advantage of this offer, and the results of this screening survey are displayed in Table 9.5. Second, the Commonwealth announced a low-cost loan program intended to provide assistance to citizens who wished to take steps to reduce the radon risks discovered in their homes. However, only about 20 applications for assistance were received during the first year of the program's availability. Third, the DER recruited a staff of 21 who were based in a town in the heart of the Reading Prong to assist people with radon measurements and to advise on how to mitigate elevated radon levels.

Finally, the Commonwealth passed legislation in 1986 establishing a $1,000,000 fund to assist selected homeowners with remedial actions.

9.2.2.2 New Jersey

The state of New Jersey started to consider the implications of the Watras discovery soon afterward, because the Reading Prong formation was known to extend throughout heavily populated areas of northwestern New Jersey. However, the initial major impetus for action in New Jersey came not from discoveries of high radon levels, but from an article that appeared in the New York Times in May 1985. This story called the public's attention to the extent and possible hazards of the Reading Prong formation. The concern fueled by this article rapidly turned radon into a political, media, and entrepreneurial issue that demanded a response from the state government.

In particular, the concern encouraged a plethora of firms to begin offering measurement and consulting services to concerned homeowners at a time when there was very little information available to the public about the nature of the problem. Thus, the initial responses of the state were to set up information services, such as a telephone hotline that could be called for assistance, and to initiate efforts to prevent irresponsible behavior by commercial firms. At the same time, the state recognized that, because of the much larger population affected, it did not have the resources to provide free measurement services commensurate to those offered in the Pennsylvania Reading Prong. Accordingly, the hallmark of the New Jersey response was to build a workable blend of public and private sector involvement, such as the state offering free confirmatory measurements to any homeowner who first received an elevated reading from a commercial testing laboratory.

New Jersey also responded with a major legislative initiative to help it lay the foundation for the most comprehensive of the state radon programs. A statewide survey of 6,000 homes is being initiated to test analytical models for predicting the distribution of indoor radon levels throughout the state. A carefully designed study of lung cancer cases is being undertaken to help assess the risks of radon. Perhaps most noteworthy, New Jersey has chosen to create and implement mandatory certification programs for measurement and mitigation firms in order to protect consumers. In addition, the state has provided itself with the authority to obtain radon measurements in private homes and protect the confidentiality of this information. Finally, under existing authorities, New Jersey also set up a low-interest loan program to assist homeowners who wish to pursue mitigation measures.

Although triggered by discoveries in Pennsylvania, the New Jersey program has since been called upon to respond to radon levels of similar magnitude. In the town of Clinton, New Jersey, a cluster of 105 homes was found in March 1986 where the basement concentrations all exceeded 4 pCi L^{-1} (150 Bq m^{-3}), the concentrations in 40 of the homes exceeded 200 pCi L^{-1} (7400 Bq m^{-3}), and those in 5 homes exceeded 1000 pCi L^{-1} (37,000 Bq m^{-3}). Although close to the Reading Prong, this cluster sits on a dolomite formation, rather than the granite of the Reading Prong, thus demonstrating that such severely elevated levels may arise in a variety of geologic settings.

9.2.2.3 Florida

The radon program in Florida began with the discovery in 1975 of elevated levels in homes built on lands reclaimed after phosphate mining activities. The focus of this effort has been on developing ways to reduce the risks in new construction rather than on mitigation of existing homes. After consultation with the USEPA, the state developed a regulation that set forth building practices that would have to be followed in new construction unless the builder demonstrated, through testing, that new homes had indoor radon levels below 4 pCi L^{-1} (0.15 Bq L^{-1}). The final rule, which was promulgated in April 1986, was initially applicable only within certain geographically defined areas involving reclaimed phosphate mining lands, although it had provisions to be extended to other areas with elevated radon levels that might be subsequently identified by the state.

However, shortly after the rule was promulgated, difficulties arose that will be instructive for the development of policies and programs for indoor radon. The map defining the geographic applicability of the new rule was very controversial. The state legislature suspended enforcement of the new rule pending completion of a radon survey of the entire state and reconsideration of the area of applicability. Florida is currently proceeding with both the survey and a study to assess how the survey data should be used in defining the applicability of the rule, and many additional challenges can be expected.

9.2.2.4 U.S. Environmental Protection Agency

As the severity of the discoveries on the Pennsylvania Reading Prong became known throughout 1985, the USEPA rapidly assembled

RISK ASSESSMENT AND POLICY

a series of initiatives to respond to the problem. Entitled the Radon Action Program, this effort sought to provide the technical assistance most needed to help Pennsylvania and other affected states respond to the immediate demands while building the partnerships necessary for the Federal government, state agencies, and the private sector to cooperate in meeting citizens' future needs. At the same time, the USEPA initiated the studies needed to assess the extent of the problem and to provide the basis for low-cost, reliable mitigation methods.

The initial results of USEPA's expedited program have included: (1) a series of radon or radon decay product measurement protocols intended to encourage consistent and reliable measurement practices; (2) a voluntary program through which commercial and government laboratories can demonstrate their proficiency in making measurements; (3) demonstration of several successful mitigation techniques in a variety of types of houses in Pennsylvania and New Jersey; (4) publication of a "Citizen's Guide to Radon" for homeowners and a companion brochure describing potential mitigation techniques; (5) a three-day training program in radon mitigation; and (6) a program to assist states in evaluating the causes of elevated radon levels in selected homes and offering recommendations for reducing those levels to homeowners.

Since taking these initiatives, the USEPA has begun programs to provide technical assistance to states that wish to conduct their own radon surveys. During the 1986–87 winter heating season, the USEPA assisted Alabama, Connecticut, Colorado, Kansas, Kentucky, Michigan, Tennessee, Rhode Island, Wisconsin, and Wyoming, and the Agency expects to assist additional states in the future. In addition, the USEPA is planning a survey to evaluate the overall distribution of indoor radon levels throughout the country.

9.2.3 Policy Issues

The roles and activities of the various governments to date and the roles they have not assumed provide a starting point for reviewing policy issues raised by the concern about indoor radon. Underlying many of the issues is a philosophical dilemma typical of issues associated with environmental carcinogens: (1) should an overall objective be to choose some acceptably small level of risk for many or all situations, after balancing various health, feasibility, or cost considerations, or (2) should greater reliance be placed on encouraging reductions of risk whenever the appropriate decision maker judges worthwhile. These two approaches need not be mutually exclusive, but they are often difficult to resolve. The first approach is more compatible with a legalistic and procedural perspective, where issues are often cast in terms of right or wrong, while the second approach may better reflect the view that the cancer risks from radon vary proportionally with the level of exposure and is more compatible with a perspective of allowing individual decision makers to make their own choices.

The second approach has been the primary thrust of the initial responses of Pennsylvania, New Jersey, and the USEPA to assemble as much information about the risks, the extent of the problem, and the potential of mitigation measures; to disseminate this information; and to provide mechanisms through which consumers can be protected from fraudulent practices and misinformation as much as possible. One reason for this emphasis has been the pervasive lack of data regarding these issues. Furthermore, because the radon contamination of concern comes entirely from naturally occurring sources, there is no external party for governments to hold accountable for any damage to citizens. As a result of these influences, the governments have tended to provide services in support of decisions by individual homeowners rather than dictating appropriate endpoints for their citizens.

In this view, the low utilization of programs like Pennsylvania's low-cost loan initiative is not an indication of failure, but merely an affirmation that relatively few homeowners believe that it is appropriate to reduce their risks at this time. And given the limited experience with the long-term effectiveness of radon mitigation measures, this might not be a surprising result. On the other hand, research conducted by the USEPA's Office of Policy Analysis suggests that government initiatives that seek to reduce health risks with information programs can result in citizen reactions that correlate poorly with the risks they face.[14] Thus, the various governments involved should carefully consider the results of their initial efforts in designing their longer-term programs.

With this background, the various types of activities now under way can be reviewed to see what policy issues they might presage.

9.2.3.1 Assessing the Health Risk

Although we would appear to know more and have better data about the effects of radon than about many other environmental pollutants, we still have much to learn about the mechanisms of cancer induction and about how the degree of risk to different elements of the population (e.g., smokers versus nonsmokers) might be quantified. The analytical tools for addressing these uncertainties include both

RISK ASSESSMENT AND POLICY

basic mechanistic research and epidemiological studies to better understand the lessons that real-world experience can teach us. There is general agreement that such research should be done, but there is also general agreement that exposures like those found in many homes in the Reading Prong are high enough to call for urgent action. Thus, the primary policy issue here is to sort out what initiatives, if any, should wait for the completion of these health risk studies.

9.2.3.2 Assessing the Extent of Elevated Levels

Some obvious questions are raised by indoor radon concentration distributions like those shown in Table 9.5. For example: Are there other areas of the country that might be like the Reading Prong in Pennsylvania? What is the national distribution of levels likely to be? These are related but quite different questions that could be addressed by indoor radon surveys of various designs. One type of survey could be designed to emphasize location of areas with particularly elevated radon concentrations. Another type of survey could focus on better estimates of national, regional, or statewide distributions of indoor radon concentrations for various comparative purposes. Important questions for policy makers are what types of objectives should receive the most attention, and whether relatively low cooperation rates (note the Pennsylvania experience) will compromise the usefulness of surveys intended to use statistical methods to characterize radon concentration distributions in various areas.

9.2.3.3 Developing Mitigation Methods

There is a clear need for reliable and inexpensive methods to lower levels judged to be unacceptably high. The initial research performed by the USEPA using homes on the Pennsylvania Reading Prong indicates that relatively modest investments (i.e., a few hundred to a few thousand dollars) should be adequate to achieve substantial reductions in seriously elevated indoor radon levels (see Chapter 8). The degree to which such research programs should be expanded and the role of government versus the private sector in furthering their objectives are policy issues of importance.

9.2.3.4 Public Information and Guidance

In addition to research, governments can develop and disseminate information about environmental and public health issues, and there may be many objectives in providing such information. For example, government may wish to motivate citizens to take certain actions by depicting a situation as alarming, or it may wish to avoid panic by suggesting that a problem is not as severe as others have portrayed it. It may wish merely to disseminate as much factual information as possible, or it may with to supplement this with recommendations about specific situations. Even a decision to provide no information is likely to have important implications because of the vital roles that governments inevitably play in affecting public perceptions.

9.2.3.5 Consumer Protection

Particularly with a potential health problem with which few citizens have any familiarity, there are rampant opportunities for the incompetent or the unscrupulous to take advantage of people. Purported measurements of radon or its decay products may be frauds, and equipment or services sold to reduce elevated radon levels may be wastes of money or may even do more damage to a home than good. In all of these situations, governments at various levels frequently are called upon to play key roles in protecting citizens' interests, and issues associated with this role are whether voluntary or mandatory programs are most effective and what types of protections should be established.

9.2.3.6 Assisting or Subsidizing Mitigation

Another key demand often made of government is that it take some financial responsibility for fixing environmental or public health hazards. This often takes the form of ensuring that the parties who cause such hazards compensate those who are damaged. However, when no one is at fault, governments are often looked to as a source of assistance. Disaster relief programs and some public health initiatives (e.g., the "swine flu" project) serve as examples of this governmental role.

The primary policy issues here are whether government should assist in funding mitigation activities other than assistance that may be a side effect of research or training purposes. If so, what should the criteria be for selecting the recipients and types of assistance: financial need, radon level, potential success of mitigation, or others?

9.2.4 Are Standards Needed?

A lot has been accomplished by the initial, generally nonregulatory, government initiatives, and there are many remaining issues associated

RISK ASSESSMENT AND POLICY

with them. However, there are some additional policy areas that have received relatively little attention to date and which may call into question the longer-term effectiveness of nonregulatory approaches. These policy areas are discussed in the following sections.

9.2.4.1 New Construction Issues

The focus of the recent activity concerning indoor radon has been to alleviate risks to the occupants of existing homes, particularly because of the magnitude of the exposures encountered in the most dangerous homes on or near the Reading Prong. However, another set of needs concerns new homes that may be built on lands with high radon availability.

Home builders would like a set of reliable, low-cost construction techniques to keep out radon where substantial amounts of it are available in surrounding soil and rock and/or measurement techniques that are also reliable and low-cost to tell them whether the particular plot of undeveloped land they are considering calls for radon reduction measures. Unfortunately, there are no measurement methods yet available that have been demonstrated to reliably predict elevated radon levels in homes based upon data collected from undeveloped land. Nor are there proven methods to predict high-radon risk areas based upon available geologic data. Thus, home builders are faced with considerable uncertainties regarding the most effective steps they can take regarding radon problems.

Beyond these technical deficiencies, however, is a potential incompatibility with the initial policy approaches currently being pursued. For these strategies basically entrust the individual homeowner with the responsibility to balance perceived costs and risks and make appropriate mitigation decisions. The home builder's motivations are usually dominated by a desire to keep his costs down. Without regulations of some kind requiring the use of radon reduction techniques in specific areas, few builders are likely to take appropriate steps. The potential motivations would be educated consumer demand or the threat of liability judgments in favor of subsequent buyers. All of these mechanisms appear to call for selection of an indoor radon or radon decay product concentration that is "acceptably low" or "safe" to guide new construction decisions.

9.2.4.2 Real Estate Transactions

Addressing the problems that indoor radon poses for the real estate industry raises a mix of technical and policy questions similar to those for the home building industry. The motivations of the parties again cause problems, because the home seller has strong incentives to avoid detection of any elevated radon levels so as not to discourage buyers. Unfortunately, the existing radon measurement protocols that are most reliable are also susceptible to being tampered with to produce artificially low readings. Furthermore, state laws often hold realtors liable for discovering and reporting defects in a house and this places the burden of dealing with this difficult situation on a group of people with little relevant technical training.

There are various approaches to this problem, such as requiring the home seller to place an appropriate amount of money in escrow to cover mitigation of any radon problems discovered by the buyer within a reasonable period of time. However, as for the new construction problem, this situation appears to call for establishment of a radon or radon decay product level that is acceptable for the transaction to proceed.

9.2.4.3 The Implications of Indoor Radon Standards

This review suggests that once enough information has been collected to:

1. assess the extent of the country that may face significant indoor radon problems;
2. better establish the costs and effectiveness of radon reduction methods; and
3. better understand, perhaps, the magnitude and incidence of the lung cancer risks posed by radon and its decay products,

it may be appropriate for indoor radon standards to be established to guide certain types of transactions.

Whether development of such standards is most effectively initiated by the federal government, by state and/or local governments, or voluntarily by private sector organizations requires further study. However, it is likely that various influences will tend to draw different standards established by different groups together over time. For example, litigation pressure may mount for judicial findings to show that compliance with standards represents "acceptable" risks—or even "no risk"—and these pressures will call into question any differences in relevant standards.

On the other hand, lifetime exposure to the lowest remedial action target currently discussed (i.e., 4 pCi L^{-1} in indoor air) equates to lifetime lung cancer risks of about one chance in a hundred[1], if current health risk estimates are correct. Many other existing environmental regulations intend to reduce individual risks to levels that are several orders of magnitude lower than this. Consequently, the potential evolution of policies regarding indoor radon may have far-reaching implications for other environmental and public health programs. Indoor radon may well become one of the most influential environmental issues of the next decade.

9.2.5 Summary

This chapter, intentionally, has identified a wide variety of policy issues associated with elevated indoor radon levels without trying to tie them together into anything like a proposed plan of action. In part, this reflects the freshness of the topic and the lack of time to evaluate the effectiveness of the steps already taken by the various States and the EPA. Also, the most useful policies may vary from region to region depending on the severity of the indoor radon levels that are discovered in different parts of the country. The three basic questions that will have to be decided in each case are: the action to be taken (be it measurement, disclosure of data, mitigation, etc.); the level(s) of society to be responsible for the action (Federal, State, or local government, private firm, or individual resident); and the quality of the information appropriate to support the action.

However, some basic tenants appear clear. First, the public health risks from indoor radon are substantial—there is virtually no scientific controversy about this, although the precise risk estimates are still unclear. The exposures in the most highly contaminated homes discovered in Pennsylvania and New Jersey are extraordinary and present such high risks that virtually everyone agrees that mitigating actions should be taken as soon as possible. Similarly, there is general agreement (see Section 9.1) that many thousands of lung cancers each year are probably attributable to indoor radon exposures. As the magnitude of this risk becomes better recognized, both private sector and government activity will inevitably increase. Initiatives to collect better information on the extent of the problem and to protect consumers when getting homes tested, mitigated, or sold are the most vital steps for the immediate future.

Glossary

Absolute risk The actual risk for a particular contaminant, independent of other background causes.

Absorbed dose (or dose) The ratio of the energy absorbed by a mass divided by the mass. For radon progeny, the mass is the mass of pertinent lung cells (units: rem or sievert).

Actinium series The naturally radioactive series starting with ^{235}U.

Actinon The isotope of radon in the actinium series.

Activated carbon A type of time-integrated sampling method which allows the detection of gamma rays from radon decay products adsorbed on carbon. Also a method for removing radon from drinking water.

Activity The rate of atomic disintegration (units: becquerel or curie).

Air changes per hour (ach) The number of air changes per hour in a building.

Alpha emission density The number of alphas per unit space emerging from a source element. For radon progeny in the lung, the emission source density is the number of alphas per unit surface area in each tube.

Alpha particle A positively charged atomic particle that is ejected from a nucleus and contains the nucleus of a helium atom or two neutrons and two protons.

Alpha track detector A type of detector that records alpha particles from decay products by radiation damage on film. After exposure, these tracks are revealed by etching with NaOH or KOH solution and counted manually or automatically under a microscope.

Anatomical dead space The region of the lung given by the tracheobronchial and nasopharyngeal regions, which normally constitute the part of the lung not responsible for gas exchange with blood.

Atomic mass number The total number of nucleons in the nucleus or the total number of protons and neutrons in the nucleus.

Atomic number The number of protons in the nucleus of a neutral atom of a nuclide.

Attached fraction Fraction of airborne radon progeny attached to aerosol particles.

Becquerel (Bq) A radioactivity of one disintegration per second.

Beta particle Electron emitted from the nucleus of a radionuclide.

Bifurcation The point at which a tube in the lung divides to form the next generation.

Bronchi The two main branches leading from the trachea to the lungs.

Bronchial epithelium The cellular lining of bronchi.

Carcinoma A malignant tumor which can affect almost any organ; tends to metastasize.

Cohort A large homogeneous group of people tested in epidemiological or socioeconomic studies.

Critical cells Those cells deemed most likely to act as essential targets for producing an effect such as cancer.

Cumulative working level months (CWLM) The sum of lifetime exposure to radon working levels expressed in total working level months.

Curie (Ci) A unit used to describe the rate of radioactive decay. One curie equals 3.7×10^{10} disintegrations per second.

Daughter Historical synonym for decay products or progeny.

Decay product A nuclide resulting from the radioactive disintegration of a radionuclide, formed either directly or as the result of successive transformations in a radioactive series. A decay product may be either radioactive or stable.

Decay series The consecutive members of a radioactive family of elements. A complete series commences with a long-lived parent such as ^{238}U and ends with a stable element such as ^{206}Pb.

Deoxyribonucleic acid (DNA) The basic instruction code which a living cell uses to direct its life processes.

Depth–Dose curve A curve describing the absorbed dose at points located at varying depths into a material such as the walls of a passageway in the human body.

GLOSSARY

Diffusion barrier Any material placed over a detector which delays, through increased diffusion time, the rate at which radon or progeny in air reach the detector.

Direct-recoil fraction Those atoms which terminate their recoil path in a pore space.

Distal lung The region of the lung farthest from the nose.

Dose The amount of the radiation or energy that a specified mass is exposed to.

Dose equivalent The product of the absorbed dose and the quality factor (units: rem or sievert).

Dose–response curve The functional relationship between dose and effect.

Drain-tile ventilation The process of using perforated pipes called drain tiles around the foundation of a house to remove radon.

Effective dose equivalent Dose equivalent weighted by a factor which measures relative sensitivity of the tissue to a radiation-induced cancer.

Effective leakage area (ELA) The hypothetical area amassed if all the actual cracks and holes in a building shell were collected in one hole with the equivalent infiltration rate.

Electrons Negatively charged particles that orbit around the nucleus of an atom.

Emanation Term used historically to describe what we now know to be the isotopes of radon. It was used by Lord Rutherford, when he was unsure of its nature.

Epidemiology The division of medical science concerned with defining and explaining the interrelationships of the host, agent, and environment in causing disease.

Equilibrium A state of balance or rest; radioactive equilibrium is the state in which the rate of formation of atoms is equal to the rate of their disintegration by radioactive decay, so that the amount of the element or isotope is constant.

Equilibrium-equivalent concentration (EC) The EC of a nonequilibrium mixture of short-lived progeny in air is that activity concentration of the parent gas in radioactive equilibrium with the concentrations of its short-lived progeny that have the same potential alpha energy concentration c_p as the nonequilibrium mixture. Units are Bqm^{-3}. (See Section 1.5.4.)

Equilibrium factor An adjustment used in converting from pCi L^{-1} to working level concentration, which takes into account the possible absence of radioactive equilibrium between radon and its progeny.

Exhalation rate The rate at which radon is released from a solid surface into the adjacent air, in Bq m^{-2} s^{-1}. It is a function of the radon production rate and the diffusion coefficient of the solid.

Exposure The amount of radiation in an environment, not necessarily indicative of absorbed energy but representative of potential health damage to the individual present. Working level months (WLM) are one way of measuring exposure.

Far-wall contribution The component of dose to cells in tube walls that arises from radiation which must first pass through the tube air to reach tissue on the "far side" from the point of decay.

Flux measurements Measurements made to determine how much radon is released into an enclosure from a source. They are used to determine the rate at which radon emanates from building materials.

Gamma radiation A ray of energy in contrast to beta and alpha particles' radiation. The properties are similar to those of X rays and other electromagnetic waves. Gamma radiation is highly penetrating but relatively low in ionizing potential.

Gamma rays Short-wavelength electromagnetic radiations.

Generation A loosely defined set of passageways in the lung characterized by roughly equal anatomical dimensions and depth into the lung.

Generational volume The sum of the volumes contained in all of the passages in a generation.

Gneiss A coarse-grained, high-grade metamorphic rock that has a distinct banding due to segregation of light and dark minerals.

Grab sample A sample indicative of the concentration of radon or progeny at one point in time.

Granular activated carbon (GAC) A unique material made of wood, coal, or other carbon base that has been carbonized (to charcoal) and activated to develop a vast internal pore structure and adsorption capacity.

Gray (Gy) A standard international unit of absorbed dose; 1 Gy = 1 J kg^{-1} = 100 rad.

Half-life The time it takes for one-half of any quantity of identical radioactive atoms to undergo decay.

Healthy-worker effect Lower mortality rate among workers than among

GLOSSARY

the general population if their health has not been adversely affected by work.

Henry's constant The ratio of the air concentration to the water concentration of a gas or volatile compound.

Igneous rock Rock formed by the cooling of liquid material; for example, basalt and granite.

Incidence A number of cases of disease or other characteristics in a given period of time (monthly or annually).

Indirect-recoil fraction Those atoms which terminate their recoil path in a solid grain but diffuse out of the solid through the recoil path, while the solid is still molten.

International system (SI) System of units commonly used in most of the world outside the United States (see Chapter 1).

In-vivo counting The measurement of the amount of radioactivity in the human body through counting of the number of gammas which emerge from that body.

Ionization chamber An instrument which detects radiation through collection of ionization products produced by the radiation in a gas-filled cavity.

Ionizing radiation Radiation which can remove electrons from atoms, creating ions.

Isotopes Nuclides of the same element which differ in number of neutrons.

Latency period The time during which disease exists without manifesting itself; a period of incubation.

LET (stopping power) The energy lost by a charged particle passing through a substance per unit length of path; related concepts are mass, atomic, molecular, and relative stopping power. Also known as linear energy transfer.

Metamorphic rocks Rocks that have been changed in their composition or texture by heat and pressure. The degree of change can be very small, such as that which occurs when shale changes into slate, or large enough to completely recrystallize the rock, as occurs in the formation of schist or gneiss.

Monodisperse Particles falling within a very narrow range of sizes so as to behave identically as far as deposition in the lung.

Mucociliary blanket The layer of mucus lining the generations of a lung above the pulmonary region and responsible for transporting deposited materials out of the lung and into the gastrointestinal (GI) tract.

Near-wall contribution   The component of dose to cells in tube walls which arises from radiation passing only through tissue (and not air). See also the entry for "Far wall."

Neutrons   Electrically neutral particles in the nucleus of an atom.

Nuclide   A species of atom characterized by the constitution of its nucleus.

Parent   A radionuclide which, upon disintegration, yields a specified nuclide—either directly or as a later member of a radioactive series.

Pegmatite   Igneous rock type that is usually found as dikes associated with granites. It can have high uranium content.

Permeability   The capacity of a solid to transmit a fluid. It is dependent on the porosity (volume of pores) as well as the extent of interconnection among pores.

Photomultiplier   A device that produces electrical pulses when struck by light pulses (such as from a scintillator).

Photons   Units of energy of electromagnetic radiations.

Plate-out   The deposition of radon progeny on indoor surfaces, causing their removal from room air.

Potential alpha activity   The total amount of alpha-particle radiation to be expected from a unit volume of air containing a mixture of radon and its progeny.

Potential alpha energy   Total energy emitted during alpha decays if a quantity of radon progeny is allowed to decay through the entirety of the decay chain (see Chapter 1).

Pressure-driven flow   Physical transport of soil gas into a structure driven by depressurization of the structure. A dominant mechanism for radon entry into structures.

Progeny   The decay products of a radioactive series decay.

Protons   Positively charged particles in the atomic nucleus. The number of protons in an atom is the atomic number of the element.

Proximal lung   The region of the lung closest to the nose.

Quality factor (Q)   The factor by which absorbed dose is to be multiplied to obtain a quantity that expresses, on a common scale, for all ionizing radiations, the irradiation incurred by exposed persons. A measure of the relative effectiveness of a radiation at producing a specific effect (primarily used for cancer). $Q = 20$ for alpha particles and $Q = 1$ for beta particles, etc.

Rad The standard unit of absorbed dose, equal to energy absorption of 100 ergs per gram (0.01 J kg^{-1}).

Radon The name of the isotope of radon in the uranium series and the generic name for all isotopes with atomic number 86.

Relative biological effect (RBE) An evaluation of the impact of a given type of radiation on tissue based on the LET of the radiation type.

Relative risk The ratio of the rate of disease in exposed to unexposed populations.

Rem A unit of dose equivalent (see Chapter 1).

RPISU Thermoluminescence radon progeny integrating sampling unit: an acoustically shielded, low-volume air sampling device that collects radon decay products on a membrane filter located near a thermoluminescence detector (TLD).

Sarcoma Cancer arising from underlying tissues (muscle, bone, connective tissue).

Schist A medium- to coarse-grained, high-grade metamorphic rock with a texture of thin, irregular plates due to abundant mica.

Scintillator A material which produces light pulses when radiation interacts with it.

Sedimentation The process by which a particle deposits on the wall of a tube in the lung due to gravitational settling.

Sensitivity The ability of a measurement system to determine the presence or quantity of radiation above background signals.

Sievert (Sv) A standard international unit for dose equivalent which equals one gray, times a quality factor, times a modifying factor; 1 Sv = 100 rem.

Spectrum A mathematical description of the relative frequencies for radiations of different energies.

Stack effect Vertical temperature differential in a structure that results in a negative pressure at the base of the structure and causes a convection pattern that draws outdoor air from the understructure.

Stochastic process A process in which the probability of an effect occurring rather than its severity is regarded as a function of dose without threshold.

Stopping power The rate at which radiation loses energy in traversing a small distance in a material, given as the ratio of the energy loss to the distance traveled.

Surface barrier working level monitor A recently invented device that uses a silicon surface barrier alpha detector instead of a TLD. Like the RPISU, it collects decay products on a membrane filter.

Symmetric branching An anatomical model which assumes that each lung passageway divides into two identical subpassages. Contrasts with asymmetric branching.

Synergism Harmoniously combined action of two or more factors such that the resulting effect is greater than the sum of the effects produced by the actions individually.

Thermal bypass An opening (usually vertical) from which warm air can leave a building.

Thermoluminescent dosimeter (TLD) A material which stores the energy absorbed from radiation passing through it, and then releases this energy as light pulses upon subsequent heating.

Thorium series The naturally radioactive series that starts with ^{232}Th.

Thoron The isotope of radon in the thorium series.

Threshold The lowest level of a stimulus above which symptoms appear, or conversely, below which no effect is observed.

Tidal volume The amount of air taken into the lung during each breath.

Transient equilibrium Relatively short-term equilibrium within a portion of a radioactive series where the parent has a long half-life only relative to successive series members. An example is the radon progeny series.

Turbulent deposition The process by which air streams fail to follow the contour of passageways and, hence, transport particles to the walls.

Unattached fraction The fraction of airborne radon progeny unassociated with aerosol particles.

Uranium Refers normally to ^{238}U, although about 0.7% is ^{235}U, the fissionable component, which is present in the natural state.

Uranium series The naturally radioactive series that starts with ^{238}U.

Weighting factor A numerical factor used to modify calculation of the equivalent dose to account for variation of LET and radiation effects with tissue type and exposure.

Working level (WL) A measure of decay product concentration which indicates the extent of alpha-particle release from short-lived products (see Chapter 1). Also a measure of the total alpha activity in air caused by the presence of attached and unattached radon progeny.

Working level month (WLM) An exposure equivalent to one working level of radon progeny for 170 hours (see Chapter 1). Also a unit of exposure to radon progeny calculated by multiplying 170 hours per month times the working level.

Working level ratio (WLR) A ratio that compares the concentration of decay products to the concentration of radon. The WLR may be computed using the formula, WLR = (WL value)(100)/radon concentration.

Appendix A

Radioactive Decay

A.1 LAW OF RADIOACTIVE DECAY

If $N(t)$ is the number of atoms present in the source at time t and each decay produces one radiation, then the number of radiations produced per unit time at time t is $-dN(t)/dt$. From this point onwards the function $N(t)$ will be given simply by N.

The constant λ is equal to the fraction of atoms decaying per unit time, so that the number of atoms decaying per unit time is simply the product of λ and the number of atoms, or

$$-dN/dt = \lambda N \tag{A.1}$$

or

$$-dN/N = \lambda\, dt$$

Integrating both sides and letting the number N at time $t = 0$ be N_0

$$-\int_0^t dN/N = \int_0^t \lambda\, dt$$

$$-\ln(N/N_0) = \lambda t \tag{A.2}$$

and

$$N/N_0 = e^{-\lambda t}$$

or

$$N = N_0 e^{-\lambda t}$$

Therefore, the number of atoms for a given radionuclide decreases exponentially with time from the initial value N_0 (assuming that the radionuclide is not also produced by the decay of another radionuclide).

The decay constant λ may be related to the time it takes for one-half of the atoms present to decay, or what is called the half-life or $T_{1/2}$. As in the case of the decay constant, $T_{1/2}$ is characteristic for each radionuclide. When $N = N_0/2$, using Equation (A.2),

$$N = N_0/2 = N_0 e^{-\lambda T_{1/2}}$$

or

$$1/2 = e^{-\lambda T_{1/2}}$$

or

$$T_{1/2} = -\ln 2/\lambda = 0.693/\lambda \qquad (A.3)$$

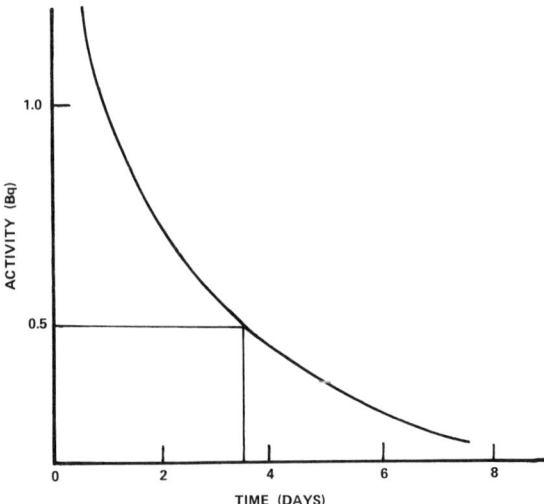

FIG. A.1. Plot of the decay of ^{222}Rn as a function of the time in days. The half-life is 3.823 d.

RADIOACTIVE DECAY

Using the decay of ^{222}Rn as an example, the activity λN as a function of time is

$$\lambda N = \lambda N_0 e^{-\lambda t} = -dN/dt$$

where $\lambda = 0.693/3.823$ d. A plot of this decay in activity is shown in Figure A.1. Note that the unit for the decay constant is inverse time.

A.2 SERIES DECAY—THE BATEMAN EQUATIONS

For the more advanced student, the following discussion shows how to calculate the activity of each of the progeny in a decay series, such as the decay series for ^{222}Rn.

In a sequential radioactive series, there are decays through several different radionuclides. Let $N_1, N_2, \ldots, N_k, \ldots, N_r$ represent the number of radionuclides in a single radioactive series where there are N_1 parent radionuclides and N_r stable end-product radionuclides. These functions, N_i, all are implicit functions of time. Then from Equation (A.1), the differential equation for the first radionuclide describing radioactive change can be generalized as follows:

$$dN_1/dt = -\lambda_1 N_1,$$

which has already been shown to have the solution $N_1 = N_{10} e^{-\lambda_1 t}$. To describe the decay of the next progeny one can observe that the number, N_2, of atoms of the second radionuclide is due to the input from the decay of the first radionuclide, N_1, with decay constant λ_1, and from the decay of the second radionuclide, N_2, with decay constant λ_2. Thus

$$dN_2/dt = \lambda_1 N_1 - \lambda_2 N_2 \tag{A.4}$$

and in general for the kth radionuclide

$$dN_k/dt = \lambda_{k-1} N_{k-1} - \lambda_k N_k \tag{A.5}$$

and

$$dN_r/dt = \lambda_{r-1} N_{r-1} \tag{A.6}$$

where $\lambda_1, \lambda_2 \ldots$ are the decay constants for each successive radionuclide in the decay series.

As an example, consider the decay from radionuclide 1 to radionuclide 2 as described by Equation (A.4). Then

$$dN_2/dt = \lambda_1 N_1 - \lambda_2 N_2 \qquad (A.7)$$

and

$$N_1 = N_{10} e^{-\lambda t}$$

N_{10} being the initial number of type 1 radionuclides. N_{20} would then be the number of type 2 radionuclides present at t equal to zero and so forth.

In general, the differential equation is

$$dN_i/dt = \lambda_{i-1} N_{i-1} - \lambda_i N_i$$

or

$$dN_i/dt + \lambda_i N_i = \lambda_{i-1} N_{i-1}$$

This is of the form specified by Bernoulli

$$dy(x)/dx + p(x)y(x) = Q(x),$$

the solution to which is

$$y(x) e^{\int p(x)\,dx} = \int e^{\int p(x)\,dx} Q(x)\,dx + C,$$

where C is a constant. Therefore

$$N_i e^{\lambda_i t} = \int e^{\lambda_i t} N_{i-1}\,dt + C_i,$$

where C_i is the initial number of atoms of the ith kind at $t = 0$.

As an alternative, one can guess that the solution to Equation (A.7) contains exponentials involving decay constants λ_1 and λ_2 as follows:

$$N_2 = a e^{-\lambda_1 t} + b e^{-\lambda_2 t} \qquad (A.8)$$

RADIOACTIVE DECAY

where a and b are constants that can be determined by substituting Equation (A.8) into Equation (A.7). This yields the following:

$$-a\lambda_1 e^{-\lambda_1 t} - b\lambda_2 e^{-\lambda_2 t} = \lambda_1 N_{10} e^{-\lambda_1 t} - a\lambda_2 e^{-\lambda_1 t} - b\lambda_2 e^{-\lambda_2 t}$$

or

$$a = [\lambda_1/(\lambda_2 - \lambda_1)]N_{10}$$

At $t = 0$, $N_2 = 0$ since no type 2 radionuclides had formed then, and from Equation (A.8),

$$a = -b$$

Thus the trial solution from Equation (A.8) is

$$N_2 = N_{10}[\lambda_1/(\lambda_2 - \lambda_1)][e^{-\lambda_1 t} - e^{-\lambda_2 t}] \tag{A.9}$$

The solutions for all members of the series was generalized by Bateman in 1910[1] as follows:

$$N_n = A_1 e^{-\lambda_1 t} + A_2 e^{-\lambda_2 t} + \cdots + A_n e^{-\lambda_n t}$$

where under the condition that $0 < j < n + 1$,

$$A_1 = \frac{N_{10} \lambda_1 \lambda_2 \cdots \lambda_{n-1}}{(\lambda_2 - \lambda_1)(\lambda_3 - \lambda_1) \cdots (\lambda_n - \lambda_1)} \tag{A.10}$$

and

$$A_j = \frac{N_{10} \lambda_1 \lambda_2 \cdots \lambda_{n-1}}{(\lambda_1 - \lambda_j)(\lambda_2 - \lambda_j) \cdots (\lambda_{j-1} - \lambda_j)(\lambda_{j+1} - \lambda_j) \cdots (\lambda_n - \lambda_j)} \tag{A.11}$$

To demonstrate how these equations can be used to calculate the activity of progeny with time, consider as an example the decay of a pure source of ^{222}Rn. Calculate the activity of ^{218}Po, ^{214}Pb, and ^{214}Bi after 60 minutes. The following table provides the half-lives, decay constants, and values of the exponentials for the three radionuclides involved:

n	Radionuclide	$T_{1/2}$	λ (min^{-1})	$e^{-\lambda t}$
1	^{222}Rn	3.823 d	1.26×10^{-4}	0.992
2	^{218}Po	3.05 min	0.227	1.22×10^{-6}
3	^{214}Pb	26.8 min	0.0259	0.211
4	^{214}Bi	19.7 min	0.0352	0.121

Using the decay equations from above:

$$N_1 = N_{10}e^{-\lambda_1 t}$$

$$= 0.992 N_{10}$$

and

$$N_1 \lambda_1 = (0.000126/\text{min})N_{10}$$

For the second radionuclide:

$$N_2 = [\lambda_1/(\lambda_2 - \lambda_1)]N_{10}e^{-\lambda_1 t} + [\lambda_1/(\lambda_2 - \lambda_1)]N_{10}e^{-\lambda_2 t}$$

$$= 5.47 \times 10^{-4} N_{10}$$

and

$$N_2 \lambda_2 = (0.000124 \text{ min}^{-1})N_{10}$$

which is approximately the same as the activity of the first radionuclide.
For the third radionuclide (assuming that $N_{20} = N_{30} = 0$),

$$N_3 = \left[\frac{\lambda_1 \lambda_2}{(\lambda_2 - \lambda_1)(\lambda_3 - \lambda_1)}\right] N_{10} e^{-\lambda_1 t}$$

$$+ \left[\frac{\lambda_1 \lambda_2}{(\lambda_1 - \lambda_2)(\lambda_3 - \lambda_2)}\right] N_{10} e^{-\lambda_2 t}$$

$$+ \left[\frac{\lambda_1 \lambda_2}{(\lambda_1 - \lambda_3)(\lambda_2 - \lambda_3)}\right] N_{10} e^{-\lambda_3 t}$$

$$= 0.00362 N_{10}$$

RADIOACTIVE DECAY

and

$$N_3\lambda_3 = (0.0000938 \text{ min}^{-1})N_{10}$$

For the fourth radionuclide (assuming that only radionuclide 1 is present at the start and the progeny atoms are not present):

$$N_4 = \left[\frac{\lambda_1\lambda_2\lambda_3}{(\lambda_2 - \lambda_1)(\lambda_3 - \lambda_1)(\lambda_4 - \lambda_1)}\right] N_{10}e^{-\lambda_1 t}$$

$$+ \left[\frac{\lambda_1\lambda_2\lambda_3}{(\lambda_1 - \lambda_2)(\lambda_3 - \lambda_2)(\lambda_4 - \lambda_2)}\right] N_{10}e^{-\lambda_2 t}$$

$$+ \left[\frac{\lambda_1\lambda_2\lambda_3}{(\lambda_1 - \lambda_3)(\lambda_2 - \lambda_3)(\lambda_4 - \lambda_3)}\right] N_{10}e^{-\lambda_3 t}$$

$$+ \left[\frac{\lambda_1\lambda_2\lambda_3}{(\lambda_1 - \lambda_4)(\lambda_2 - \lambda_4)(\lambda_3 - \lambda_4)}\right] N_{10}e^{-\lambda_4 t}$$

and

$$N_4\lambda_4 = (0.0000620 \text{ min}^{-1})N_{10}$$

A.3 EQUILIBRIUM

Two general situations can occur for the decay of a radioactive series with the first two decaying to a stable third. The two situations [refer to Equation (A.9)] are for the parent longer-lived than the progeny ($\lambda_1 < \lambda_2$) and the progeny longer-lived than the parent ($\lambda_1 > \lambda_2$).

If the parent is longer-lived than the progeny ($\lambda_1 < \lambda_2$), a state of radioactive equilibrium is reached where the ratio of the decay rates of parent and progeny are constant. This can be seen by using Equation (A.9) for a large time t. After sufficient time, $e^{-\lambda_2 t}$ is negligible compared to $e^{-\lambda_1 t}$ and

$$N_2 = [\lambda_1/(\lambda_2 - \lambda_1)]N_{10}e^{-\lambda_1 t}$$

but since $N_1 = N_{10}e^{-\lambda_1 t}$

$$N_2/N_1 = \lambda_1/(\lambda_2 - \lambda_1) \tag{A.12}$$

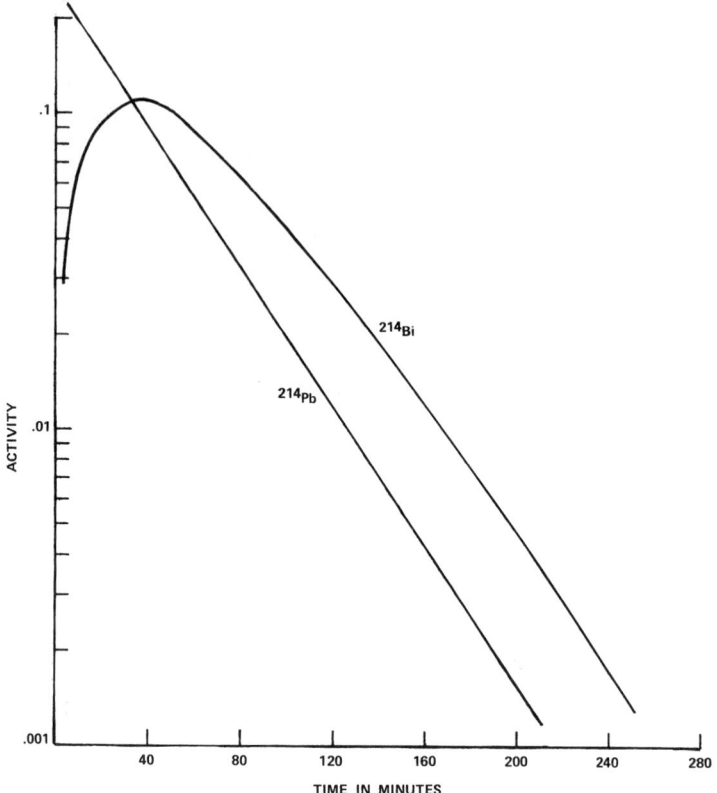

FIG. A.2. Plot of the decay of ^{214}Pb and its immediate progeny ^{214}Bi. As can be seen, in time the slopes of the decay curves are approaching the same value. The relative abundance remains constant and the two species are said to be in equilibrium.

This situation is called transient equilibrium and is shown graphically in Figure A.2. For the limiting case of Equation (A.12) where $\lambda_1 \ll \lambda_2$,

$$N_2/N_1 = \lambda_1/\lambda_2$$

or

$$\lambda_1 N_1 = \lambda_2 N_2$$

which is called secular equilibrium. In the case of secular equilibrium, the parent and progeny eventually decay with equal activities.

If the parent is shorter-lived than the progeny ($\lambda_1 > \lambda_2$), no equilibrium is attained at any time.

REFERENCE

1. H. Bateman, The solution of a system of differential equations occurring in the theory of radioactive transformations, *Proceedings of the Cambridge Philosophical Society 15*, 423–427 (1910).

Appendix B

Conversion Factors

$$1 \text{ Bq} = 27 \text{ pCi}$$

$$1 \text{ pCi L}^{-1} = 37 \text{ Bq m}^{-3}$$

$$1 \text{ J h m}^{-3} = 285 \text{ WLM}$$

$$1 \text{ WLM} = 0.0035 \text{ J h m}^{-3}$$

$$1 \text{ WL} = 2.08 \times 10^{-5} \text{ J m}^{-3}$$

$$1 \text{ WL} = 51.6 \text{ WLM y}^{-1}$$

$$1 \text{ WLM y}^{-1} = 4.03 \times 10^{-7} \text{ J m}^{-3}$$

References

CHAPTER 1

1. P. R. Fields, L. Stein, and M. H. Zirin, Radon flouride, *J. Am. Chem. Soc.* **84**, 4164–4165 (1962).
2. L. Stein, The chemistry of radon, *Radiochim. Acta* **32**, 163–171 (1983).
3. United Nations Scientific Committee on the Effects of Atomic Radiation, *Ionizing Radiation: Sources and Biological Effects,* United Nations Press, New York (1982).
4. E. Nussbaum, Radon Solubility in Body Tissues and in Fatty Acids, AEC Research and Development Report, UR-503, University of Rochester, Rochester, NY (1957).
5. A. Nesmith and A. Long, Arduous march toward standardization, *Smithsonian 1983* (May), 176–194.
6. J. Q. Adams, *Report on Weights and Measures,* Gales and Seaton, Washington, DC (1821).
7. National Council on Radiation Protection and Measurements, SI Units in Radiation Protection and Measurements, NCRP Report No. 82, Bethesda, MD (1985).
8. E. Landa, *History of Geophysics,* Vol. 3, American Geophysical Union, Washington, DC (in press).
9. R. D. Evans, Fundamentals of radioactivity and its instrumentation, *Adv. Biol. Med. Phys.* **1**, 151–221 (1948).
10. A summary of this meeting is given in "Radiation Exposure of Uranium Miners," Hearings before the Subcommittee on Research, Development and Radiation of the Joint Committee on Atomic Energy, Congress of the United States, Part 2, Additional Backup and Reference Material to the Hearings Held May–August 1967, U.S. Government Printing Office, Washington, DC.
11. D. A. Holaday, D. E. Fushing, R. D. Coleman, P. F. Fushing, R. D.

Coleman, P. F. Woorich, H. L. Kusetz, W. F. Blane, and W. F. Gafafer, Control of Radon and Daughters in Uranium Mines and Calculations on Biologic Effects, *U.S. Department of Health, Education and Welfare, Public Health Series* (1957).

12. National Academy of Sciences, *Radiation Exposure of Uranium Miners*, National Academy of Engineering, NAS-NRC, Washington, DC (1968).
13. E. Rutherford, Radioactive change, *Phil. Mag. 1903*, 576–591.
14. R. E. Evans, Engineers' guide to the elementary behavior of radon daughters, *Health Phys. 38*, 1173–1197 (1968).
15. W. Hofmann, F. Steinhausler, and E. Pohl, Age-, sex- and weight-dependent dose patterns due to inhaled natural radionuclides, in: *Natural Radiation Environment III* (T. F. Gesell and W. M. Lowder, eds.), Technical Information Center, U.S. Department of Energy, Washington, DC (1980).
16. F. T. Cross, N. I. Harley, and W. Hofmann, Health effects and risks From ^{222}Rn in drinking water, *Health Phys. 48*, 649–670 (1985).
17. National Council on Radiation Protection and Measurements, Evaluation of Occupational and Environmental Exposures to Radon and Radon Daughters in the United States, NCRP Report No. 78, Bethesda, MD (1984).
18. Charged particle tracks in solids and liquids, *Proceedings of the Second L. H. Gray Conference*, published by the Institute of Physics and the Philosophical Society, London (1970).
19. K. Z. Morgan, Rolf M. Sievert: The pioneer in the field of radiation protection, *Health Phys. 31*, 263–264 (1976).
20. International Commission on Radiological Protection, *Recommendations of the International Commission on Radiological Protection*, ICRP Publication 26, Pergamon Press, New York (1977).
21. International Commission on Radiological Protection, *Limits for Intakes of Radionuclides by Workers*, ICRP Publication 30, Pergamon Press, New York (1979).
22. International Commission on Radiological Protection, Statement and Recommendation of the 1980 Brighton Meeting of the ICRP, *Annals of the ICRP*, Vol. 2, No. 3/4, Pergamon Press, New York (1979).
23. International Commission on Radiological Protection, *Limits for Inhalation of Radon Daughters by Workers*, ICRP Publication 32, Pergamon Press, New York (1981).

CHAPTER 2

1. R. D. Evans and C. Goodman, Determination of the thoron content of air and its bearing on lung cancer hazards of industry, *J. Ind. Hyg. Toxicol. 22*, 89–99 (1940); note by N. S. Stannard, *Health Phys. 38*, 919 (1980).

2. E. Lorenz, Radioactivity and lung cancer: A critical review of lung cancer in the miners of Schneeberg and Joachimsthal, *J. Natl. Cancer Inst.* 5, 1–15 (1944); Agricola, *De Re Metallica* (translated by Herbert and Lou Hoover), Dover Publications, New York (1950); B. M. Fried, *Bronchogenic Carcinoma and Adenoma*, The Williams and Wilkins Co., Baltimore (1948).
3. D. Wilson, *Rutherford, Simple Genius*, MIT Press, Cambridge, MA (1983).
4. M. Uhlig, Über den Schneeberger Lungenkrebs, *Virchows Arch. Pathol. Anat. Physiol.* 230, 76–98 (1921).
5. A. Arnstein, Über der Sogenannten Schneeberger Lungenkrebs, *Verh. Dtsch. Ges. Pathol.* 16, 332–342 (1913).
6. A. Romer, *The Restless Atom: The Awakening of Nuclear Physics*, Dover Publications, Inc., New York (1982); A. Romer, ed., *The Discovery of Radioactivity and Transmutation*, Dover Publications, Inc., New York (1964); A. Romer, ed., *Radiochemistry and the Discovery of Isotopes*, Dover Publications, Inc., New York (1970); *Encyclopaedia Britannica*, 1911 edition, S. V. "radioactivity," written by Ernest Rutherford, and the 1973 edition, S. V. "radioactivity," written by Norman Feather, one of Rutherford's students; E. Rutherford, *Radioactive Substances and Their Radiations*, Cambridge University Press, G. P. Putnam's Sons, New York (1913); Marie Curie, Sc.D. thesis, published in *The Chemical News* 88, 85–86, 97–99, 134–135, 145–147, 159–160, 169–171, 175–177, 187–188, 199–201, 211–212, 223–224, 235–236, 247–249, 259–261, 271–272 (1903); L. Badash, *Radioactivity in America, Growth and Decay of a Science*, Johns Hopkins University Press, Baltimore (1979).
7. H. Becquerel, Sur les radiations émises par phosphorescence, *C. R. Acad. Sci.* 122, 420–421 (1896); a translation appears in the *The Discovery of Radioactivity and Transmutation* (A. Romer, ed.), Dover Publications, Inc., New York (1964).
8. H. Becquerel, Emission de radiations nouvelles par l'uranium métallique, *C. R. Acad. Sci.* 122, 1086–1088 (1896).
9. M. Curie, P. Curie, and M. G. Bémont, Sur une nouvelle substance fortement radio-active, contenue dans la pitchblende, *C. R. Acad. Sci.* 127, 1215–1217 (1898).
10. M. E. Weeks, *Discovery of the Elements*, 7th Ed., Journal of Chemical Education, Easton, PA (1968); L. F. Miller, Electroscopes and methods of radioactive measurements, *Quarterly of the Colorado School of Mines* 9, 1–15 (1914).
11. A. Romer, *The Restless Atom*, Dover Publications, Inc., New York (1982).
12. M. Curie, Recherches sur les substances radioactives, *Ann. Chim. Phys.* 30, 199–203 (1903).
13. E. Rutherford, A radioactive substance emitted from thorium compounds, *The London, Edinburgh and Dublin Philosophical Magazine and Journal of Science* 49, 1–14 (1900).
14. E. Rutherford, Radioactivity produced in substances by the action of thorium compounds, *The London, Edinburgh and Dublin Philosophical Magazine and Journal of Science* 49, 161–192 (1900).

15. A. S. Eve, *Rutherford,* Cambridge University Press, Cambridge (1939); N. Feather, *Lord Rutherford,* Blackie, Glasgow (1940).
16. P. Curie and A. Debierne, Sur la radio-activité et les gaz activés par le radium, *C. R. Acad. Sci. 132,* 768–770 (1901).
17. E. Dorn, Versuche Über Sekundarstrahlen und Radiumstrahlen, *Abhandlungen der Naturforschenden Gesellschaft zu Halle 22,* 37–43 (1900).
18. E. Rutherford and H. T. Brooks, The new gas from radium, *Transactions of the Royal Society of Canada 7,* 21–25 (1901).
19. E. Rutherford and F. Soddy, The radioactivity of thorium compounds, I. An investigation of the radioactive emanation, *J. Chem. Soc., Transactions 81,* 321–350 (1902).
20. E. Rutherford, The succession of changes in radioactive bodies, *Philosophical Transactions of the Royal Society of London 204A,* 169–219 (1904).
21. J. Elster and H. Geitel, Über Eine Fernere der Naturlichen und der Durch Becquerelstrahlen Abnorm Leitend Gemachten Luft, *Physik. Z. 2,* 590–593 (1901).
22. E. Rutherford and F. Soddy, The radioactivity of thorium compounds, II. The cause and nature of radioactivity, *J. Chem. Soc., Transactions 81,* 837–860 (1902).
23. T. J. Trenn, *The Self Splitting Atom,* Taylor and Francis, Ltd., London (1977).
24. A. Debierne, Sur les gaz produits par l'actinium, *C. R. Acad. Sci. 141,* 383–385 (1905); F. O. Giesel, Über den Emanationskorper (Emanium) *Ber. 37,* 1696–1699, 3963–3936 (1904).
25. E. Rutherford and F. Soddy, The radioactivity of uranium, *The London, Edinburgh and Dublin Philosophical Magazine and Journal of Science 5,* 441–445 (1903); E. Rutherford and F. Soddy, A comparative study of the radioactivity of radium and thorium, *ibid. 5,* 445–457 (1903); E. Rutherford and F. Soddy, Condensation of the radioactive emanations, *ibid. 5,* 561–576 (1903).
26. W. Ramsay and J. W. Collie, The Spectrum of radium, *Proc. Roy. Soc. 73,* 470–476 (1904).
27. J. J. Thomson, Experiments on induced-radioactivity in air, and on the electrical conductivity produced in gases when they pass through water, *The London, Edinburgh and Dublin Philosophical Magazine and Journal of Science 4,* 352–367 (1902).
28. F. R. Von Traubenberg, Über die Gultigkeit des Daltonschen Resp. Henryschen Gesetzes Bei der Absorption der Emanation des Frieburger Leitungswassers und der Radiumemanation Durch Verschiedene Flussigkeiten, *Physik. Z. 5,* 130–134 (1904).
29. H. A. Bumstead and L. P. Wheeler, On the properties of a radioactive gas found in the soil and water near New Haven, *Am. J. Sci. 167,* 97–111 (1904).
30. L. Badash, The suicidal success of radiochemistry, *The British Journal for the History of Science 12,* 245–256 (1979).

REFERENCES—CHAPTER 2

31. K. Fajans, Über Eine Beziehung Zwischen der Art Einer Radioactiven Umwandlung und dem Elektrochemischen Verhalten der Betreffenden Radioelemente, *Physik. Z. 14*, 131–136 (1913); Die Stellung der Radioelemente im Periodischen System, *ibid. 14*, 136–142 (1913).
32. R. D. Evans, Determination of small quantities of radon and thoron, *Phys. Rev. 39*, 1014–1020 (1932).
33. R. D. Evans, Apparatus for the determination minute quantities of radium, radon, and thorium in solids, liquids and gases, *Rev. Sci. Instrum. 6*, 99–112 (1935).
34. L. F. Curtiss and F. J. Davis, A counting method for the determination of small amounts of radium and of radon, *J. Nat. Bur. Stand. 31*, 181–195 (1943).
35. H. F. Lucas, Improved low-level alapha-scintillation counter for radon, *Rev. Sci. Instrum. 28*, 680–683 (1957).
36. H. Jansen and P. Schultzer, Experimental investigations into internal radium emanation therapy, I. Emanatorium experiments with rats, *Acta Radiol. 6*, 631–646 (1926).
37. P. Schultzer, Experimental investigations into internal radium emanation therapy, II. On the cause of the effect on rats of continuous emanation treatment, *Acta Radiol. 6*, 647–657 (1926).
38. J. H. Harley, Sampling and measurement of airbone daughter products of radon, *Nucleonics 11*, 12–15 (1953).
39. A. Pirchan and H. Sikl, Cancer of the lung in the miners of Jachymov (Joachimsthal), report of cases observed in 1929–1930, *Am. J. Cancer 16*, 681–722 (1932).
40. E. Lorenz, Radioactivity and lung cancer: a critical review of lung cancer in the miners of Schneeberg and Joachimsthal, *J. Natl. Cancer Inst. 5*, 1–15 (1944).
41. F. B. Flynn, Dangers of internal radium therapy, *Am. J. Phys. Therap. 9*, 65–66 (1932).
42. H. Schlundt, W. McGavock, Jr., and M. Brown, Dangers in refining radioactive substances, *J. Ind. Hyg. 13*, 117–134 (1931).
43. T. H. Oddie, A method for the routine purification of radon, *Br. J. Radiol. 10*, 348–359 (1937).
44. R. D. Evans, Inception of standards for internal emitters, radon and radium, *Health Phys. 41*, 437–448 (1981).
45. S. C. Lind, *The Chemical Effects of Alpha Particles and Electrons*, 2nd Ed., American Chemical Society Monograph Series, The Chemical Catalog Company, New York (1928).
46. M. C. McPherson, *Time Bomb*, E. P. Dutton, New York (1986).
47. C. T. Hess, private communication, Physics Department, University of Maine, Orono.
48. Michael Lafavore, Warning! This house contains radon, *Readers Digest* (June 1986); M. LaFavore, The radon report, *New Shelter* (January 1986).
49. H. L. Kusnetz, Radon daughters in mine atmospheres, a field method for determining concentrations, *Am. Ind. Hyg. Assoc. Q. 17*, 85–88 (1956).

50. E. Landa, Colorado radium: Mining, processing and usage in medicine, science and industry, *Colorado School of Mines Quarterly, 82*(2) (in press).
51. W. C. Stevenson, Preliminary clinical report on a new and economical method of radium therapy by means of emanation needles, *Br. Med. J. 2*, 9–10 (1914).
52. N. Goldstein, Radon seed implants, *Arch. Dermatol. 111*, 757–759 (1975).
53. E. P. Hendricks, B. D. Massey, E. F. Nation, C. A. Gallup, B. D. Massey, Jr., and J. W. Edwards, Radon in treatment of infiltrating carcinoma of urinary bladder, *Urology 5*, 465–469 (1975).
54. B. S. Hilaris, J. H. Kim, and N. Tokita, Low energy radio-nuclides for permanent interstitial implantation, *Am. J. Roentgenol. 126*, 171–178 (1976).
55. N. Beheshiti and N. Javid, Oral tissue and irradiation therapy, *Isr. J. Dent. Med. 27*, 31–35 (1978).
56. E. R. Landa, The first nuclear industry, *Sci. Am. 247*, 180–193 (1982).
57. For more details on the use of radon in earthquake prediction see: *Earthquake Prediction, Proceedings of the International Symposium on Earthquake Prediction*, 1984, United Nations Educational, Scientific and Cultural Organization, 7 Place de Fontenoy, 75700 Paris, France (1984); June 10, 1980 issue of the *J. Geophys. Res.*; May 1981 issue of *Geophys. Res. Lett.*; and such journals as *Earthquake Prediction Research* and the *Journal of Seismological Research*.
58. B. S. Amin and Rama, A search for correlation between seismicity and radon anomaly in hot springs, *Proc. Indian Acad. Sci. 91*, 15–19 (1982).
59. S. C. Liu, J. R. McAfee, and R. J. Cicerone, Radon 222 and tropospheric vertical transport, *J. Geophys. Res. 89*, 7291–7297 (1984).
60. C. Rangarajan, S. Gopalakrishnan, V. R. Chandrasekanan, and C. D. Eapen, The relative concentrations of radon daughter products in surface air and the significance of their ratios, *J. Geophys. Res. 80*, 845–848 (1975).
61. G. Lambert, A. Buisson, J. Sanak, and B. Ardouin, Modification of the atmospheric polonium 210 to lead 210 ratio by volcanic emission, *J. Geophys. Res. 84*, 6980–6986 (1979).
62. J. Sanak, G. Lambert, and B. Ardouin, Lead-210 in the atmosphere, in: *Natural Radiation Environment III* (T. F. Gesell and W. M. Lowder, eds.), pp. 445–467, Technical Information Center, U. S. Department of Energy, Washington, DC (1980).
63. R. E. Larson and D. J. Bressan, Radon-222 as an indicator of continental air masses and air mass boundaries over ocean areas, in: *Natural Radiation Environment III* (T. F. Gesell and W. M. Lowder, eds.), pp. 308–326, Technical Information Center, U. S. Department of Energy, Washington, DC (1980).
64. Rama, Using natural radon for delineating monsoon circulation, *J. Geophys. Res. 75*, 2227–2229 (1970).
65. *Encyclopedia of Science and Technology*, Vol. 11, p. 373, McGraw-Hill, New York (1982).
66. R. L. Kathren, *Radioactivity in the Environment: Sources, Distribution and Surveillance*, Harwood Academic Publishers, New York (1984).

67. *Nostrums and Quackery, Articles on the Nostrum Evil and Quackery*, Reprinted from the *Journal of the American Medical Association*, Press of the American Medical Association, Chicago (1936).
68. R. L. Kathren, Historical development of radiation measurement and protection, in: *Handbook of Radiation Measurement and Protection* (A. Brodsky, ed.), CRC press, Boca Raton, FL (1978), pp. 13–52.
69. E. R. Landa, C. L. Miller, and R. F. Brich, Radioactive and nonradioactive solutes in drinking water from radon-charging devices (submitted for publication).
70. The Denver Radium Service, *Radium Therapeutics and Methods of Administration For the General Practitioner*, The Denver Radium Service, Denver, CO (1930).
71. J. Schubert and R. E. Lapp, *Radiation: What It Is and How It Affects You*, Viking Press (1957); E. W. Robinson, The Use of Radium in Consumer Products, U.S. Public Health Service Report MORP 68–5, Washington, DC (1968).
72. W. A. Jennings and S. Russ, *Radon: Its Techniquue and Use*, Published for the Middlesex Hospital Press by John Murray, Albemarle Street, London, W1 (1948).
73. Report of Council on Physical Medicine and Rehabilitation, Alphatron radon ointment not acceptable, *J. Am. Med. Assoc. 140*, 667 (1949).
74. *New York Times*, Sunday, November 25, 1984, p. 25.
75. W. V. Lewis, *Arthritis and Radioactivity, A Story of Montana's Free Enterprise Uranium-Radon Mine*, The Christopher Publishing House, Boston (1955).
76. P. Curtis and E. Laborde, Sur la réactivité des gaz qui se dégagent de leau des sources thermales, *C. R. Acad. Sci. 138*, 1150–1153 (1904).
77. I. Uznov, R. Steinhausler, and E. Pohl, Carcinogenic risk of exposure to radon daughters associated with radon spas, *Health Phys. 41*, 807–813 (1981).
78. W. Chruschielewski, T. Domanski, and W. Orzechowski, Concentrations of radon and its progeny in the rooms of Polish spas, *Health Phys. 45*, 421–424 (1983).
79. S. Kimura and T. Komae, Applications of environmental radon-222 to some cases of water circulation, pp. 581–599, *The Natural Environment III* (T. F. Gesell and W. M. Lowder, eds.), Technical Information Center, U.S. Department of Energy, Washington, DC (1980).
80. A. B. Tanner, Physical and chemical controls of distribution of radium-226 and radon-222 in ground water near the Great Salt Lake, Utah, in *The Natural Radiation Environment* (J. A. S. Adams and W. M. Lowder, eds.), pp. 253–274, University of Chicago Press, Chicago (1964).
81. W. S. Broecker, An application of natural radon to problems of ocean circulation, in: Symposium on Diffusion in Ocean and Freshwater, Lamont Geological Observatory, Palisades, NY (1964).
82. A. B. Tanner, Radon migration in the ground: A supplementary review, in: *Natural Radiation Environment III* (T. F. Gesell and W. M. Lowder,

eds.), pp. 5–56, Technical Information Center, U.S. Department of Energy, Washington, DC (1980).
83. J. W. Card, K. Bell, G. M. Derham, and S. R. S. Shah, Radon decay product measurements in radiometric uranium exploration: Implications for petroleum exploration, *Oil Gas J. 1985* (June 24) 114–118.
84. A. B. Tanner, Radon migration in the ground: A review, in: *The Natural Radiation Environment* (J. A. S. Adams and W. M. Lowder, eds.), pp. 161–190, University of Chicago Press, Chicago (1964).

CHAPTER 3

1. T. F. Gesell, Background atmospheric radon concentrations outdoors and indoors: A review, *Health Phys. 43*, 277–289 (1983).
2. "Grand Junction Remedial Action Criteria," Code of Federal Regulations, Title 10, Part 712 (10CFR712).
3. A. C. George, Instruments and methods for measuring indoor radon and radon progeny concentrations, in: *Indoor Radon,* pp. 87–101, Air Pollution Control Association, Pittsburgh, PA (1986).
4. H. F. Lucas, Jr., Alpha scintillation radon counting, in: *Workshop on Methods for Measuring Radiation in and around Uranium Mills* (E. D. Harward, ed.), Vol. 3, pp. 69–95, Atomic Industrial Forum, Bethesda, MD (1977).
5. H. F. Lucas, Jr., Improved low-level alpha scintillation counters for radon, *Rev. Sci. Instrum. 28*, 680–685 (1957).
6. A. C. George, Scintillation flasks for the determination of low level concentrations of radon, in: *Proceedings of the Ninth Midyear Health Physics Symposium,* Colorado Chapter of the Health Physics Society, Denver, CO (1976), pp. 112–115.
7. C. W. Sill, An integrating air sampler for determination of Rn-222, *Health Phys. 16*, 371–377 (1969).
8. J. W. Thomas and R. J. Countess, Continuous radon monitor, *Health Phys. 36*, 734–738 (1979).
9. W. W. Nazaroff, F. J. Offerman, and A. W. Robb, Automated system for measuring air exchange rate and radon concentration in houses, *Health Phys. 45*, 525–539 (1983).
10. J. Harley (ed.), HASL Procedures Manual, Health and Safety Laboratory Report HASL-300, New York (1972).
11. M. E. Wrenn, Design of a continuous digital output environmental radon monitor, *IEEE Trans. Nucl. Sci. 22*, 645–648 (1975).
12. M. E. Wrenn and H. B. Spitz, The design and application of a continuous digital readout radon measuring instrument, in: *Workshop on Methods for Measuring Radiation in and around Uranium Mills* (E. D. Harward, ed.), Vol. 3, pp. 119–130, Atomic Industrial Forum, Bethesda, MD (1977).

REFERENCES—CHAPTER 3

13. R. Rolle, Rapid working level monitoring, *Health Phys.* 22, 233–238 (1972).
14. A. C. George and A. J. Breslin, Measurement of environmental radon with integrating instruments, in: *Workshop on Methods for Measuring Radiation in and around Uranium Mills* (E. D. Harward, ed.), Vol. 3, Atomic Industrial Forum, Bethesda, MD (1977), pp. 105–115.
15. J. W. Thomas and P. C. LeClare, A study of the two filter method for Rn-222, *Health Phys.* 18, 113–122 (1970).
16. A. C. George, A cumulative environmental radon monitor, in: *Proceedings of the Ninth Midyear Health Physics Symposium*, pp. 116–120, Colorado Chapter of the Health Physics Society, Denver, CO (1976).
17. A. C. George, Passive integrated measurement of indoor radon using activated carbon, *Health Phys.* 46, 867–872 (1984).
18. B. L. Cohen and E. S. Cohen, Theory and practice of radon monitoring with charcoal adsorption, *Health Phys.* 45, 501–508 (1983).
19. E. L. Geiger, Radon film badge, *Health Phys.* 13, 407–411 (1967).
20. H. W. Alter and R. L. Fleischer, Passive integrating radon monitor for environmental monitoring, *Health Phys.* 40, 693–702 (1981).
21. M. Urban and E. Piesch, Low level environmental radon dosimetry with a passive track etch device, *Radiat. Prot. Dosim.* 1, 97 (1982).
22. H. L. Kusnetz, Radon daughters in mine atmospheres. A field method for determining concentrations, *Am. Ind. Hyg. Assoc. J.* 17, 85 (1956).
23. E. G. Tsviglou, H. E. Ayers, and D. A. Holaday, Occurrence of nonequilibrium atmospheric mixtures of radon and daughters, *Nucleonics* 11, 40 (1953).
24. C. Rangarajan and S. Gopalakrishnan, The estimation of the relative concentrations of short-lived radon daughters by gamma measurements, *Health Phys.* 24, 433–436 (1973).
25. L. B. Lockhart, R. L. Patterson, and C. R. Hosler, Determination of Radon Concentration in Air through Measurement of Its Solid Decay Products, Report 6229, U.S. Naval Research Lab, Washington, DC (1965).
26. D. E. Martz, D. F. Holleman, D. E. McCurdy, and K. J. Schiager, Analysis of atmospheric concentrations of RaA, RaB and RaC by alpha spectroscopy, *Health Phys.* 17, 131–138 (1969).
27. J. Bigu, R. Raz, K. Golden, and P. Dominquez, A Computer Based Continuous Monitor for the Determination of the Short Lived Decay Products of Radon and Thoron, Division Report MPR/MRL 83 (OP) J, Canada Centre for Mineral and Energy Technologies, Energy, Mines and Resources of Canada, Elliott Lake, Ontario (1983).
28. J. A. Auxier, K. Becker, E. M. Robinson, D. R. Johnson, R. H. Boyett, and C. H. Abner, A new radon progeny personnel dosimeter, *Health Phys.* 21, 126–128 (1971).
29. D. B. Lovett, Track etch detectors for alpha exposure estimation, *Health Phys.* 16, 623–628 (1969).
30. K. Becker, Alpha particle registration in plastics and its applications for radon and neutron personnel dosimetry, *Health Phys.* 16, 113–123 (1969).

31. O. White, Jr., Environmental Measurements Lab, Report HASL TM 71-17, New York (1971).
32. K. J. Schiager, Integrating radon progeny air sampler, *Am. Ind. Hyg. Assoc. J. 35,* 165 (1974).
33. A. J. Breslin, S. F. Guggenheim, A. C. George, and R. T. Graveson, A Working Level Dosimeter for Uranium Miners, Report EML-333, U.S. Department of Energy, New York (1977).
34. F. S. Guggenheim, A. C. George, R. T. Graveson, and A. J. Breslin, A time-integrating environmental radon daughter monitor, *Health Phys. 36,* 452–455 (1979).
35. O. White, Jr., USAEC Health and Safety Laboratory (now the Environmental Measurements Lab) Report HASL TM 69–23A, New York (1969).
36. J. Bigu and R. Kaldenbach, Theory, operation and performance of a time-integrating continuous radon/thoron daughter working level monitor, *Radiat. Prot. Dosim. 9,* 19 (1984).
37. M. Eisenbud, In-vivo measurement of Pb-210 as an indicator of cumulative radon daughter exposure in uranium mines, *Health Phys. 16,* 637–646 (1969).
38. H. L. Fisher, Jr., A model for estimating the inhalation exposure to radon-222 and daughter products from the accumulated lead-210 body burden, *Health Phys. 16,* 597–616 (1969).
39. R. F. Bell and J. C. Gilliland, Urinary lead-210 as an index of mine radon exposure, in: *Radiological Health and Safety in the Mining and Milling of Nuclear Materials,* Vol. 2, pp. 411–412, International Atomic Energy Agency, Vienna (1964).
40. J. Michel and W. S. Moore, Sources and Behavior of Natural Radioactivity in Fall Line Aquifers near Larvette, S.C., Water Resources Research Institute Report No. 83, Clemson University, Clemson, SC (1980).
41. H. M. Pritchard and T. F. Gesell, An estimate of population exposure due to radon in public water supplies in the area of Houston, Texas, *Health Phys. 41,* 599 (1981).
42. M. Asikainen, State of disequilibrium between ^{238}U, ^{234}U, ^{226}Ra, and ^{222}Rn in groundwater from bedrock, *Geochim. Cosmochim. Acta 45,* 201–206 (1981).
43. H. F. Lucas, A fast and accurate survey technique for both radon-222 and radium-226, in: *Natural Radiation Environment* (J. A. S. Adams and W. M. Lowder, eds.), University of Chicago Press, Chicago (1964).
44. H. M. Prichard, T. F. Gesell, and C. R. Meyer, Liquid scintillation analyses for radium-226 and radon-222 in potable waters, in: *Liquid Scintillation Counting, Recent Applications and Development, Volume 1, Physical Aspects* (C.-T. Peng, D. L. Horrocks, and E. L. Alpen, eds.), Academic Press, New York, pp. 347–355 (1980).
45. H. M. Pritchard and T. F. Gesell, Rapid measurement of ^{222}Rn concentration in water with a commercial liquid scintillation counter, *Health Phys. 33,* 577–581 (1977).

46. Research Planning Institute, Inc., Statistical Analysis of Analytical Methods for Radionuclide in Drinking Water, Report to U.S. Environmental Protection Agency, Office of Drinking Water, Washington, DC (1986).

CHAPTER 4

1. S. P. Clark, Jr., *Handbook of Physical Constants*, Geol. Soc. Am. Memoir 97, New York (1966).
2. J. B. Mertie, Jr., Monazite deposits of the southeastern United States, *U.S. Geol. Survey Circular 237*, 31 (1953).
3. A. B. Tanner, in: *Natural Radiation Environment* (J. A. S. Adams and W. M. Lowder, eds.), pp. 161–190, University of Chicago Press, Chicago (1964).
4. A. B. Tanner, in: *Natural Radiation Environment III* (T. F. Gesell and W. J. Lowder, eds.), pp. 5–56, U.S. Department of Energy Report CONF-780422, Washington, DC (1980).
5. G. Lambert, P. Bristeau, and G. Polian, Evidence of little migration of radon within rock grains, *C. R. Acad. Sci., Ser. D*, 274, 3333–3336 (1972).
6. G. Lambert and P. Bristeau, Migration of radon atoms implanted in crystals by recoil energy, *J. Phys. (Paris), Colloq. C5*, 137–138 (1973).
7. A. V. Nero and W. W. Nazaroff, Characterizing the source of radon indoors, *Radiat. Prot. Dosim. 7*, 23–39 (1984).
8. A. B. Tanner, in: *Proc. Air Pollution Control Association International Specialty Conference*, pp. 1–12, Air Pollution Control Association Pub. SP-54, Pittsburgh, PA (1986).
9. Rama and W. S. Moore, Mechanism of transport of U-Th series radioisotopes from solids into ground water, *Geochim. Cosmochim. Acta 48*, 395–399 (1984).
10. Rama and W. S. Moore, Nanoporosity in natural minerals, unpublished manuscript, Department of Geology, University of South Carolina, Columbia (1986).
11. P. M. C. Barretto, Emanation Characteristics of Rocks, Soils, and Rn-222 Ion Effect on the U-Pb System Discordance, Ph.D. thesis, Rice University, Houston, TX (1973).
12. K. Megumi and T. Mamuro, Emanation and exhalation of radon and thoron gases from soil particles, *J. Geophys. Res. 79*, 3357–3360 (1974).
13. K. Megumi and T. Mamuro, Concentration of uranium series nuclides in soil particles in relation to their size, *J. Geophys. Res. 82*, 353–356 (1977).
14. E. Stranden, A. K. Kolstad, and B. Lind, Radon exhalation: Moisture and temperature dependence, *Health Phys. 47*, 480–484 (1984).
15. W. Jacobi and K. Andre, The vertical distribution of radon-222, radon-

220, and their decay products in the atmosphere, *J. Geophys. Res.* **68**, 3799–3814 (1963).
16. Rama, Monsoon circulation from observation of natural radon, *Earth Planet. Sci. Lett.* **6**, 56–60 (1969).
17. R. E. Larson and D. J. Bressan, in: *Natural Radiation Environment III* (T. F. Gesell and W. M. Lowder, eds.), pp. 308–326, U.S. Department of Energy Report CONF-780422, Washington, DC (1980).
18. T. F. Gesell, Background atmospheric ^{222}Rn concentrations outdoors and indoors: A review, *Health Phys.* **45**, 289–302 (1983).
19. United Nations Scientific Committee on the Effects of Atomic Radiation, Sources and Effects of Ionizing Radiation, Report to the General Assembly with Annexes, United Nations, New York (1977).
20. H. M. Pritchard and T. F. Gessell, Radon in the environment, in: *Advances in Radiation Biology*, Vol. II, John Lett (ed.), pp. 391–428, Academic Press, New York (1984).
21. G. Akerblom, P. Andersson, and B. Clavensjo, Soil gas radon—a source for indoor radon daughters, *Radiat. Prot. Dosim.* **7**, 49–54 (1984).
22. A. Hesselborn, Radon in Soil Gas: A Study of Methods and Instruments for Determining Radon Concentration in the Ground, Geological Survey of Sweden, Series C, No. 803, Stockholm (1985), p. 58.
23. J. Otten, Indoor radon—Geologic controls in Pacific Northwest (abs.), in: *Proc. Geological Society of American, Southeast Section Meeting* (1987).
24. T. E. Myrick, B. A. Borven, and F. F. Haywood, Determination of concentrations of selected radionuclides in surface soils in the United States, *Health Phys.* **45**, 631–642 (1983).
25. B. A. Moed, W. W. Nazaroff, A. V. Nero, M. B. Schwehr, and A. Van Heuvelen, Identifying Areas with Potential for High Indoor Radon Levels: Analysis of the National Airborne Radiometric Reconnaissance Data for California and the Pacific Northwest, Lawrence Berkeley Laboratory Report (LBL-16955), University of California, Berkeley (1984), p. 70.
26. G. V. Akerblom, in: Proc. International Meeting on Radon—Radon Progeny Measurements, U.S. EPA Report 520/5-83/021 (1982), pp. 171–185.
27. W. W. Nazaroff and A. V. Nero, Transport of Radon from Soil into Residences, Lawrence Berkeley Laboratory Report (LBL-16823), University of California, Berkeley (1984).
28. W. W. Nazaroff, S. R. Lewis, S. M. Doyle, B. A. Moed, and A. V. Nero, Experiments in Pollutant Transport from Soil into Residential Basements by Pressure-Drive Air Flow, Lawrence Berkeley Laboratory Report (LBL-18374), University of California, Berkeley (1986), p. 25.
29. W. E. Clements and M. H. Wilkening, Atmospheric pressure effects on ^{222}Rn transport across the earth-air interface, *J. Geophys. Res.* **79**, 5025–5029 (1974).
30. G. A. Swedjemark, Radon and Its Decay Products in Housing, Department of Radiation Physics, University of Stockholm Report ISBN91-7146-637-7 (1985).

31. C. Wilson, in: Proc. International Meeting on Radon—Radon Progeny Measurements, U.S. EPA Report 520/5-83/021 (1982), pp. 209–233.
32. G. Akerblom, Investigation and Mapping of Radon Risk Areas, Swedish Geological Co. Report: IRAP 86036, Lulea, Sweden (1986), p. 15.
33. U.S. Environmental Protection Agency, Areas with Potentially High Radon Levels FACT Sheet, U.S. EPA, Office of Radiation Programs, Washington, DC (1986).
34. E. P. Wagner, Radon monitoring in northern communities, *Chronic Diseases in Canada 4*, 6–7 (1983).
35. J. K. Cochran, in: *Uranium Series Disequilibrium: Applications to Environmental Problems* (M. Ivanovitch and R. S. Harmon, eds.), pp. 384–430, Clarendon Press, Oxford, England (1982).
36. R. J. Elsinger and W. S. Moore, Gas exchange in the Pee Dee River based on ^{222}Rn evasion, *Geophys. Res. Lett. 10*, 443–446 (1983).
37. S. Emerson, Gas exchange rates in small Canadian shield lakes, *Limnol. Oceanogr. 20*, 754–761 (1975).
38. W. S. Broecker, in: *Symposium on Diffusion in Oceans and Fresh Water* (T. Ichiye, ed.), pp. 116–145, Lamont Doherty Geological Observatory, Columbia University, Palisades, NY (1965).
39. T. R. Horton, Nationwide Occurrence of Radon and Other Natural Radioactivity in Public Water Supplies, U.S. EPA Report 520/5-85-008, U.S. EPA Eastern Environmental Radiation Facility, Montgomery, AL (1985), p. 208.
40. C. T. Hess, J. Michel, T. R. Horton, H. M. Pritchard, and W. A. Coniglio, The occurrence of radioactivity in public water supplies in the United States, *Health Phys. 48*, 553–586 (1985).
41. D. P. Loomis, The relationship between water system size and ^{222}Rn concentration in North Carolina public water supplies, *Health Phys. 52*, 69–71 (1987).
42. C. T. Hess, C. V. Weiffenbach, and S. A. Norton, Environmental radon and cancer correlations in Maine, *Health Phys. 45*, 339–348 (1983).
43. D. P. Loomis, Radon-222 concentration and aquifer lithology in North Carolina, *Ground Water Monitoring Review 7*, 33–39 (1987).
44. J. Michel and W. S. Moore, ^{228}Ra and ^{226}Ra Content of Groundwater in Fall Line Aquifers near Leesville, SC, S.C. Water Resources Research Inst. Report 83, Clemson University, Clemson, SC (1980), p. 50.
45. M. Aastrup, Natural Activities of Uranium, Radium, and Radon in Groundwater, Geological Survey of Sweden, SKBF-KBS-TR-81-08 (1981), p. 23
46. J. Michel and C. D. Pollman, A Model for the Occurrence of ^{228}Ra in Groundwater, Environmental Science and Engineering Report No. 81-227-270, Gainesville, FL (1982), p. 50.
47. R. S. Livley and G. B. Morey, in: *Isotope Studies of Hydrologic Processes* (E. C. Perry, Jr. and C. W. Montgomery, eds.), pp. 91–108, Northern Illinois University Press, DeKalb, IL (1982).

48. M. K. Sasser and J. E. Watson, An evaluation of the radon concentrations of North Carolina groundwater, *Health Phys. 34,* 667–671 (1978).
49. J. Michel and W. S. Moore, ^{228}Ra and ^{226}Ra content of groundwater in Fall Line acquifers, *Health Phys. 38,* 663–671 (1980).
50. King, J. Michel, and W. S. Moore, Grondwater geochemistry of ^{228}Ra, ^{226}Ra, and ^{222}Rn, *Geochim. Cosmochim. Acta, 46* 1173–1182 (1982).
51. W. Cline, S. Adamovitz, C. Blackman, and B. Kahn, Radium and uranium concentrations in Georgia community water supplies, *Health Phys. 43,* 1–12 (1983).
52. C. R. Cothern, W. L. Lappenbusch, and J. Michel, Drinking water contribution to natural background radiation, *Health Phys. 50,* 33–47 (1986).
53. R. Collé, R. J. Rubin, L. I. Knob, and J. M. R. Hutchinson, Radon Transport through and Exhalation from Building Materials: A Review and Assessment, U.S. Department of Commerce, National Bureau of Standards, Washington, DC, Technical Note 1139 (1981).
54. N. Jonassen and J. P. McLaughlin, in: *Natural Radiation Environment III* (T. F. Gesell and W. J. Lowder, eds.), pp. 1211–1224, U.S. Department of Energy Report CONF-780422, Washington, DC (1980).
55. G. V. Akerblom and C. Wilson, Radon gas—a radiation hazard from radioactive bedrock and building materials, *Bull. Int. Assoc. Eng. Geol. 23,* 51–61 (1981).
56. E. Stranden, Assessment of the radiological impact of using fly ash in cement, *Health Phys. 44,* 145–153 (1982).
57. B. Kahn, G. G. Eicholz, and F. J. Clarke, Assessment of the Critical Populations at Risk Due to Radiation Exposure in Structures, Georgia Institute of Technology Report, Atlanta, GA (1979).
58. J. G. Ingersoll, A survey of radionuclide contents and radon emanation rates in building materials used in the United States, *Health Phys. 45,* 363–368 (1983).
59. A. C. George, E. O. Knutson, and H. Franklin, Radon and radon daughter measurements in solar buildings, *Health Phys. 45,* 413–420 (1983).
60. T. F. Gesell, Occupational radiation exposure due to ^{222}Rn in natural gas and natural gas products, *Health Phys. 29,* 681–687 (1975).
61. U.S. Environmental Protection Agency, Preliminary Findings: Radon Daughter Levels in Structures Constructed on Reclaimed Florida Phosphate Lands, U.S. EPA Tech. Note ORP/CDS-75-4 (1975).
62. R. J. Guimond, Jr., W. H. Ellett, J. E. Fitzgerald, Jr., S. T. Windham, and P. A. Curry, Indoor Radiation Exposure Due to Radium-226 in Florida Phosphate Lands, U.S. EPA Report 520/4-78-013 (1979).
63. C. E. Roessler, G. S. Roessler, and W. E. Bolch, Indoor radon progeny exposure in the Florida phosphate mining region: A review, *Health Phys. 45,* 389–396 (1983).
64. C. W. Roessler, R. Kantz, W. E. Bolch, Jr., and J. A. Wethington, Jr., in: *Natural Radiation Environment III* (T. F. Gesell and W. M. Lowder, eds.),

pp. 1476–1493, U.S. Department of Energy Report CONF-78-0422, Washington, DC (1980).
65. J. N. Hartley and H. D. Freeman, Radon Flux Measurements in Gardinier and Royster Phosphogypsum Piles near Tampa and Mulberry, Florida, U.S. EPA Report 520/5-85-029 (1985), p. 29.
66. A. V. Nero, R. G. Sextro, S. M. Doyle, B. A. Moed, W. W. Nazaroff, K. L. Revzan, and M. B. Schwehr, Characterizing the sources, ranges, and environmental influences of radon-222 and its decay products, *The Science of the Total Environment 45*, 233–244 (1985).
67. T. F. Gesell and H. M. Pritchard, in: *Natural Radiation Environment III* (T. F. Gesell and W. M. Lowder, eds.), pp. 1347-1363, U.S. Department of Energy Report CONF-780477, Washington, DC (1980).
68. T. F. Gesell and H. M. Pritchard, The technologically enhanced natural radiation environment, *Health Phys. 28*, 361–366 (1975).
69. W. W. Nazaroff, S. M. Doyle, A. V. Nero, and R. G. Sextro, Potable water as a source of airborne radon-222 in U.S. dwellings: A review and assessment, *Health Phys. 52*, 281–295 (1987).

CHAPTER 5

1. United Nations Scientific Committee on the Effects of Atomic Radiation, *Sources and Effects of Ionizing Radiation*, United Nations, New York (1977).
2. D. Holaday, History of the exposure of miners to radon, *Health Phys. 16*, 547 (1969).
3. R. Pugh, in: *Sources and Effects of Ionizing Radiation*, United Nations, New York (1977).
4. A. George and A. J. Breslin, Deposition of radon daughters in humans exposed to uranium mine atmospheres, *Health Phys. 50*, 605–618 (1986).
5. S. Kumazawa, D. R. Nelson, and A. C. B. Richardson, Occupational Exposure to Ionizing Radiation in the United States, U.S. Environmental Protection Agency, Report EPA 520/1-84-005, Washington, DC (1984).
6. U.S. Environmental Protection Agency, Radiological Impact Caused by Emissions of Radionuclides into Air in the United States, Report EPA 520/7-79-006, Washington, DC (1979).
7. K. Neilson, Prediction of Net Radon Emission from a Model Open Pit Uranium Mine, Battelle Northwest, Report PNL 2889/NUREG-CR-0628, Richland, WA (1979).
8. W. E. Kisieleski (ed.), Radon Release and Dispersion from an Open Pit Uranium Mine, Argonne National Laboratory, Report NUREG/CR-1583 (1980).
9. R. L. Rock and D. K. Walker, *Controlling Employee Exposure to Alpha Radiation in Underground Uranium Mines*, Vol. 1., U.S. Bureau of Mines (1970).

10. M. V. Morgan and J. M. Samet, Radon daughters exposures of New Mexico U miners, *Health Phys. 50*, 656–662 (1986).
11. U.S. Environmental Protection Agency, Background Information Document—Standard for Radon-222 Emissions for Underground Uranium Mines, EPA 520/1-85-010 (1985).
12. U.S. Nuclear Regulatory Commission, Draft Environmental Statement related to operation of Highland Uranium Solution Mining Project, Report NUREG-0407, Washington, DC (1978).
13. U.S. Atomic Energy Commission, Environmental Survey of the Uranium Fuel Cycle, Report WASH-1284, Washington, DC (1974).
14. W. E. Clements, S. Barr, and M. L. Marple, Uranium mill tailings piles as sources of atmospheric radon-222, in: *Natural Radiation Environment III* (T. F. Gesell and W. M. Lowder, eds.), pp. 1559–1583, U.S. Department of Energy, Washington, DC (1980).
15. W. A. Goldsmith, F. F. Haywood, and R. W. Leggett, Transport of radon which diffuses from uranium mill tailings, in: *Natural Radiation Environment III* (T. F. Gesell and W. M. Lowder, eds.), pp. 1584–1600, U.S. Department of Energy, Washington, DC (1980).
16. U.S. Environmental Protection Agency, Regulatory Impact Analysis of Final Environmental Standards for Uranium Mill Tailings at Active Sites, Report 520/1-83-010, Washington, DC (1983).
17. "Use of Uranium Mill Tailings for Construction Purposes," in: *Hearings before the Subcommittee on Raw Materials,* Joint Committee on Atomic Energy, 92nd U.S. Congress, Washington, DC (1971).
18. K. J. Schiager and H. G. Olson, "Radon Progeny Exposure Control in Buildings," in: *Hearings before the Subcommittee on Raw Materials,* Joint Committee on Atomic Energy, 92nd U.S. Congress, Washington, DC (1971).
19. W. E. Bolch, N. Desai, C. E. Roessler, and R. S. Kautz, Determinants of radon flux from complex media: Virgin and reclaimed lands in Florida phosphate region, in: *Natural Radiation Environment III* (T. F. Gesell and W. M. Lowder, eds.), pp. 1673–1681, U.S. Department of Energy, Washington, DC (1980).
20. C. E. Roessler, R. Kautz, W. E. Bolch, and J. A. Wethington, The effect of mining and land reclamation on the radiological characteristics of the terrestrial environment of Florida's phosphate region, in: *Natural Radiation Environment III* (T. F. Gesell and W. M. Lowder, eds.), pp. 1476–1493, U.S. Department of Energy, Washington, DC (1980).
21. R. J. Guimond and S. T. Windham, Radioactivity Distribution in Phosphate Products, By-products, Effluents and Wastes, U.S. Environmental Protection Agency, Technical Note ORP/CSD-75-3, Washington, DC (1975).
22. R. J. Guimond, Preliminary Findings: Radon Daughter Levels in Structures Constructed on Reclaimed Florida Phosphate Land, U.S. Environmental Protection Agency, Technical Note ORP/CSD-75-4, Washington DC (1975).
23. G. G. Eichholz, J. P. Ambrose, and M. O. Skowroski, Evaluation of

REFERENCES—CHAPTER 5

potential radon exposure from development of Georgia phosphate deposits, *Min. Eng. 38*(3), 195–196 (1986).

24. C. C. Travis, A. P. Watson, L. M. McDowell-Boyer, S. J. Cotter, M. L. Randolph, and D. E. Fields, A Radiological Assessment of Radon-222 Released from Uranium Mills and Other Natural and Technologically Enhanced Sources, Report, Oak Ridge National Laboratory NUREG/CR-0573 (1979).
25. U.S. Environmental Protection Agency, Radiological Impact Caused by Emissions of Radionuclides into Air in the United States, EPA520/7-79-006, Washington, DC (1979).
26. J. Faucett Associates, Regulatory Impact Analysis of Emission Standards for Elemental Phosphorus Plants, U.S. Environmental Protection Agency, Report EPA 520/1-84-025, Washington, DC (1984).
27. R. J. Guimond, W. H. Ellett, J. E. Fitzgerald, S. T. Windham, and P. A. Cuny, Indoor Radiation Exposure Due to Radium-222 in Florida Phosphate Lands, U.S. Environmental Protection Agency, Report EPA 520/4-78-013, Washington, DC (1979).
28. C. E. Roessler, G. S. Roessler, and W. E. Bolch, Indoor radon progeny exposure in the Florida phosphate mining region: A review, *Health Phys. 45*, 389–396 (1983).
29. G. G. Eichholz, F. J. Clarke, and B. Kahn, Radiation exposure from building materials, in: *Natural Radiation Environment III* (T. F. Gesell and W. M. Lowder, eds.), pp. 1331–1346. U.S. Department of Energy, Washington, DC (1980).
30. B. Kahn, G. G. Eichholz, and F. J. Clarke, Search for building materials as sources of elevated radiation dose, *Health Phys. 45*, 349–361 (1983).
31. H. L. Leira, E. Lund, and T. Refseth, Mortality and cancer incidence in a small cohort of miners exposed to low levels of radiation, *Health Phys. 50*, 189–194 (1986).
32. V. E. Archer, Health concerns in uranium mining and milling. *J. Occup. Med. 23*, 502–505 (1981).
33. Conference of Radiation Control Program Directors Inc., Natural Radioactivity Contamination Problems, U.S. Environmental Protection Agency, Report EPA 520/4-77-015, Washington, DC (1978).
34. J. L. Horwood, Preliminary Investigation of Airborne Thoron Decay Products in two Uranium Mines at Elliott Lake, Ontario, CANMET Centre for Mineral and Energy Technology, Report MRP/MSL 79-1, Ottawa, Ontario, Canada (1979).
35. V. J. Tennery, E. S. Bomar, W. D. Bond, L. E. Morse, H. P. Meyer, J. E. Till, and M. G. Yalcintas, Environmental Assessment of Alternate FBR Fuels: Radiological Assessment of Airborne Releases from Thorium Mining and Milling, Oak Ridge National Laboratory, Report ORNL/TM-6474 (1978).
36. C. J. Barton, R. E. Moore, and P. S. Rohwer, Contribution of radon in natural gas to the dose from airborne radon daughters in homes, in: *Noble*

Gases (Proc. Las Vegas Symp.) R. E. Stanley and A. A. Moghissi, eds., U.S. Environmental Protection Agency, CONF-730915 (1975).
37. T. F. Gesell, Some radiological health aspects of radon-222 in liquefied petroleum gas, in: Noble Gases (Proc. Las Vegas Symp.) (R. E. Stanley and A. A. Moghissi, eds.), U.S. Environmental Protection Agency, CONF-730915 (1975).
38. R. H. Johnson, D. E. Bernhardt, N. S. Nelson, and H. W. Calley, Assessment of Potential Radiological Health Effects from Radon in Natural Gas, U.S. Environmental Protection Agency, Report EPA-520/1-73-004 (1973).
39. T. P. Barton and P. I. Ziemer, The effects of particle size and moisture content on the emanation of Rn from coal ash, *Health Phys.* 50, 581–588 (1986).
40. National Council on Radiation Protection and Measurements, Natural Background Radiation in the United States, NCRP Report No. 45, Bethesda, MD (1975).
41. C. C. Travis and S. J. Cotter, Radon-222 releases associated with cultivation of agricultural land, *Trans. Am. Nucl. Soc.* 32, 112 (1979).
42. I. Kobal, B. Sunodis, and M. Skofljanec, Radon-222 air concentrations in the Slovenian Karst caves of Yugoslavia, *Health Phys.* 50, 830–834 (1986).
43. M. H. Wilkening and D. E. Watkins, Air exchange and ^{222}Rn concentrations in the Carlsbad Caverns, *Health Phys.* 31, 139–145 (1976).
44. H. E. Moore, S. E. Poet, and E. A. Martell, Size distribution and origin of lead-210, bismuth-210 and polonium-210 on airborne particles in the troposphere, in: *Natural Radiation Environment III* (T. F. Gesell and W. M. Lowder, eds.), pp. 415–429, U.S. Department of Energy, Washington, DC (1980).
45. J. A. Auxier, Respiratory exposure in buildings due to radon progeny, *Health Phys.* 31, 119–125 (1976).
46. National Council on Radiation Protection and Measurements, Exposure from the Uranium Series with Emphasis on Radon and its Daughters, NCRP Report No. 77, Bethesda, MD (1984).
47. D. C. Kocher, Effects of Man's Residence inside Building Structures on Radiation Doses from Routine Releases of Radionuclides to the Atmosphere, Oak Ridge National Laboratory, Report ORNL/TM-6526 (1978).
48. J. B. Hursh, D. A. Morken, T. P. Davis, and A. Lovaas, The fate of radon ingested by man, *Health Phys.* 11, 465–476 (1965).
49. K. D. Cliff, Measurements of radon-222 concentrations in dwellings in Great Britain, in: *Natural Radiation Environment III* (T. F. Gesell and W. M. Lowder, eds.), pp. 1260–1271, U.S. Department of Energy, Washington, DC (1980).
50. W. A. Goldsmith, J. W. Poston, P. T. Perdue, and M. O. Gibson, Radon-222 and progeny measurements in "typical" east Tennessee residences, *Health Phys.* 45, 81–88 (1983).

REFERENCES—CHAPTER 5

51. G. A. Swedjemark, Radon in dwellings in Sweden, in: *Natural Radiation Environment III* (T. F. Gesell and W. M. Lowder, eds.), pp. 1237–1259, U.S. Department of Energy, Washington, DC (1980).
52. A. C. George and A. J. Breslin, The distribution of ambient radon and radon daughters in residential buildings in the New Jersey-New York area, in: *Natural Radiation Environment III* (T. F. Gesell and W. M. Lowder, eds.), pp. 1272–1292, U.S. Department of Energy, Washington, DC (1980).
53. T. F. Gesell and H. M. Prichard, The contribution of radon in tap water to indoor radon Concentrations, in: *Natural Radiation Environment III* (T. F. Gesell and W. M. Lowder, eds.), pp. 1347–1363, U.S. Department of Energy, Washington, DC (1983).
54. A. V. Nero and W. W. Nazaroff, Characterizing the source of radon indoors, *Radiat. Prot. Dosim.* 7, 23–39 (1985).
55. J. Rundo, F. Markun, and N. J. Plondke, Observation of high concentration of radon in certain houses, *Health Phys.* 36, 729–730 (1979).
56. B. L. Cohen, A national survey of ^{222}Rn in U.S. homes and correlating factors, *Health Phys.* 51, 175–183 (1986).
57. U.S. Radiation Policy Council, Report of the Task Force on Radon in Structures, Report RPC-80-002, Washington, DC (1980).
58. C. R. Cothern, W. L. Lappenbusch, and J. Michel, Drinking water contribution to natural background radiation, *Health Phys.* 50, 33–47 (1986).
59. M. Suomela and H. Kahlos, Radiation exposure following ingestion of radon-222 rich water, *Health Phys.* 23, 641–652 (1972).
60. F. T. Cross, N. H. Harley, and W. Hofmann, Health effects and risks from ^{222}Rn in drinking water, *Health Phys.* 48, 649–670 (1985).
61. C. T. Hess, J. Michel, T. R. Horton, H. P. Prichard, and W. A. Coniglio, The occurrence of radioactivity in public water supplies in the United States, *Health Phys.* 48, 553–586 (1985).
62. M. V. J. Culot, K. J. Schiager, and H. G. Olson, Prediction of increased gamma fields after application of a radon barrier on concrete surfaces, *Health Phys.* 30, 471–478 (1976).
63. D. W. Moeller, D. W. Underhill, and G. V. Gulezian, Population dose equivalent from naturally occurring radionuclides in building materials, in: *Natural Radiation Environment III* (T. F. Gesell and W. M. Lowder, eds.), pp. 1424–1443, U.S. Department of Energy, Washington, DC (1980).
64. N. Jonassen and J. P. McLaughlin, Exhalation of radon-222 from building materials and walls, in: *Natural Radiation Environment III* (T. F. Gesell and W. M. Lowder, eds.), pp. 1211–1224, U.S. Department of Energy, Washington, DC (1980).
65. G. G. Eichholz, M. D. Matheny, and B. Kahn, Control of radon emanation from building materials by surface coatings, *Health Phys.* 39, 301–304 (1980).

CHAPTER 6

1. International Commission on Radiation Units and Measurements, Radiation Quantities and Units, ICRU Report 33, Bethesda, MD (1980).
2. H. H. Rossi, The role of microdosimetry in radiobiology, *Radiat. Environ. Biophys. 17*, 29 (1979).
3. International Commission on Radiation Units and Measurements, Microdosimetry, ICRU Report 36, Bethesda, MD (1983).
4. E. Polig, Hit probabilities for cellular targets by bone surface seeking alpha emitters, *Phys. Med. Biol. 26*, 369–377 (1981).
5. D. J. Crawford-Brown, Age dependent hit probabilities for lung cancer induction following exhalation of ingested radon, *Proc. Second Workshop on Lung Dosimetry*, Cambridge, England, September 1985 (to be published).
6. J. A. Simmonds and S. R. Richards, Microdosimetry of alpha irradiated lung, *Health Phys. 46*, 607 (1984).
7. W. Hofmann, Microdosimetry of plutonium in lungs, *Health Phys.* Suppl. 1, *44*, 419–429 (1983).
8. D. R. Fisher, In search of the relevant lung dose, in: *Current Concepts in Lung Dosimetry* (D. R. Fisher, ed.) CONF 820492, pp. 29–37, National Technical Information Center, U.S. Department of Energy, Washington, DC (1983).
9. ICRP Task Group on Lung Dynamics, Deposition and Retention models for internal dosimetry of the human respiratory tract, *Health Phys. 12*, 173–207 (1966).
10. A. VanAs and I. Webster, The morphology of mucus in mammalian pulmonary airways, *Environ. Res. 7*, 1–12 (1974).
11. G. A. Laurenzi, The mucociliary stream, *J. Med., 15*, 175–176 (1973).
12. M. A. Sleigh, *The Biology of Cilia and Flagella*, Pergamon Press, London (1962).
13. J. Sade, N. Eliezer, A. Silberg, and A. C. Nevo, The role of mucus in transport by cilia, *Am. Rev. Resp. Dis. 102*, 48–52 (1970).
14. G. E. Angus and W. M. Turbeck, Number of alveoli in the human lung, *J. Appl. Phys. 32*, 483–483 (1972).
15. K. Horsfield, Quantitative morphology and structure: Functional correlations in the lung, in: *The Lung: Structure, Function and Disease*, W. M. Thurlbeck and M. R. Abell, eds.), p. 151, Williams and Wilkins Co., Baltimore, MD (1978).
16. E. R. Weibel, The cell population of the normal lung, in: *Lung Cells in Disease* (A. Bouhuys, ed.), p. 3, North Holland Publishing Co., Amsterdam (1976).
17. E. R. Weibel, *Morphometry of the Human Lung*, Academic Press, New York (1963).
18. C. N. Davies, A formalized anatomy of the human respiratory tract, in:

International Symposium on Inhaled Particles and Vapors (C. N. Davies, ed.), p. 82, Pergamon Press, Oxford (1960).
19. H. D. Landhahl, On the removal of airborne droplets by the human respiratory tract: I. The lung, *Bull. Math. Biophys. 12*, 43 (1950).
20. W. Findeisen, Über das Absetzen Kleiner, in der Luft Suspendierter Teilchen in der Menshlichen Lunger bei der Atmung, *Pfluegers Arch. J. Ges. Physiol. 236*, 367 (1935).
21. K. Horsfield, Models of the human bronchial tree, *J. Appl. Phys. 31*, 207–217 (1971).
22. H. C. Yeh and M. Schum, Models of human lung airways and their application to inhaled particle deposition, *Bull. Math. Biol. 42*, 461 (1980).
23. O. G. Raabe, H. C. Yeh, G. M. Schum, and R. F. Phalen, Tracheobronchial Geometry: Human, Dog, Rat, Hamster, Lovelace Foundation Report, LF-53, Albuquerque, NM (1976).
24. A. Hislop, D. C. F. Muir, M. Jacobson, G. Simon, and L. Reid, Postnatal growth and functions of the pre-acinar airways, *Thorax 27*, 265–274 (1972).
25. W. Hofmann, F. Steinhausler, and E. Pohl, Dose calculations for the respiratory tract from inhaled natural radioactive nuclides as a function of age, *Health Phys. 37*, 517–532 (1979).
26. International Commission on Radiological Protection, *Report of the Task Group on Reference Man*, ICRP Publication 23, Pergamon Press, New York (1974).
27. D. J. Crawford-Brown, A Generalized Age Dependent Lung Model with Applications to Radiation Standards, Oak Ridge National Laboratory Report, NUREG/CR-1955, Oak Ridge, TN (1981).
28. D. J. Crawford-Brown, Identifying critical human subpopulations by age groups: Radioactivity and the lung, *Phys. Med. and Biol. 27*, 539–552 (1982).
29. D. J. Crawford-Brown and K. F. Eckerman, Modifications of the ICRP Task Group lung model to reflect age dependence, *Radiat. Prot. Dosim. 2*, 209–220 (1983).
30. R. A. Millikan, *The Electron, Protons, Photons, Neutrons, Mesotrons and Cosmic Rays*, 2nd Ed., University of Chicago Press, Chicago (1974).
31. P. G. Gormley and M. Kennedy, Diffusion from a stream flowing through a cylindrical tube, *Proc. R. Ir. Acad. 52*, 163 (1949).
32. C. N. Davies, Diffusion and sedimentation of aerosol particles from Poiseville flow in pipes, *J. Aerosol Sci. 4*, 317–328 (1973).
33. D. V. Ingham, Diffusion of aerosols from a stream flowing through a circular tube, *J. Aerosol Sci. 6*, 125–132 (1975).
34. J. W. Thomas, Particle loss in sampling conduits, in: *Assessment of Airborne Radioactivity*, p. 701, International Atomic Energy Agency, Vienna (1967).
35. R. E. Pattle, *Inhaled Particles and Vapours*, p. 70, Pergamon Press, Oxford (1961).
36. H. D. Landahl and S. Black, Filtration of airborne particulates through the human nose, *J. Ind. Hyg. Toxicol. 29*, 269 (1947).
37. M. Lippmann, Deposition and clearance of inhaled particles in the human nose, *Ann. Otol. Rhinol. Laryngol. 79*, 519–528 (1970).

38. J. Heyder, Total deposition of aerosol particles in the human respiratory tract for nose and mouth breathing, *J. Aerosol Sci. 6*, 311–328 (1975).
39. F. A. Fry, Charge distribution on polystyrene aerosols and deposition in the human nose, *J. Aerosol Sci. 1*, 135–146 (1970).
40. R. F. Hounam, A. Black, and M. Walsh, The deposition of aerosol particles in the naseopharyngeal region of the human respiratory tract, *J. Aerosol Sci. 2*, 47–61 (1971).
41. T. T. Mercer, The deposition model of the Task Group on Lung Dynamics: A comparison with recent experimental data, *Health Phys. 29*, 673–680 (1975).
42. A. George and A. J. Breslin, Deposition of radon daughters in humans exposed to uranium mine atmospheres, *Health Phys. 17*, 115–124 (1969).
43. W. L. Dennis, (comments on discussion) in: *Inhaled Particles and Vapors* (C. N. Davies, ed.), p. 88, Pergamon Press, Oxford (1961).
44. W. Stahlhofen, J. Gebhart and J. Heyder, Experimental determination of the regional deposition of aerosol particles in the human respiratory tract, *J. Am. Ind. Hyg. Assoc. 41*, 385–398 (1980).
45. H. Landahl and R. Herrmann, Sampling of liquid aerosols by wires, cylinders and slides and the efficiency of impaction of the droplets, *J. Colloid Sci. 4*, 103 (1949).
46. J. Johnstone, I. Isles, and D. Muir, Inertial deposition of particles in the lung, *J. Aerosol Sci. 4*, 269–270 (1973).
47. K. Takahashe and H. Ito, A Computational Model for Regional Deposition of Aerosol Particles in the Human Lung, Technical Report of the Institute of Atomic Energy 17, Kyoto University, Kyoto, Japan (1976).
48. H. Yeh, Use of a heat transfer analogy for a mathematical model of respiratory tract deposition, *Bull. Math. Biol. 36*, 105 (1974).
49. International Commission on Radiological Protection, *The Metabolism of Compounds of Plutonium and Other Actinides,* ICRP Publication 19, Pergamon Press, Oxford (1972).
50. M. Friedman, F. D. Scott, D. O. Poole, R. Dougherty, G. A. Chapman, H. Watson, and M. A. Sackner, A new roentgenographic method for estimating mucus velocity in airways, *Am. Rev. Resp. Dis. 115*, 67–72 (1977).
51. D. Yeates, N. Aspin, H. Levison, M. T. Jones, and A. C. Bryan, Mucociliary tracheal transport in man, *J. Appl. Phys. 39* (1975).
52. A. D. Barclay, K. J. Franklin, and R. G. Macbeth, Roentogenographic studies of the excretion of dusts from the lungs, *Am. J. Roentgenol. 39*, 673–686 (1938).
53. G. Gamsu, R. M. Weintraub, and J. A. F. Nadel, Clearance of tantalum from airways of different caliber in man evaluated by a roentgenographic method, *Am. Rev. Resp. Dis. 107*, 214–224 (1973).
54. B. Altshuler, N. Nelson, and M. Kuschner, Estimation of lung tissue dose from the inhalation of radon and daughters, *Health. Phys. 10*, 1137–1161 (1964).
55. A. K. M. Haque and A. J. L. Collinson, Radiation dose to the respiratory

system due to radon and its daughter products, *Health Phys.* 13, 431–443 (1967).
56. G. A. Laurenzi, The mucociliary stream, *J. Occup. Med.* 15, 175–176 (1973).
57. W. Whaling, The energy loss of charged particles in matter, in: *Handbuch der Physik* (S. Flugge, ed.), p. 193, Springer, Berlin (1958).
58. H. Bichsel, Charged particle interactions, in: *Radiation Dosimetry* (F. H. Attix and W. C. Roesch, eds.), p. 158, Academic Press, New York (1968).
59. E. Rotondi, Energy loss of alpha particles in tissue, *Radiat. Res.* 33, 1–9 (1968).
60. P. J. Walsh, Stopping power and range of alpha particles, *Health Phys.* 19, 312–316 (1970).
61. A. K. M. Haque, Energy expended by alpha particles in lung tissue, *Br. J. Appl. Phys.* 17, 905 (1966).
62. N. H. Harley and B. S. Pasternack, Alpha absorption measurements applied to lung dose from radon daughters, *Health Phys.* 23, 771–782 (1972).
63. D. J. Crawford, Radiological Risk of Actinon (^{219}Rn), Oak Ridge National Laboratory Report ORNL/TM-7977, Oak Ridge, TN (1980).
64. D. E. Lea, *Actions of Radiations on Living Cells*, Cambridge University Press, London (1955).
65. W. Jacobi, The dose to the human respiratory tract by inhalation of short-lived ^{222}Rn and ^{220}Rn decay products, *Health Phys.* 10, 1163–1174 (1964).
66. V. N. Kirichenko, Experimental studies of the short-lived daughters of radon in the respiratory tract, *Gig. Sanit.* 2, 52 (1970).
67. National Council on Radiation Protection and Measurement, *Evaluation of Occupational and Environmental Exposures to Radon and Radon Daughters in the United States*, NCRP Report 78, Bethesda, MD (1984).
68. H. Goldziecher, Über Baselzellen Wucherungea der Bronchial Schlemeit," *Zentralbl. Allg. Path. Pathol. Anat.* 29, 506 (1918).
69. P. Kotin, D. Courington, and H. L. Falk, Pathogenesis of cancer in ciliated mucus secreting epithelium, *Am. Rev. Resp. Dis.* 93, 115–124 (1966).
70. S. Hattori, M. Matsuda, R. Tateishi, H. Nishihara, and T. Harai, Oat cell carcinoma of the lung, *Cancer 30*, 1014–1024 (1972).
71. E. M. McDowell and B. F. Trump, Histogenesis of preneoplastic and neoplastic lesions in tracheobronchial epithelium, *Surv. Synth. Pathol. Res.* 2, 235–242 (1983).
72. J. Horacek, V. Placek, and J. Sevc, Histologic types of bronchogenic cancer in relation to different conditions of radiation exposure, *Cancer* 40, 832–835 (1977).
73. R. M. Gastineau, P. J. Walsh, and N. Underwood, Thickness of bronchial epithelium with relation to exposure to radon, *Health Phys.* 23, 857–860 (1972).
74. O. G. Raabe, Deposition and Clearance of Inhaled Aerosols, Laboratory for Energy Related Health Research Report UCD 472–503, University of California, Davis (1979).

75. N. H. Harley and B. S. Pasternack, Environmental radon daughter alpha dose factors in a five-lobed human lung, *Health Phys. 42*, 789–799 (1982).
76. A. C. Chamberlain and E. D. Dyson, The dose to the trachea and bronchi from the decay products of radon and thoron, *Br. J. Radiol. 29*, 317–325 (1956).
77. P. J. Walsh, Radiation dose to the respiratory tract of uranium miners—A review of the literature, *Environ. Res. 3*, 14–36 (1970).
78. W. Hofmann and F. Steinhausler, Dose calculations for infants and youths due to the inhalation of radon and its decay products, in: *Proceedings of DECUS Europe Symposium*, London, pp. 315–320 (1977).
79. R. Schlesinder and M. Lippmann, Particle deposition in the trachea, in vivo and hollow casts, *Thorax 31*, 678–684 (1976).
80. G. A. Ferron, Deposition of polydisperse aerosols in two glass models representing the upper human airways, *J. Aerosol Sci. 8*, 409–427 (1977).
81. P. Hammill, Particle deposition due to turbulent diffusion in the upper respiratory system, *Health Phys. 36*, 355–369 (1979).
82. T. Martonen, personal communication (1986).
83. D. J. Crawford-Brown, On a theory of age dependence in the incidence of lung carcinomas following inhalation of a radioactive atmosphere, in: *Current Topics in Lung Dosimetry* (D. Fisher, ed.), pp. 178–188, CONF-820492, Batelle Northwest Laboratory, Richland, WA (1983).
84. T. L. Chan and M. Lippmann, Experimental measurements and empirical modelling of the regional deposition of inhaled particles in humans, *J. Am. Ind. Hyg. Assoc. 41*, 399–409 (1980).
85. G. Giacomelli-Maltoni, C. Melandri, V. Prodi, and G. Tarroni, Deposition efficiency of monodisperse particles in the human respiratory tract, *J. Am. Ind. Hyg. Assoc. 33*, 603–610 (1972).
86. F. Shanty, Deposition of Ultrafine Aerosols in the Respiratory Tract of Human Volunteers, Doctoral dissertation, School of Hygiene and Public Health, Johns Hopkins University, Baltimore (1974).
87. B. Altshuler, L. Yarmus, E. Palmes, and N. Nelson, Aerosol deposition in the human respiratory tract, *AMA Arch. Ind. Health 15*, 293–303 (1957).
88. N. Foord, A. Black, and M. Walsh, Regional Deposition of 2.5–7.5 μm Diameter Inhaled Particles in Healthy Male Non-Smokers, AERE Harwell Report, ML-76-2892, Great Britain (1976).
89. M. Lippmann and R. Albert, The effect of particle size on the regional deposition of inhaled aerosols in the human respiratory tract, *J. Am. Ind. Hyg. Assoc. 30*, 257–275 (1969).
90. F. T. Cross, N. H. Harley, and W. Hofmann, Health effects and risks from Rn-222 in drinking water, *Health Phys. 48*, 649 (1985).
91. D. J. Crawford-Brown, Age dependent lung doses from ingested ^{222}Rn in drinking water, submitted to *Health Phys. 52*, 149–156 (1987).
92. I. O. Anderson and I. Nilsson, Exposure following ingestion of water containing Rn-222, in: *Assessment of Radioactivity in Man*, p. 317, International Atomic Energy Agency, Vienna (1964).

93. W. vonDobeln and B. Lindell, Some aspects of Rn-222 contamination following ingestion, *Arkiv für Fysik 27*, 531 (1964).
94. J. B. Hursh, D. A. Morken, T. P. Davis, and A. Lovaas, The fate of Rn-222 ingested by man, *Health Phys. 11*, 465–476 (1965).
95. M. Suomela and H. Kohlos, Studies on the elimination radiation and the radiation exposure following ingestion of Rn-222 rich water, *Health Phys. 23*, 641–652 (1972).

CHAPTER 7

1. International Commission on Radiological Protection (ICRP), *Biological Effects of Inhaled Radionuclides*, ICRP Publication 31, Pergamon Press, New York (1980).
2. R. W. Hornung and T. J. Meinhardt, Quantitative Risk Assessment of Lung Cancer in U.S. Uranium Miners, National Institute for Occupational Safety and Health (NIOSH), Cincinnati, OH (1986).
3. H. G. Muller, Radiation damage to the genetic material, *Sci. Prog. 7*, 93–493 (1951).
4. C. O. Nordling, A new theory of the cancer inducing mechanism, *Br. J. Cancer 7*, 68–72 (1953).
5. P. Armitage and R. Doll, in: *Proceedings of the Fourth Berkeley Symposium on Mathematical Statistics and Probability*, Vol. 4, pp. 19–38, University of California Press, Berkeley and Los Angeles (1961).
6. S. H. Moolgavkar and A. G. Knudson, Jr., Mutation and cancer: A model for human carcinogenesis, *J. Natl. Cancer Inst. 66*, 1037–1052 (1981).
7. A. G. Knudson, Jr., Hereditary cancer, oncogenes, and antioncogenes, *Cancer Res. 45*, 1437–1443 (1985).
8. V. R. Potter, A new protocol and its rationale for the study of initiation and promotion of carcinogenesis, *Carcinogenesis 2*, 1375–1379 (1981).
9. H. R. Hennings, R. Shores, M. L. Wenk, E. F. Spangler, R. Tarone, and S. H. Yuspa, Malignant conversion of mouse skin tumors is increased by tumor initiators and unaffected by tumor promoters, *Nature 304*, 67–69 (1983).
10. K. G. Vohra, K. C. Pillai, U. C. Mishra, and S. Sadasivan, in: *Natural Radiation Environment—Proceedings of the Second Special Symposium on Natural Radiation Environment* (K. G. Vohra, K. C. Pillai, U. C. Mishra, and S. Sadasivan, eds.), John Wiley and Sons, New York (1982).
11. W. Hofmann, R. Katz, and Z. Chunxiang, Lung cancer risk at low doses of α particles, *Health Phys. 51*, 457–468 (1986).
12. E. G. Letourneau and D. T. Wigle, in: *Proceedings of the Specialist Meeting on Assessment of Radon and Daughter Exposure and Related Biological Effects*

(G. F. Clemente, A. V. Nero, F. Steinhausler, and M. E. Wrenn, eds.), pp. 239–251, R. D. Press, Salt Lake City, UT (1982).
13. I. Uzunov, F. Steinhausler, and E. Pohl, Carcinogenic risk of exposure to radon daughters associated with radon spas, *Health Phys.* 41, 807–813 (1981).
14. W. Chruschielewski, T. Domanski, and W. Orzechowski, Concentrations of radon and its progeny in the rooms of Polish spas, *Health Phys.* 45, 421–424 (1983).
15. G. Agricola, *De Re Metallica Basil* (1597). [Translated by H. C. Hoover and L. H. Hoover, Dover Publications, New York (1950).]
16. H. Q. Woodard, Radiation Carcinogenesis in Man: A Critical Review, U.S. DOE Report EML-380, National Technical Information Service, Springfield, VA (1980).
17. J. G. Batsakis, *Tumors of the Head and Neck*, p. 124, Williams and Wilkins, Baltimore, MD (1976).
18. E. McDowell, J. S. McLaughlin, D. K. Merenyl, R. F. Kiefer, C. C. Harris, and B. F. Trump, The respiratory epithelium histogenesis of lung carcinoma in the human, *J. Natl. Cancer Inst.* 2, 587–606 (1978).
19. M. J. Evans, in: *Proceedings of the 37th Annual Meeting of the American College of Veterinary Pathologists and the 21st Annual Meeting of the American Society for Veterinary Clinical Pathology/Toxicologic Pathology*, pp. 193–202, American College of Veterinary Pathologists, Kennett Square, PA (1986).
20. G. Saccomanno, V. E. Archer, O. Auerbach, M. Kuschner, M. Egger, S. Wood, and R. Mick, in: *Proceedings of the International Conference on Radiation Hazards in Mining: Control, Measurement and Medical Aspects* (M. Gomez, ed.), pp. 675–679, Kingsport Press, Inc., Kingsport, TN (1981).
21. G. Saccomanno, V. E. Archer, R. P. Saunders, L. A. James, and P. A. Beckler, Lung cancer of uranium miners on the Colorado Plateau, *Health Phys.* 10, 1195–1201 (1964).
22. G. Saccomanno, V. E. Archer, O. Auerbach, M. Kuschner, R. P. Saunders, and M. G. Klein, Histologic types of lung cancer among uranium miners, *Cancer* 27, 515–523 (1971).
23. J. Horacek, V. Placek, and J. Sevc, Histologic types of bronchogenic cancer in relation to different conditions of radiation exposure, *Cancer* 40, 832–835 (1977).
24. V. E. Archer, E. P. Radford, and O. Axelson, in: *Conference/Workshop on Lung Cancer Epidemiology and Industrial Applications of Sputum Cytology*, pp. 324–367, Colorado School of Mines Press, Golden, CO (1979).
25. National Institute for Occupational Safety and Health/National Institute of Environmental Health Sciences, Radon Daughter Exposure and Respiratory Cancer—Quantitative and Temporal Aspects, Joint Monograph No. 1, National Technical Information Service, Springfield, VA (1971).
26. National Academy of Sciences, *The Effects on Populations of Exposure to Low Levels of Ionizing Radiation* (BEIR-III Report), National Academy Press, Washington, DC (1980).

REFERENCES—CHAPTER 7

27. United Nations Scientific Committee on the Effects of Atomic Radiation, Sources and Effects of Ionizing Radiation, United Nations, New York (1977).
28. R. D. Evans, J. H. Harley, W. Jacobi, A. S. McLean, W. A. Mills, and C. G. Stewart, Estimate of risk from environmental exposure to radon-222 and its decay products, *Nature 290,* 98–100 (1981).
29. National Council on Radiation Protection and Measurements (NCRP), Evaluation of Occupational and Environmental Exposures to Radon and Radon Daughters in the United States, NCRP Report No. 78, National Council on Radiation Protection and Measurements, Bethesda, MD (1984).
30. J. Muller, W. C. Wheeler, J. F. Gentleman, G. Suramji, and R. A. Kusiak, Study of Mortality of Ontario Mines 1955–1977, Part I, Ontario Ministry of Labour, Toronto (1983).
31. J. Muller, W. C. Wheeler, J. F. Gentleman G. Suramji, and R. A. Kusiak, in: *Occupational Radiation Safety in Mining* (H. Stocker, ed.), Vol. I, pp. 335–343, Canadian Nuclear Association, Toronto (1985).
32. International Commission on Radiological Protection (ICRP), Exposure and Lung Cancer Risk of the General Public from Inhaled Radon Daughters, Report of a Task Group of the ICRP (Draft February 1985; in press).
33. N. Harley, J. M. Samet, F. T. Cross, T. Hess, J. Muller, and D. Thomas, Contribution of radon and radon daughters to respiratory cancer, *Environ. Health Perspect. 70,* 17–21 (1986).
34. D. C. Thomas and K. G. McNeill, Risk Estimates for the Health Effects of Alpha Radiation, Research Report, INFO-0081, Atomic Energy Control Board of Canada, Ottawa (1982).
35. A. S. Whittemore and A. McMillan, Lung cancer mortality among U.S. uranium miners: A reappraisal, *J. Natl. Cancer. Inst. 71,* 489–499 (1983).
36. International Commission on Radiological Protection (ICRP), *Limits for Inhalation of Radon Daughters by Workers,* ICRP Publication 32, Pergamon Press, New York (1981).
37. Environmental Protection Agency (EPA), Final Environmental Impact Statement of Standards for Control of Byproduct Materials from Uranium Ore Processing (40CFR192), Vol. I, EPA 520/1-83-008-20, Washington, DC (1983).
38. A. F. Cohen and B. L. Cohen, Tests of the linearity assumption in the dose–effect relationship for radiation-induced cancer, *Health Phys. 38,* 53–69 (1980).
39. United States Department of Health and Human Services (USDHHS), The Health Consequences of Smoking—Cancer: A Report of the Surgeon General, DHHS Publication No. (PHS)82-50179 (1982).
40. World Health Organization (WHO), Smoking and Its Effects on Health, WHO Technical Report Series No. 568, WHO, Geneva (1975).
41. World Health Organization (WHO), Controlling the Smoking Epidemic, WHO Technical Report Series No. 636, WHO, Geneva (1979).

42. F. E. Lundin, J. W. Loyd, E. M. Smith, V. E. Archer, and D. A. Holaday, Mortality of uranium miners in relation to radiation exposure, hard-rock mining and cigarette smoking—1950 through September 1967, *Health Phys. 16*, 571–578 (1969).
43. E. P. Radford, in: *Quantification of Occupational Cancer* (R. Peto and M. Schneiderman, eds.), Banbury Report 9, pp. 151–163, Cold Spring Harbor Laboratory, Cold Spring Harbor, NY (1981).
44. E. P. Radford and K. G. St. Clair Renard, Lung cancer in Swedish iron miners exposed to low doses of radon daughters, *N. Engl. J. Med. 310*, 1485–1494 (1984).
45. O. Axelson and L. Sundell, Mining, lung cancer and smoking, *Scand. J. Work Environ. Health 4*, 46–52 (1978).
46. Federal Radiation Council, *Guidance for the Control of Radiation Hazards in Uranium Mining*, Staff Report No. 8 (Revised), U.S. Government Printing Office, Washington, DC (1967).
47. V. E. Archer, in: *Carcinogenesis: A Comprehensive Survey* (M. J. Mass *et al.*, eds.), Vol. 8, pp. 23–37, Raven Press, New York (1985).
48. A. E. Reif, Synergism in carcinogenesis, *J. Natl. Cancer Inst. 73*, 25–39 (1984).
49. J. W. Horm and A. J. Asire, Changes in lung cancer incidence mortality rates among Americans, *J. Natl. Cancer Inst. 69*, 833 (1982).
50. D. L. Preston, H. Kato, K. J. Kopecky, and S. Frujita, Lifespan Study Report 10, Part 1, Cancer Mortality among A-Bomb Survivors in Hiroshima and Nagasaki, 1950–82, RERF Technical Report 1-86, Radiation Effects Research Foundation, Hiroshima, Japan (1986).
51. Nuclear Energy Agency (NEA), Dosimetry Aspects of Exposure to Radon and Thoron Daughter Products, NEA (OECD) Report, Paris (1983).
52. R. O. McClellan, B. B. Boecker, R. G. Cuddihy, W. C. Griffith, F. F. Hahn, A. Muggenburg, B. R. Scott, and F. A. Seiler, in: *Proceedings of the Symposium on the Control of Exposure of the Public to Ionizing Radiation*, April 27–29, 1981, Reston, VA, pp. 28–39, National Council on Radiation Protection and Measurements, Bethesda, MD (1982).
53. R. C. Nair, J. D. Abbatt, G. R. Howe, H. B. Newcombe, and S. E. Frost, in: *Occupational Radiation Safety in Mining* (H. Stocker, ed.), Vol. 1, pp. 354–364, Canadian Nuclear Association, Toronto (1985).
54. American Cancer Society (ACS), *1986 Facts and Figures*, ACS, New York (1986).
55. A. S. Whittemore, The age distribution of human cancer for carcinogenic exposures of varying intensity, *Am. J. Epidemiol. 106*, 418–432 (1977).
56. N. E. Day and C. C. Brown, Multistage models and primary prevention of cancer, *J. Natl. Cancer Inst. 64*, 977–989 (1980).
57. N. H. Harley and B. S. Pasternack, A model for predicting lung cancer risks induced by environmental levels of radon daughters, *Health Phys. 40*, 307–316 (1981).
58. J. Lafuma, M. Marin, J. Beaumatin, and R. Masse, in: *Proceedings of the*

Third International Symposium on Radiological Protection—Advances in Theory and Practice, pp. 82–86, ICRP, Pergamon Press, Oxford (1982).
59. H. Jansen and P. Schultzer, Experimental investigations into the internal radium emanation therapy. I. Emanatorium experiments with rats, *Acta Radiol. 6,* 631 (1926).
60. J. Read and J. C. Mottram, The 'tolerance concentration' of radon in the atmosphere, *Br. J. Radiol. 12,* 54 (1939).
61. M. L. Jackson, The Biological Effects of Inhaled Radon, Master's thesis, Massachusetts Institute of Technology, Cambridge (1940).
62. C. R. Richmond and B. B. Boecker, Experimental Studies, Final Report of Subgroup I.B, Interagency Uranium Mining Radiation Review Group, Environmental Protection Agency, Rockville, MD (1971).
63. Senes Consultants, Ltd., Assessment of the Scientific Basis for Existing Federal Limitations on Radiation Exposure to Underground Uranium Miners, Report prepared for the American Mining Congress, Senes Consultants, Ltd., Willowdale, Ontario (1984).
64. F. T. Cross, A Review of Radon Inhalation Studies in Animals with Reference to Epidemiological Data, Research Report for Senes Consultants, Ltd., Ontario, Canada, Battelle, Pacific Northwest Laboratories, Richland, WA (1984).
65. W. F. Bale, Hazards associated with radon and thoron, Memo dated March 14, 1951, Division of Biology and Medicine, Atomic Energy Commission, Washington, DC. [Also found in *Health Phys. 38,* 1061–1066 (1951).]
66. S. J. Harris, Radon levels in mines in New York State, *Arch. Ind. Hyg. Occup. Med. 10,* 54–60 (1954).
67. D. A. Morken, Acute toxicity of radon, *AMA Arch. Ind. Health 12,* 435 (1955).
68. J. Shapiro, An Evaluation of the Pulmonary Radiation Dosage from Radon and Its Daughter Products, Project Report UR-298 to the U.S. Atomic Energy Commission, University of Rochester, Rochester, NY (1954).
69. S. H. Cohn, R. K. Skow, and J. K. Gong, Radon inhalation studies in rats, *Arch. Ind. Hyg. Occup. Med. 7,* 508–515 (1953).
70. D. A. Morken, in: *Noble Gases* (R. E. Stanley and A. A. Moghissi, eds.), pp. 469–471, CONF-730915, National Technical Information Service, Springfield, VA (1973a).
71. D. A. Morken, in: *Noble Gases* (R. E. Stanley and A. A. Moghissi, eds.), pp. 501–506, CONF-730915, National Technical Information Service, Springfield, VA (1973b).
72. D. A. Morken and J. K. Scott, Effects on Mice of Continual Exposure to Radon and Its Decay Products on Dust, Project Report UR-669 to the U.S. Atomic Energy Commission, University of Rochester, Rochester, NY, National Technical Information Service, Springfield, VA (1966).
73. R. F. Palmer, P. O. Jackson, J. C. Gaven, and B. O. Stuart, Dosimetric studies of inhaled radon daughters in dogs, in: *Pacific Northwest Laboratory*

Annual Report for 1975 to the ERDA Division of Biomedical and Environmental Research, BNWL-2000, Part 1, pp. 53–54, National Technical Information Service, Springfield, VA (1976).

74. A. E. Desrosiers, A. Kennedy, and J. B. Little, ^{222}Rn daughter dosimetry in the Syrian Golden hamster lung, Health Phys. 35, 607–623 (1978).

75. N. H. Harley and B. S. Pasternack, Alpha absorption measurements applied to lung dose from radon daughters, Health Phys. 23, 771–782 (1972).

76. N. H. Harley and B. S. Pasternack, Environmental radon daughter alpha dose factors in a five-lobed human lung, Health Phys. 42, 789–799 (1982).

77. W. Hofmann, Cellular lung dosimetry for inhaled radon decay products as a base for radiation induced lung cancer risk assessment. I. Calculation of mean cellular doses, Radiat. Environ. Biophys. 20, 95–112 (1982).

78. W. Jacobi and K. Eisfeld, Dose to Tissues and Effective Dose Equivalent by Inhalation of Radon-222, Radon-220 and Their Short-Lived Daughters, Gesellschaft für Strahlen-und Umweltforschung MBH, Report GSF S-626, Institut für Strahlenschutz, München-Nürnberg (1980).

79. A. C. James, W. Jacobi, and F. Steinhausler, F., in: *Proceedings of the International Conference on Radiation Hazards in Mining: Control, Measurement and Medical Aspects* (M. Gomez, ed.), 42–54, Kingsport Press, Inc., Kingsport, TN (1981).

80. J. Chameaud, R. Perraud, J. Lafuma, R. Masse, and J. Pradel, in: *Experimental Lung Cancer. Carcinogenesis and Bioassays* (E. Karbe and J. F. Park, eds.), pp. 411–421, Springer-Verlag, New York (1974).

81. J. Chameaud, R. Perraud, J. Chretien, R. Masse, and J. Lafuma, in: *Pulmonary Toxicology of Respirable Particles* (C. L. Sanders, F. T. Cross, G. E. Dagle, and J. A. Mahaffey, eds.), pp. 551–557, CONF-791002, National Technical Information Service, Springfield, VA (1980).

82. J. Chameaud, R. Perraud, R. Masse, J. C. Nenot, and J. Lafuma, in: *Biological and Environmental Effects of Low Level Radiation*, Vol. II, pp. 223–228, IAEA STI/PUB/409, International Atomic Energy Agency, Vienna (1976).

83. J. Chameaud, R. Perraud, J. Chretien, R. Masse, and J. Lafuma, in: *Late Biological Effects of Ionizing Radiation*, Vol. II, pp. 429–436, IAEA SIT/PUB/489, International Atomic Energy Agency, Vienna (1978).

84. J. Chameaud, R. Perraud, R. Masse, and J. Lafuma, in: *Proceedings of the International Conference on Radiation Hazards in Mining: Control, Measurement and Medical Aspects* (M. Gomez, ed.), pp. 222–235, Kingsport Press, Inc., Kingsport, TN (1981).

85. J. Chameaud, R. Perraud, J. Lafuma, and R. Masse, in: *Proceedings of the Specialist Meeting on Assessment of Radon and Daughter Exposure and Related Biological Effects* (G. F. Clemente, A. V. Nero, F. Steinhausler, and M. E. Wrenn, eds.), pp. 198–209, R. D. Press, Salt Lake City, UT (1982).

86. J. Chameaud, R. Masse, and J. Lafuma, Influence of radon daughter exposure at low doses on occurrence of lung cancer in rats, Radiat. Prot. Dosim. 7, 385–388 (1984).

87. J. Chameaud, R. Masse, M. Morin, and J. Lafuma, in: *Proceedings of the International Conference on Occupational Radiation Safety in Mining* (H. Stocker, ed.), pp. 350–353, Canadian Nuclear Association, Toronto (1984).
88. F. T. Cross, R. F. Palmer, R. E. Filipy, R. H. Busch, and B. O. Stuart, Study of the Combined Effects of Smoking and Inhalation of Uranium Ore Dust, Radon Daughters and Diesel Oil Exhaust Fumes in Hamsters and Dogs, PNL-2744, Pacific Northwest Laboratory, Richland, WA, National Technical Information Service, Springfield, VA (1978).
89. F. T. Cross, R. F. Palmer, R. H. Busch, R. E. Filipy, and B. O. Stuart, Development of lesions in Syrian Golden hamsters following exposure to radon daughters and uranium ore dust, *Health Phys.* 41, 135–153 (1981).
90. F. T. Cross, R. F. Palmer, R. E. Filipy, G. E. Dagle, and B. O. Stuart, Carcinogenic effects of radon daughters, uranium ore dust and cigarette smoke in beagle dogs, *Health Phys.* 42, 33–52 (1982).
91. F. T. Cross, R. F. Palmer, R. H. Busch, and R. L. Buschbom, in: *Proceedings of the Specialist Meeting on Assessment of Radon and Daughter Exposure and Related Biological Effects* (G. F. Clemente, A. V. Nero, F. Steinhausler, and M. E. Wrenn, eds.), pp. 189–197, R. D. Press, Salt Lake City, UT (1982).
92. F. T. Cross, R. L. Buschbom, G. E. Dagle, R. E. Filipy, P. O. Jackson, S. M. Loscutoff, and R. F. Palmer, Inhalation hazards to uranium miners, in: *Pacific Northwest Laboratory Annual Report for 1982 to the DOE Office of Energy Research*, PNL-4600, Part 1, pp. 77–80, National Technical Information Service, Springfield, VA (1983).
93. F. T. Cross, R. L. Buschbom, G. E. Dagle, P. O. Jackson, R. F. Palmer, and H. A. Ragan, Inhalation hazards to uranium miners, in: *Pacific Northwest Laboratory Annual Report for 1983 to the DOE Office of Energy Research*, PNL-5000, Part 1, pp. 41–44, National Technical Information Service, Springfield, VA (1984).
94. F. T. Cross, R. F. Palmer, G. E. Dagle, R. H. Busch, and R. L. Buschbom, Influence of radon daughter exposure rate, unattachment fraction, and disequilibrium on occurrence of lung tumors, *Radiat. Prot. Dosim.* 7, 381–384 (1984).
95. F. T. Cross, R. F. Palmer, R. H. Busch, G. E. Dagle, R. E. Filipy, and H. A. Ragan, pp. 608–623 in: *Life-Span Radiation Effects Studies in Animals: What Can They Tell Us?* (R. C. Thompson and J. A. Mahaffey, eds.), 22nd Hanford Life Sciences Symposium, September 27–29, 1983, Richland, WA, CONF-830951 National Technical Information Service, Springfield, VA (1987).
96. B. O. Stuart, R. F. Palmer, R. E. Filipy, and J. Gaven, Inhaled radon daughters and uranium ore dust in rodents, in: *Pacific Northwest Laboratory Annual Report for 1977 to the DOE Assistant Secretary for Environment*, PNL-2500, Part 1, pp. 3.70–3.72, National Technical Information Service, Springfield, VA (1978).
97. R. B. Schlesinger and M. Lippman, Selective particle deposition and bronchogenic carcinoma, *Environ. Res.* 15, 424–431 (1978).

98. V. E. Archer, in: *Proceedings, Workshop of Dosimetry for Radon and Radon Daughters*, pp. 23–25, ORNL-53481, Oak Ridge National Laboratory, Oak Ridge, TN, National Technical Information Service, Springfield, VA (1978).
99. National Academy of Sciences, *The Effects on Populations of Exposure to Low Levels of Ionizing Radiation* (BEIR-I Report), National Academy Press, Washington, DC (1972).
100. R. G. Gray, J. Lafuma, S. E. Parish, and R. Peto, pp. 592–607 in: *Life-Span Radiation Effects Studies in Animals: What Can They Tell Us?* (R. C. Thompson and J. A. Mahaffey, eds.), 22nd Hanford Life Sciences Symposium, September 27–29, 1983, Richland, WA, CONF-830951, National Technical Information Service, Springfield, VA (1987).
101. E. Kunz, J. Sevc, V. Placek, and J. Horacek, Lung cancer in man in relation to different time distribution of radiation exposure, *Health Phys. 36*, 699–706 (1979).
102. V. E. Archer, B. E. Carrol, H. P. Brinton, and G. Saccomanno, in: *Radiological Health and Safety in Mining and Milling of Nuclear Materials*, Vol. 1, pp. 21–36, IAEA Symposium, August 26–31, 1963, International Atomic Energy Agency, Vienna (1964).
103. R. F. Palmer, B. O. Stuart, and R. E. Filipy, in: *Noble Gases* (R. E. Stanley and A. A. Moghissi, eds.), pp. 507–519, CONF-730915, National Technical Information Service, Springfield, VA (1973).
104. B. Altshuler, N. Nelson, and M. Kuschner, Estimation of lung tissue dose from the inhalation of radon and daughters, *Health Phys. 10*, 1137–1161 (1964).
105. R. Masse, in: *Pulmonary Toxicology of Respirable Particles* (C. L. Sanders, F. T. Cross, G. E. Dagle, and J. A. Mahaffey, eds.), pp. 498–521, CONF-791002, National Technical Information Service, Springfield, VA (1980).
106. W. Jacobi, in: *Personal Dosimetry and Area Monitoring Suitable for Radon and Daughter Products*, Proceedings of NEA Specialist Meeting, pp. 33–48, Nuclear Energy Agency, OECD, Paris (1977).
107. K. D. Cliff, B. L. Davis, and J. A. Riessland, Little danger from radon, *Nature 279*, 12 (1979).

CHAPTER 8

1. D. C. Sanchez and D. B. Henschel, Radon Reduction Techniques for Detached Houses—Technical Guidance, U.S. Environmental Protection Agency Report EPA/625/5-86/019, Center for Environmental Research Information, Cincinnati, OH (June 1986).
2. U.S. Environmental Protection Agency and U.S. Department of Health and Human Services, A Citizen's Guide to Radon: What It Is and What

to Do About It, U.S. Environmental Protection Agency Pamphlet Number OPA86-004, Office of Public Awareness, Washington, DC (August 1986).
3. U.S. Environmental Protection Agency, Radon Reduction Methods: A Homeowner's Guide, U.S. Environmental Protection Agency Booklet Number OPA-86-005, Office of Public Awareness, Washington, DC (August 1986).
4. American Society of Heating, Refrigerating, and Air-Conditioning Engineers, Inc., *ASHRAE Handbook 1985 Fundamentals,* ASHRAE, Atlanta, GA (1985).
5. U.S. Department of Commerce, *Statistical Abstract of the United States 1982–1983,* Bureau of the Census, Washington, DC (1982).
6. American Society of Heating, Refrigerating, and Air-Conditioning Engineers, Inc., *Ventilation for Acceptable Indoor Air Quality,* ASHRAE Standard 62-1981, ASHRAE, Atlanta, GA (1981).
7. G. W. Reid, P. Lassovszky, and S. Hathaway, Treatment, waste management and cost for removal of radioactivity from drinking water, *Health Phys. 48,* 671 (1985).

CHAPTER 9

1. U.S. Environmental Protection Agency and U.S. Department of Health and Human Services, A Citizen's Guide to Radon: What It Is and What To Do About It, U.S. Environmental Protection Agency Pamphlet Number OPA86-004, Office of Public Awareness, Washington, DC (August 1986).
2. *Evaluation of Occupational and Environmental Exposures to Radon and Radon Daughters in the United States,* National Council on Radiation Protection and Measurements, NCRP Report No. 78 (May 1984).
3. *Exposure from the Uranium Series with Emphasis on Radon and Its Daughters,* National Council on Radiation Protection and Measurements, NCRP Report No. 77 (March 1984).
4. *Safety and Health Standards, Underground Metal and Nonmetal Mines, Underground Radiation,* Mine Safety and Health Administration, 30 CFR 57.5037-57.5047, 42 CFR 29418 (June 1977).
5. R. D. Evans, J. H. Harley, W. Jacobi, A. S. McLean, W. A. Mills, and C. G. Stewart, Estimate of risk from environmental exposure to radon-222 and its decay products, *Nature 290,* 98–100 (1981).
6. W. Hofmann, R. Katz, and C. Zhang, Lung cancer at low doses of α-particles, *Health Phys. 51,* 457–468 (1986).
7. A. V. Nero, M. B. Schwehr, W. W. Nazaroff, and K. L. Revzan, Distribution of airborne radon-222 in U.S. homes, *Science 234,* 992–997 (1986).
8. American Cancer Society, *1986 Cancer Facts and Figures,* 86-500M-No. 5008-LE, New York, NY.

9. J. L. Repace and A. H. Lowrey, A quantitative estimate of nonsmoker's lung cancer risk from passive smoking, *Environmental International 11*, 3-22 (1985).
10. M. E. Ginevan and W. A. Mills, Assessing the risk of radon exposure: The influence of cigarette smoking, *Health Phys. 51*, 163–174 (1986).
11. G. Saccomanno, C. Yale, W. Dixon, O. Auerbach, and G. C. Huth, An epidemiological analysis of the relationship between exposure to radon progeny, smoking and bronchogenic carcinoma in the U-mining population of the Colorado Plateau—1960–1980, *Health Phys. 50*, 605–618 (1986).
12. E. A. Martell, α-Radiation dose at bronchial bifurcations of smokers from indoor exposure to radon progeny, *Proceed. Natl. Acad. Sci. U.S.A. 80*, 1285–1289 (1983).
13. E. A. Martell, Critique of current lung dosimetry models for radon progeny exposure, *Radon and Its Decay Products: Occurrence, Properties, and Health Effect* (P. K. Hopke, ed.), American Chemical Society, Symposium Series No. 331 (1987).
14. R. F. Johnson and R. A. Luken, Radon risk information and voluntary information protection: Evidence from a natural experiment, *Risk Analysis 7*, 97–107 (1987).

General Reading List

1. William W. Nazaroff and Anthony V. Nero (eds.), *Radon and Its Decay Products in Indoor Air*, Wiley (in press).
2. Special Issue on Indoor Radon, *Health Phys.* (A. V. Nero and W. M. Lowder, guest editors) (August 1983).
3. National Council on Radiation Protection and Measurement, Exposures from the Uranium Series with Emphasis on Radon and Its Daughters, NCRP Report No. 77, Bethesda, MD 20814 (1984).
4. National Council on Radiation Protection and Measurement, Evaluation of Occupational and Environmental Exposures to Radon and Radon Daughters in the United States, NCRP Report No. 78, Bethesda, MD 20814 (1984).
5. U.S. Environmental Protection Agency and U.S. Department of Health and Human Services, A Citizen's Guide to Radon: What It Is and What to Do About It, U.S. Environmental Protection Agency Pamphlet Number OPA-86-004, Office of Public Awareness, Washington, D.C. (1986).
6. U.S. Environmental Protection Agency, Radon Reduction Methods: A Homeowner's Guide, U.S. Environmental Protection Agency Booklet Number OPA-86-005, Washington, D.C. (1986).
7. D. C. Sanchez and D. B. Henschel. Radon Reduction Techniques for Detached Houses: Technical Guidance, U.S. Environmental Protection Agency Report EPA/625/5-86/019/, Center for Environmental Research Information, Cincinnati, OH (June 1986).

About the Authors and Editors

Judith E. Cook
Office of Research and Development
Mail Drop 60
U.S. Environmental Protection Agency
Research Triangle Park, NC 27711

Cook holds a Bachelor of Arts degree in English from Mary Washington College of the University of Virginia. Her professional background includes ten years as a management analyst with the U.S. Environmental Protection Agency, and her primary interest is in technical writing in the area of environmental protection research.

C. Richard Cothern
Office of the Administrator (A101F)
U.S. Environmental Protection Agency
Washington, DC 20460

Cothern holds a Ph.D. in nuclear physics and has research experience in nuclear and environmental physics as well as surface chemistry. He has 17 years of teaching experience at the Universities of Manitoba and Dayton. He has been at the U.S. Environmental Protection Agency since 1978 and was involved in developing the regulations for radionuclides in drinking water and providing quantitative risk assessments for drinking water contaminants. He is currently the Executive Secretary of the Environmental Health Committee of EPA's Science Advisory Board. He is an author and co-editor of this book.

Douglas J. Crawford-Brown
Department of Environmental Science and Engineering
School of Public Health
University of North Carolina, Chapel Hill, NC 27514

Crawford-Brown is an Assistant Professor in the Department of Environmental Sciences and Engineering at the University of North Carolina at Chapel Hill. His main research areas lie in the theory of metabolism, dosimetry, and radiobiology of radionuclides and in the logical and ethical foundations of risk analysis. His most recent work examines the logical and empirical uncertainties underlying radiobiology, particularly as these uncertainties impact on the process of decision in a societal setting.

Fred T. Cross
Biology and Chemistry Department
Battelle Pacific Northwest Laboratory
P.O. Box 999
Richland, WA 99352

Cross is involved with the work of many organizations and agencies on the uranium miner problem, radiation dosimetry of radon, the evaluation and derivation of exposure standards, the risk analysis of persons exposed to radon, and health effects studies in animals. He is a staff scientist at Battelle in the Biology and Chemistry Department. Over the past 20 years he has been involved with, among others, the following organizations: NIOSH, USEPA, NCRP, ICRP, NAS, DOE, Bureau of Mines, Atomic Energy Control Board of Canada, and the American Mining Congress.

Daniel J. Egan, Jr.
Office of Radiation Programs (ANR-460)
U.S. Environmental Protection Agency
Washington, DC 20460

Egan holds a Bachelor of Science and a Master of Engineering in nuclear engineering from Renselaer Polytechnic Institute, and he also has a Master of Science in energy systems from George Washington University. He was on active duty with the Air Force from 1969 until 1976, working primarily on nuclear safety and spacecraft technology development. He has been with the U.S. Environmental Protection Agency since 1976 and was project officer for the high-level radioactive waste standards that were promulgated in August 1985. Since then he has helped manage EPA's Radon Action Program, currently heading up the Problem Assessment Team.

Geoffery G. Eichholz
Nuclear Engineering and Health Physics Program
Georgia Institute of Technology
Atlanta, GA 30332

Eichholz is the Regent's Professor of Nuclear Engineering at the Georgia Institute of Technology. He has published widely on the subject of radon as well as other related nuclear topics.

Jacqueline Michel
Research Planning Institute, Inc.
925 Gervais Street
Columbia, SC 29201

Michel, with a Ph.D. in geology from the University of South Carolina, has focused on study of the behavior of natural radioactivity in soils, rocks, and water. She has published more than a dozen articles and reports on the distribution of radionuclides in groundwater, with emphasis on the relationship between aquifer type and radionuclide concentrations. She is Director of the Environmental Technology Division of RPI International, Inc.

William A. Mills
Oak Ridge Associated Universities
Suite 700
1019 19th Street, N. W.
Washington, DC 20036

Mills has more than 35 years of experience in radiation research and in addressing science and policy issues associated with radiation protection. He became interested in the environmental radon issue during the mid-1960s with the development of the U.S. Surgeon General's recommendations on radiation's remedial action levels for measurement of uranium mill tailings wastes in Grand Junction, Colorado. He is currently employed by Oak Ridge Associated Universities to serve as the Senior Technical Advisor to the federal Committee on Interagency Radiation Research and Policy Coordination, Office of Science and Technology Policy.

James E. Smith, Jr.
Center for Research Information
Office of Research and Development
U.S. Environmental Protection Agency
26 West St. Clair Street
Cincinnati, OH 45268

Smith is a co-editor of this book, holds an Sc.D. in civil and environmental engineering, has research experience in residuals waste management, and has consulted for the EPA in water and wastewater treatment. He has worked for the World Health Organization and consulted with the Pan American Health Organization and the United Nations Development Program. He has been at the U.S. Environmental Protection Agency since 1968 and is presently involved in preparing design manuals, guidance handbooks, and technical assistance programs in water treatment and hazardous waste management.

Index

Absolute risk, 222, 297
Absorbed dose, 26–27, 297
Actinium series, 9, 11, 25, 297
Activated charcoal, 65, 69, 70, 297
Activity, 14–16, 297
Activity concentration, 15–16
Acute early effects, 215
Aeration, 269
Aerosol, 157
Age dependence, 182–185
Agricola, 32, 33
Air cleaner, 251, 269–270
Air-to-air heat exchanger, 258
Alpha particle spectroscopy, 71
Alpha recoil, 86–87
Alpha track, 65, 68, 69, 70, 72, 297
Alveoli, 179–180
Anatomical dead space, 298
Anatomical models of the lung, 182
Anatomy of the lung, 176–180
Animal studies, 229–244
Arkosic sediments, 114
Arnstein, 32, 33
Atmospheric transport, 53, 54
Attached fraction, 135, 156–157, 298

Basal cells, 179, 200, 219
Basalt, 83
Basement houses, 253
Bateman equations, 309–313

Becquerel (unit), 14, 298
Becquerel, Henri, 14, 32, 34, 42
Bemont, Gustove, 35
Bergkrankheit, 33, 219
Bifurcating tubes, 178, 185, 298
Boltwood, Bertram, 45
Branching, 180
Brick, 121
Bronchial passages, 178–179, 219
Bronchiolar adenocarcinomas, 236
Bronchiolaveolar carcinomas, 236
Bronchioles, 179
Brooks, Harriet, 32, 41
Bumstead, Henry, 44

Calibration, 75–77
Carbonate aquifer, 113
Charcoal canisters, 67
Cigarette smoke, 218, 235, 236–237, 281, 282
Coal fired power plant, 153–154
Collie, 44
Compagnie General des Matieres Nucleaires Laboratory (COGEMA)—France, 230, 234–237
Compounds, 1
Concrete, 120
Concrete blocks, 259
Construction issues, 293

Consumer protection, 292
Continuous sampling, 61, 69
Coprecipitation, 90–91
Crawl space, 252, 253
Critical cells, 175, 176, 199–201, 298
Curie (Ci), 13, 14, 298
Curie, Marie, 14, 32, 35, 36, 39, 40, 43
Curie, Pierre, 14, 32, 42, 43, 45

Debierne, Andre, 32, 44, 45
Decay constant, 15
Decay equations, 309–313
Demarcay, Eugene, 41
Denver Radium Service, 55
Deposition in the lung, 185–189
Depressuration, 254, 268
Depth-dose curves, 176, 199
Diesel engine exhaust, 238
Diffusion, 188–189
Diffusion constant, 110
Dilution, 250
Diorites, 83
Dogs, 237
Dorn, 32, 41
Dose equivalent, 27–29, 299
Dosimetry, 174

Earthquake prediction, 51–53
Effective dose equivalent, 28–29, 299
Electrometer, 37–38
Electroscope, 6, 37
Eman, 15
Emanation, 85
  coefficient of, 86
Emanatorium, 57
Emphysema, 245–246
Energy density, 175
Energy-efficient house, 255
Epidemiology, 244
Epidermoid carcinoma, 219–220, 236
Epithelial cells, 178
Equilibrium factor, 19–20, 299
Equilibrium–equivalent
  concentration, 18, 299

Erz Mountains, 33
Evans, Robley, 47
Eve, 45
Exposure chamber, 76, 77

Fajans, Kasimir, 32, 46
Faults, 85
Fermi, Enrico, 48
Footings, 260
Fractionation, 218
Fractures, 85

Gas, natural, 151–153
Geiger counter, 45
Geisel, Freidrich, 32, 44
Geitel, 45
Grab samples, 61
Granites, 83, 99, 105
Granular activated carbon, 269, 300
Gray (Gy), 27, 300
Group displacement laws, 45–46
Gypsum, 121

Hahn, Otto, 45, 48
Hartung, 33
Healthy worker effect, 222, 300
Heat recovery ventilators, 258
Hesse, 33
Histopathological changes, 233
Hit probabilities, 175
Hyperplastic lesions, 232–233

Igneous rocks, 83, 84, 99, 105, 111–113, 301
Ingested radon, 210–212
Inhalatorium, 57
Initiator, 216–217
International system (SI), 13, 301
Ionization, 6
Ionization chamber, 64, 65, 301
Isotopes, 4

Jachymov, 33
Joachimsthal, 33
Joly, John, 50

INDEX

Kusnetz, 71
   Tsviglou method, 71, 72, 73

Life table, 222, 277
Lifespan shortening, 238–239
Liquid scintillation, 69, 80
Lucas cell, 47, 63, 80
Lung cancer risk, 281–282
Lung compartments, 177
Lung models, 175

Mache, 15
Macroscopic scale, 83
Measurement
   of radon alone, 63–70
   of radon progeny in air, 59–63
   of radon progeny in the body, 74–75
   of radon in water, 77–80
Medical applications, 50–51
Mesoscopic scale, 85
Metamorphic rock, 84, 85–86, 111–113, 301
Metaplastic lesions, 232–233
Meyer, 45
Microdosimetry, 175
Microscopic scale, 85
Milling activities, 136–151
Miner exposure, 32–33
Mining activities, 136–151
Mitigation, 240–272
Monozite, 85–86
Mucin, 179
Mucociliary, 175, 189–192, 301
Multistage model, 216–218

Nanopores, 89–90
Nasopharynx, 177, 216, 242
National Airborne Radiometric Reconnaissance Program (NARR), 106
National Council on Radiation Protection (NCRP), 13, 203, 274
National Uranium Resource Evaluation (NURE), 106

Natural gas, 123
Negative pressure, 257
Neptunium series, 8

Oat cell carcinoma, 219
Overburden, 149
Owens, Robert, 39

Pacific Northwest Laboratory, 230, 237–243
Passive alpha track, 72, 73
Passive smoke, 274, 281
Pegmatites, 105
Phillips, Charles, 45
Phosphate, 99, 115, 123–126
Phosphate mining and milling, 147–150
Policy, 283–295
Polonium, 35, 132
Potential alpha energy, 16–18, 20, 302
   concentration, 18
   exposure, 25–26
Pressure driven flow, 102
Pressure neutralization, 256
Preventing radon entry, 258–269
Private wells, 254
Progeny, 5
Promoter, 216–217
Properties, 1
Pulmonary, 177, 216
Pulmonary emphysema, 238–239
Pulmonary fibrosis, 215–216, 238–239, 240
Pulmonary lymph nodes, 215

Rad, 13, 303
Radioactive series, 6–11
Radiotorium, 57
Radium Institute of Vienna, 33
Radon
   in the atmosphere, 93–97, 154–157
   in building materials, 118–122, 255
   charging device, 55–56

Radon (*continued*)
  in fuels, 123, 126, 151–154
  history, 31–50
  in indoor air, 126, 130, 157–172
  medical applications, 50–51
  in the oceans, 109
  seeds, 50–51
  in soils, 98–108
  spa workers, 219
  uses, 50–58
  in water, 108–111, 115–118, 166–172
Rainfall, 104
Ramsay, 44
Reading Prong, 48–50, 111–112, 163–164, 285–288
Real estate, 293
Recoil, 88
Reducing radon entry, 250–251
Relative risk, 222, 303
Rem, 13, 303
Respiratory carcinoma, 238–239
Risk assessment, 273, 283
Rochester, University of, 230, 232–234
Roentgen, Wilhelm, 34
Royds, 41
Rutherford (unit), 15
Rutherford, Ernest, 11, 21, 32, 35, 37, 39–46

SI units, 13, 301
Sandstone, 85, 99, 105
Schmidt, Gerhard, 35
Schneeberg, 32–33
Schweider, 45
Scintillation cell, 65, 66, 74
Sedimentary aquifers, 113–115
Sedimentation in the lung, 186
Shales, 99, 105
Sievert (Sv), 27, 303
Slab house, 253
Small cell carcinoma, 219
Smoke tracer test, 264
Smoking, 225–226

Soddy, Frederick, 11, 21, 35, 42, 45
Soil gas, 249
Solubility, 2, 3, 93
Solution mining, 142
Specific energy distribution, 175
Squamous carcinomas, 241
Stack effect, 249
Stevenson, Walter, 50
Stopping power, 176, 198
Strassman, F., 48
Sump, 252, 253
Surface barrier detector, 74
Surface mining, 138
Sweden, 105–107

Tailings piles, 144–146
Thermoluminescent dosimeter (TLD), 66, 72, 74, 304
Thomson, J.J., 32, 35, 44
Thorium series, 8, 10, 25, 39–41, 135, 304
Thoron, 151
Tobacco leaves, 132–133
Tracheobronchial, 177, 216, 242
Transmutation, 43
Transport mechanisms, 86–96
Tsviglou, 71
Two-filter method, 65, 66
Two-stage model of carcinogenesis, 217

Unattached fraction, 73, 135, 156–157, 243, 304
Underground caves, 155–160
Underground miners, 219–220
Units, 12
Uranium
  in earth, 84
  exploration, 58
  mobilization, 85
  series, 7, 10, 304

Ventilation
  drain tile, 260–263, 299
  forced air, 257–258

INDEX

Ventilation (*continued*)
    general, 255–258
    natural, 256–257
    sub-slab, 263–265
    wall, 265–268
Volcanic rocks, 105
Von Traubenberg, H.F.R., 44

Wheeler, Lyndie, 44
Working level (WL), 13, 20–25, 305

Working level concentration, 59, 71, 73
Working level month (WLM), 26, 305
Working level ratio, 162, 305
Wrenn chamber, 66

Zircons, 90